Nuptiality in
Sub-Saharan Africa

international union for the scientific study of population

The International Union for the Scientific Study of Population Problems was set up in 1928, with Dr Raymond Pearl as President. At that time the Union's main purpose was to promote international scientific co-operation to study the various aspects of population problems, through national committees and through its members themselves. In 1947 the International Union for the Scientific Study of Population (IUSSP) was reconstituted into its present form.

It expanded its activities to:
- stimulate research on population
- develop interest in demographic matters among governments, national and international organizations, scientific bodies, and the general public
- foster relations between people involved in population studies
- disseminate scientific knowledge on population

The principal ways through which the IUSSP currently achieves its aims are:
- organization of worldwide or regional conferences
- operations of Scientific Committees under the auspices of the Council
- organization of training courses
- publication of conference proceedings and committee reports.

Demography can be defined by its field of study and its analytical methods. Accordingly, it can be regarded as the scientific study of human populations primarily with respect to their size, their structure, and their development. For reasons which are related to the history of the discipline, the demographic method is essentially inductive: progress in knowledge results from the improvement of observation, the sophistication of measurement methods, and the search for regularities and stable factors leading to the formulation of explanatory models. In conclusion, the three objectives of demographic analysis are to describe, measure, and analyse.

International Studies in Demography is the outcome of an agreement concluded by the IUSSP and the Oxford University Press. The joint series is expected to reflect the broad range of the Union's activities and, in the first instance, will be based on the seminars organized by the Union. The Editorial Board of the series is comprised of:

Nuptiality in Sub-Saharan Africa

Contemporary Anthropological and Demographic Perspectives

Edited by
Caroline Bledsoe
and
Gilles Pison

CLARENDON PRESS · OXFORD
1994

Oxford University Press, Walton Street, Oxford OX2 6DP

Oxford New York
Athens Auckland Bangkok Bombay
Calcutta Cape Town Dar es Salaam Delhi
Florence Hong Kong Istanbul Karachi
Kuala Lumpur Madras Madrid Melbourne
Mexico City Nairobi Paris Singapore
Taipei Tokyo Toronto
and associated companies in
Berlin Ibadan

Oxford is a trade mark of Oxford University Press

Published in the United States
by Oxford University Press Inc., New York

British Library Cataloguing in Publication Data
Data available

Library of Congress Cataloging in Publication Data
Nuptiality in Sub-Saharan Africa : contemporary anthropological and demographic perspectives / edited by Caroline Bledsoe and Gilles Pison.
p. cm. — (International studies in demography)
Papers from a conference held in Paris in 1988 under the auspices of the International Union for the Scientific Study of Population.
Includes bibliographical references.
1. Marriage—Africa, Sub-Saharan—Congresses. 2. Marriage customs and rites—Africa, Sub-Saharan—Congresses. 3. Sex customs—Africa, Sub-Saharan—Congresses. 4. Polygamy—Africa, Sub-Saharan—Congresses. 5. Africa, Sub-Saharan—Social life and customs—Congresses. I. Bledsoe, Caroline H. II. Pison, Gilles. III. Series.
GN645.N82 1994 93–48892 306.81'0967—dc20
ISBN 0–19–828761–5

1 3 5 7 9 10 8 6 4 2

Typeset by Graphicraft Typesetters Ltd., Hong Kong
Printed in Great Britain
on acid-free paper by
Bookcraft (Bath) Ltd, Midsomer Norton, Avon

Acknowledgements

This book represents some of the papers first presented at a 1988 conference on 'Nuptiality in Sub-Saharan Africa'. The conference was organized in Paris by the Anthropology and Demography Committee (chaired by Gilles Pison), under the auspices of the International Union for the Scientific Study of Population. We would like to thank the other members of the Anthropology and Demography Committee who worked with us to arrange the seminar, select the papers, and offer comments on them: Monica Das Gupta, Tim Dyson, Peter Aaby, and Valerie Hull. Particular thanks for support of the project go to Bruno Remiche, Allan Hill, Susan Greenhalgh, and members of the Union's Publications Committee.

We are grateful also to Meg Greene for her skilful translations of five of the papers, to John Massad for his insightful suggestions regarding statistical analyses, and to Enid Barker and Anna Zaranko at Oxford University Press.

Contents

List of Contributors

CAROLINE BLEDSOE, Department of Anthropology, Northwestern University, Evanston, Illinois

JOHN CALDWELL, Health Transition Centre, National Centre for Epidemiology and Population Health, The Australian National University, Canberra, Australia

PAT CALDWELL, Health Transition Centre, National Centre for Epidemiology and Population Health, The Australian National University, Canberra

MICHEL CARAËL, Institut de Sociologie, Université Libre de Bruxelles, Brussels

CATHERINE ENEL, Laboratoire d'Anthropologie, Musée de l'Homme, Paris

ANASTASIA GAGE, Institute for Resource Development, Demographic and Health Surveys, Calverton, Maryland

JANE GUYER, Department of Anthropology, Northwestern University, Evanston, Illinois

IMAN NGONDO A PITSHANDENGE, Département de Démographie, Université de Kinshasa

UCHE C. ISIUGO-ABANIHE, Department of Sociology, University of Ibadan

WAMBUI WA KARANJA, United Nations Economic Commission for Africa, Addis Ababa

GEORGIA KAUFMANN, Institute for Development Studies, University of Sussex

MONIQUE LEFEBVRE, Institut National d'Études Démographiques, Paris

RON LESTHAEGHE, Interuniversity Programme in Demography, Centrum Sociologie, Vrije Universiteit Brussel, Brussels

THÉRÈSE LOCOH, Centre Français sur la Population et le Développement, Paris

KRISTIN MANN, Department of History, Emory University, Atlanta, Georgia

DOMINIQUE MEEKERS, Population Research Institute, Department of Sociology, The Pennsylvania State University, University Park, Pennsylvania

MARC PILON, Unité de Recherches Dynamique des Peuplements Humains, Institut Français de Recherche Scientifique pour le Développement en Coopération (ORSTOM), Bondy Cedes

GILLES PISON, Musée National d'Histoire Naturelle, Laboratoire d'Anthropologie, Musée de l'Homme, Paris

JOHAN SURKYN, Interuniversity Programme in Demography, Centrum Sociologie, Vrije Universiteit Brussel, Brussels

ÉTIENNE VAN DE WALLE, Population Studies Center, University of Pennsylvania, Philadelphia, Pennsylvania

List of Figures

List of Tables

Introduction

CAROLINE BLEDSOE AND GILLES PISON

Anthropology and demography potentially have much to contribute to each other. Anthropology has drawn increasing interest in recent years as the scope of demographic attention has expanded to the developing world, traditionally the scholarly turf of anthropologists. Questions about forms of social organization and modes of production in the societies from which population data is drawn have spurred efforts to create variables that more accurately reflect the cultural and social make-up of these societies. An earlier IUSSP-sponsored volume, *Micro-Approaches to Demographic Research* (Caldwell *et al.* (eds.), 1988), highlighted some potential methodological contributions of anthropology to demographic research. The chapters in this book cover similar ground. But they move on to the task of incorporating some theoretical contributions from anthropology into population science: contributions that explain empirical findings and stretch their applications to broader contexts beyond specific locations in time and space. In attempting to integrate some bodies of diverse scholarship into a viable socio-cultural demography, the studies bring new insights to bear on a lively topic of long-standing interest in African ethnography: the dynamics of conjugal union and their implications for fertility change.

Sub-Saharan Africa is a world area where interpretations of demographic data on marriage and fertility call for fresh collective insight. To be sure, marriage in this part of the world has absorbed anthropological attention for a very long time. Some of the most important concepts of socio-cultural anthropology, particularly those pertaining to marriage and kinship, were forged and refined in this world region (see, e.g., the contributors to Radcliffe-Brown and Forde (eds.), 1950). The role of economic transactions in transferring legal rights in a woman's labour and her sexual and reproductive services have particularly preoccupied anthropological studies, as has polygyny. Understanding the dynamics of marriage in Africa has also become increasingly important to demographers as a vital area for comparative international research. And new cadres of African social scientists from a wide range of disciplines are taking hard looks at increasing problems of bearing and raising children in an economic recession that spans the continent. This confluence of events and analytical interests makes it imperative for scholars to share information across international as well as disciplinary boundaries.

To stimulate research in such domains as well as to encourage collaboration between demographers and anthropologists, the International Union for the Scientific Study of Population, through its committee on Anthropology and Demography,[1] organized a seminar in Paris in 1988 on the theme, 'Nuptiality in Sub-Saharan Africa: Ongoing Changes and their Impact on Fertility'. Specialists from anthropology, demography, and related social sciences convened to examine current transformations in African conjugal unions. Though they were encouraged to assess the implications of these changes for reproduction, the emphasis was on nuptiality itself. And although the emphasis was on topical breadth and strength, some degree of regional coverage was sought; in the end, however, studies of West African contexts predominated.

Although this volume can be read as a document on the complexity of African demography, its distinctiveness lies in its concerted effort to incorporate recent theoretical perspectives in anthropology into the interpretation of demographic data. Its most unique contribution lies in an explicit reconceptualization of marriage, arguably the most interesting, yet least studied, of the core demographic variables. Nuptiality commands attention because it establishes—or reflects—the social contexts that surround reproduction. In Africa, marriage is often less a discrete event than a process sometimes lasting over years: a process involving transfers of wealth, the inception of sexual relations, the birth of children, etc. In many areas of the subcontinent, marriage is becoming even more attenuated under pressures from economic forces, urbanization, and education. Confronting recent views of the fluidity and negotiability of marital statuses, the contributions in this book offer new ways of thinking about nuptiality and its implications for fertility.

Marriage in the Proximate Determinants Framework

Of all the developing world areas where thirty years ago analysts argued that fertility would never decline, only Africa maintains persistent high fertility with, as yet, signs of decline in only a few countries.[2] Women can be divorced and men humiliated for not producing children. Adults want children for farm and household labour and for future support, and they hope that some of their children will raise the family's status by acquiring formal schooling, urban employment, and political connections. Yearly growth rates from 2 per cent to almost 4 per cent make this the fastest growing population in the world. Scholars who previously studied fertility transitions in other parts of the world are now turning to Africa, poised to watch how and when declines might occur.

Most students of population characterize African reproductive regimes as those governed by what Louis Henry (1961) called 'natural fertility'. That is, because the demand for children is high and few people appear to be interested in reducing numbers of children, women do not alter their fertility behaviour appreciably from one birth to the next. Among such populations, research on the biological proximate determinants of fertility (Bongaarts, 1983) comprises a critical domain of inquiry. One such topic is the practices that affect child spacing: coital frequency, the length of breast-feeding and postpartum abstinence are examples. (Collections on the socio-cultural underpinnings of fertility in Africa are those edited by Page and Lesthaeghe, 1981; Lesthaeghe, 1989; and van de Walle, forthcoming.)

Marriage is another locus of research attention in the natural fertility framework. Some scholars hold that certain features of marriage for women—age at first marriage, rates of divorce or widowhood, rates of remarriage, time spent in union, and proportions of never-married women—have played key roles in regulating fertility in the course of human history (see especially Wrigley, 1969). Many of the chapters represented here begin from this natural fertility paradigm, which treats marriage as a regulator of fertility. But they go on to examine some of the social and economic factors that belie simplistic applications of the model to Africa. Some examine links between fertility and factors such as age at first marriage, multiple partner unions, and variant residence arrangements. But most concentrate on conjugal union itself: its myriad manifestations, the pressures that produce them, and the social contexts they set for fertility and childbearing under conditions of change.

Age at first marriage is a focal point for fertility interests in 'natural fertility' populations because it usually marks the onset of regular sexual relations for women and, thus, entry into regular childbearing. A population in which women marry early will likely have higher fertility rates than one in which women marry late. Age at first marriage, a point of much attention in this volume, appears to have played a major regulatory role in the history of European fertility. When privately held arable land fell in short supply in some areas, fewer marriages were contracted and people who married did so at older ages, producing a temporary reduction in births. Later, when more land became available, more marriages were contracted and by younger partners, producing a rise in births (Wrigley, 1969).

Age at first marriage is a promising area of inquiry for predicting future trends in African fertility as well. Because of relatively low levels of contraception and a continued high value placed on children, van de Walle and Meekers (Chapter 2) predict that for many African countries fertility may decline through a rise in ages at marriage before it does so through altered childbearing patterns within marriage. But their caution, that analysing marriage presents numerous problems, suggests that the marriage–fertility relationship may take quite a different trajectory in Africa than it did in Europe. The collective control of land in much of the African subcontinent, for example, may undercut the kinds of homeostatic conjugal mechanisms reported in Europe. The readiness with which women in some areas contract new forms of union undermines the strength of fertility inferences drawn from marriage and divorce figures. And the possibility that one man can have multiple wives means that in theory economic pressures could result in lower numbers of polygynous households, rather than rising ages at marriage.

Age at marriage, then, is by no means linked to fertility in mechanical ways. In fact, a theme of lively demographic debate centres on why marriage, which is generally considered a crucial event for gauging the onset of childbearing, is only tenuously connected to childbearing in many African countries. Even if we set aside the difficulties of ascertaining ages and dates (e.g. van de Walle, 1968), we quickly run up against the problem of how to define marriage—and thus how to assess whether people are married and when they first married. Many forms of union in Africa are not well described by classical survey categories of celibate, married, widowed, and divorced. Earlier approaches assumed marriage across the continent to be a single, undifferentiated category, wherein people were married or unmarried, and all marriages had the same social significance. However, many African societies recognize a range of union types: those endorsed by legal/statutory ceremonies, consensual arrangements, and various religious or customary procedures. Ngondo a Pitshandenge (Chapter 6), for example, describes the conflicts that stem from the residue of multiple systems of legislation—traditional, colonial, and post-colonial—under which people might marry. Guyer (Chapter 11) adds that many changes in nuptiality during the past century that stemmed from shifts in systems of authority and reform movements sought to regularize certain marital types and ban others; but instead of clarifying the meaning of marriage, they actually made it more ambiguous by adding more criteria.

To complicate matters further, there are multiple potential interpretations within these types. For customary unions, a group may recognize a union only if bridewealth (the money or goods given by the man's family to that of the woman in exchange for rights in her labour and fertility) has been paid, although parties may disagree later over how much was actually paid. Further, the elements that comprise unions can be ephemeral and subject to later denial. If relations deteriorate between two families that consider themselves linked by a marriage, then the money that the man's family originally handed over as bridewealth may suddenly be recalled as payments on a past loan; the food and drinks that were consumed together are now remembered as casual tokens of hospitality rather than as elements of a marriage ritual.

One of the more serious obstacles to establishing when marriage begins grows out of new views of marriage among many African groups as a long, ambiguous process rather than a discrete event established by clear legal, ritual, and economic transactions (see e.g. Comaroff and Roberts, 1977; see also Guyer, Chapter 11, for a discussion of this intellectual development in anthropology). The process transpires through a sequence of events that might include the exchange of symbolic tokens, making instalments on bridewealth payments, establishing a joint residence, or even the birth of a child. This process may extend over a period of months or even years. At some point, cohabitation and sexual relations begin and children may be born, but these events do not necessarily coincide with, or follow predictably from, certain marriage rituals or economic transfers. These events can best be seen as 'conjugal testing', wherein the partners and their families build mutual confidence in each other by taking incremental steps towards a union. If one partnership does not appear to be working, a young woman—sometimes with her family's implicit permission—may test out potential relationships with several young men before establishing a more stable one. In the meantime she may have borne a child or two by different partners.

If marriage is a process, it may not be clear when, in this series of events, a woman is 'married'. Both Locoh and Guyer (Chapters 10 and 11, respectively) point out that in some contexts a woman who has a child by a man considers herself married. Discerning a predictable order for the marriage process might give us a foothold. However, van de Walle and Meekers (Chapter 2) show that the order of the process differs from group to group; and chapters by Karanja and Mann (Chapters 9 and 8, respectively) describe a form of union, 'outside marriage', in which conjugal transactions seldom conclude in a legal marriage. Further, marriage in many areas tends to be less a linear process than a constant mix and movement: as Nici Nelson, a conference participant, pointed out, people may slip back and forth among kinds of union with little fanfare. Still, because much is at stake in marriage—property, children, labour, and political alliance—disputes may arise far into the future (especially when inheritance time arrives) about whether a marriage really occurred.

Another problem in trying to establish the start of a marriage, as many of the authors represented here acknowledge, is that a certain conjugal transaction may have meaning only in the historical context of a particular relationship. An exchange of kola nuts may signify the initiation of courtship in one situation; in the next household, it may signify the conclusion of a marriage. There are similar fallacies in trying to pin down and measure phenomena that are inherently ambiguous and fluid. This task is made all the more elusive because most scholars identify a current trend towards yet more informality and fluidity in conjugal relations. External ideological influences, including demands from conservative Christian elements to distinguish once and for all between 'married' and 'unmarried' people, ironically appear less to have constrained choices or eliminated ambiguity than to have added yet more idioms for rationalizing action.

Given these complications of definition and ordering, it should not be surprising that some African countries show only weak associations between age at first marriage and age at beginning childbearing (e.g. Gyepi-Garbrah, 1985). Moreover, the fluidity that has long characterized African conjugal practices appears to be increasing dramatically in the contemporary era for various reasons: diminished familial sanctions, young women's desires to remain single while they continue schooling, and young women's difficulties in obtaining contraception. As a result, many African countries are witnessing rising rates of reported premarital fertility, particularly in urban areas.

For some authors, these facts mandate abandoning the search for clear conjugal categories that determine behaviour, and confronting instead the fact that marital statuses are not only fluid but subject to constant renegotiation. For many of the case studies that move beyond proximate determinants questions of biological reproduction, their findings suggest that in contemporary Africa the major impact on fertility or 'completed family size' is the outcome of strategic attempts by individuals to shape their familial obligations and sources of support by social means.

The Social Demography of Marriage

These lines of query suggest that besides treating marriage as having a possible role as a proximate determinant of fertility, we can also view it as an important indicator of the social conditions under which families join and births occur: conditions that can have considerable importance for the well-being of women and children. Since the status of children born into one form of union or another may vary, for example, how a union is formed and regarded by society may also affect the likelihood of abortion, sexually transmitted diseases, or child survival. The topics that loom large for Africa are legitimacy and multiple unions.

The Question of Legitimacy. Since marriage may consist of a long process of conjugal testing, childbearing does not necessarily wait until everyone agrees that a marriage is concluded. The logical question to ask at this point is whether a child born 'premaritally'—that is, before the union is concluded—is less welcome than one born into an established union. (See Adegbola, 1987, for a related discussion.)

To distinguish a premarital birth from a marital birth, we would need to know precisely when marriage occurred, a task that poses considerable difficulty, given the problems of ambiguous conjugal transactions, differing forms of legitimation, and the processual nature of marriage. Yet before wading into a problem of such complexity, we need to recognize that in much of Africa, premarital births are not necessarily illegitimate, in the sense of lacking formal recognition. Contemporary Westerners tend to see a child's legitimacy as a question of timing—whether the child was born into an extant legal union (see, however, Laslett's (1980: 8) distinction between 'pre-nuptial' and 'extra-marital' births in historical Europe). Most African societies, by contrast, tend to see a child's legitimacy as a function of paternal recognition and support; hence, the mother's marital status at the time of her delivery is at best a rough proxy for reflecting a child's legitimacy. If the entire conjugal process is proceeding smoothly, a birth that occurs before a young woman's union is fully recognized does not necessarily represent an embarrassing case of illegitimacy. It may instead be a welcome event that, by proving her fertility, nudges the union further along the continuum toward a legitimate marriage. Only if the union is dissolved—or if the man dies before it is fully concluded, and inheritance is suddenly at stake—will questions of 'illegitimacy' be likely to arise.

Multiple Unions. Closely related to the question of a union's legitimacy in contemporary Africa is that of multiple unions, for the legitimacy status of one union may affect that of another.

Sub-Saharan Africa is the only major world region in which polygyny is still widely practised, despite the welter of legal and religious codes that colonial as well as modern states have invoked to control or eliminate it. The possible effects of polygyny on fertility have drawn considerable research attention. From the perspective of a proximate determinates approach, the fact that every woman in a polygynous society is in theory marriageable almost from puberty loosens whatever demographic constraints monogamy might place on ages at first marriage, time spent in union, and proportions of women married. As a result, women in polygynous societies often marry quite young and, upon widowhood or divorce, remarry quickly. Overall, however, the relationship of polygyny to fertility remains unclear. While polygyny may increase fertility by absorbing more women in marriage, it may reduce the number of an individual woman's sexual encounters. And in areas where outside affairs are common, polygyny may have no perceptible effect on fertility. (For a sample of discussions, see Pison, 1986; Borgerhoff Mulder, 1989; Garenne and van de Walle, 1989; and Goldman and Pebley, 1989.)

To be sure, some of the institutional supports of polygyny are falling away, especially among urban élites whose patterns of union formation take a monogamous trajectory. Yielding to social and legal pressure, many urban men take only one legal partner at a time. Young women with education and independent incomes have considerable independence to select a partner rather than bowing to family decree. And they find it easier to divorce or separate if their unions prove unsatisfactory. Such shifts also tend to accompany an increase in age at marriage for girls.

These changes might appear to benefit women more than men. Besides giving women more choice in marriage, they should, in theory, erode further the demographic bases of polygyny. But it is not clear that these changes are as deep as surface transformations would suggest. Many of the new marriage forms that outwardly resemble monogamy actually follow patterns of *de facto* polygyny: 'serial monogamy' (Comaroff and Roberts, 1977), 'outside marriage' (Lacombe, 1983; Clignet, 1987; Karanja, 1987), 'sugar daddy' relationships (Dinan, 1983), and so on. Recognizing that such forms remain far removed from Western conjugal unions, van de Walle and Kekovole (1984), among others, rejected Goode's (1970) evolutionary prediction of Africa turning toward Western-style marriages. (See also the contributors to Parkin and Nyamwaya (eds.), 1987.) Even 'free choice' of partner may be less free than it appears. Although women seem to have more freedom to choose within a certain range of possible partners, growing class structures, combined with monogamous ethics, make inter-class marriage less probable. If we can treat these unions as 'signs', to use Ariès's (1980: 645) term, of varying degrees of legitimacy that ultimately affect the resources that women and their children receive, it becomes clear that while élite women benefit from support from the Church or the legal system against potential competitors for their husband's resources, these same institutions exclude other women from equal support. (Similar short-falls may be experienced by 'polyandrous mothers', the other example of women in multiple unions that this volume examines; see Chapter 11.)

New Points of Juncture between Anthropology and Demography

At the outset, the methods of demography and socio-cultural anthropology appear to pose formidable barriers to meaningful discourse. Despite their overlapping interests, practitioners of the two disciplines often find certain areas of each other's discourse impenetrable. More seriously, they often have quite different conventions for framing research questions. Demographic studies of marriage and reproduction tend to look for factors that affect biological determinants of fertility: cohabitation, consummation, time in union, post-partum abstinence, child spacing, breast-feeding, and terminal abstinence. Many socio-cultural anthropologists resist demographers' attempts to boil down behavioural phenomena into discrete variables and trace them to ultimate causes. They shy away from what they interpret as narrow, quantitative interests in a small set of topics, and from research strategies that appear to stress numerical significance and comparability at the expense of culturally informed notions of meaning and causality. The latter, anthropologists insist, can only be gleaned from 'holistic' studies that encompass a very wide socio-cultural context. Anthropological studies of marriage typically involve 'thick description' (Geertz, 1973): broad, diffuse descriptions of group structure and function. Patterns of kinship, residence, bridewealth, inheritance, descent, alliance, divorce, and the transference of legal rights in women are considered basic to the anthropological study of marriage in Africa. To demographic sceptics, such studies produce very long descriptions of very small groups: works that ultimately have limited generalizability.

These two polar types, of course, are overdrawn. As in any interdisciplinary endeavour, there are inevitably overlaps—so much so that despite their differences, anthropologists and demographers sometimes share shortcomings. Guyer (Chapter 11) suggests, for example, that although anthropologists and demographers alike have viewed fertility in Africa as a source of future security and labour, disciplinary blinkers may be hindering both fields at a crucial moment in African economic and demographic history. The need to build lateral ties to other groups is leading women to place less emphasis on children's future returns and more on their immediate potentials for creating links to men's resource networks. As a result, conjugal ties are becoming less a means to fertility than desired ends in themselves.

The task of bridging disciplinary gaps to study nuptiality in Africa is further complicated by the fact that the explanatory paradigms guiding research are changing as quickly as their subject matter. Models of acculturation, evolution, and diffusion that were popular in the social sciences resurface occasionally to fuel speculations about the motors of population change, particularly studies of how contraception is adopted. David Parkin, a conference participant (1988), drew on his earlier collaboration with David Nyamwaya (Parkin and Nyamwaya (eds.), 1987) to review several key anthropological approaches to African nuptiality and link them to the present demographic concerns. He reported anthropologists' increasing defections from normative explanations for conjugal patterns, and their movement instead toward 'interactional' approaches that emphasize process, strategy, and the negotiability of conjugal and filial relations. Parkin singled out four major areas that African scholars themselves identified as critical for fresh perspectives on marriage and fertility: polygynous marital templates, children's custodial negotiability, the meaning of children for both men and women, and the control of women's sexuality—themes that have obvious relevance for demographic questions, and that appear with surprising frequency in the chapters that follow.

Parkin's observations highlight two features of demographic categories that are relevant to our present purposes. The first is that these categories represent cultural notions—our own or those of the people whose lives we study. Hence, to call a woman 'married' is to assign her a status either through our own notions of marriage or through those of the group with whom she identifies. The discussion above should leave no doubt that both of these strategies have shortcomings: the former inevitably builds ethnocentric assumptions into the categories, while the latter precludes cross-cultural comparability.

The second fact about marital categories that concerns us is that cultural norms are attached to them. In interpretations of survey findings, these norms are often treated implicitly as rules or traditions that people are obliged to follow. That is, a woman who is considered married is expected to behave in certain ways because of societal norms attached to the category of 'marriage', and to refrain from behaviours deemed inappropriate to married women. The problem with such approaches, even those that employ local categories, is that they present rigid views of behaviour. They assume that people seek to clarify which categories they belong to and hence which rules they should follow. Issues of ambiguity—for example, who is actually married and which of several mutually contradictory norms a person is supposed to follow—are troublesome but eventually dismissible.

Increasingly, as Parkin's observations suggest, social scientists reject the jural determinacy of mechanistic, rule-centred categories and explanations, and ask instead how individuals seek to 'construct' their own realities. People may try, for instance, to construct reality socially by trying to construe statuses such as 'married' or 'outside wife' so as to incorporate—or exclude—more individuals than biology or formal rules alone would produce. (See Burnham, 1987, for a lucid application of this general approach to marriage in Africa. See also Kopytoff, 1987, on processes of assimilating strangers and newcomers into local lineage groups and Comaroff and Roberts, 1981, on the strategic 'management of meaning' of marriage in southern Africa.) Whereas analysts usually try to remove ambiguity, local people may resist being pigeon-holed into tidy social categories, not because they fail to understand the categories but because ambiguity may represent a strategy for keeping options open. As many of the papers here suggest, fulfilling particular responsibilities to spouses, children, and affines helps define marriages and make them reality, rather than the other way around. The same would apply to children's legitimacy: adults may attempt to validate or undermine children's claims to rights by construing the parents' union as being of one kind or another.

Two strands of recent theoretical innovation in the social sciences that hold special promise for interpreting African nuptiality are the 'processual' or 'constructionist' approaches. Such approaches do not accept at face value the empirical veracity of particular social labels such as 'married'. Instead, they examine the process by which such labels come to be applied to particular individuals. Placing the emphasis on process also draws attention to the fact that social statuses are frequently subject to dispute—is a woman married legally, or only a casual girlfriend?—and to efforts to negotiate their meaning. These perspectives dovetail with what students of African social organization have been insisting for some time: marriage is less a single event in time than a process built up out of visits, transfers of wealth and symbolic tokens, the births of children, etc. Hence, a birth may be a means to bolster a marriage, rather than its goal. Such an approach also implies that conjugal status is not frozen in time or place. Instead it is characterized by enormous fluidity: individuals move back and forth between marital statuses as various socio-economic pressures come to bear on them. This approach is even more appropriate to contemporary Africa: pressures to become educated or to migrate for work before marriage attenuate further the marriage process, creating even more ambiguity among marital statuses.

The Papers

Reviewing some of the major patterns of marriage in Africa that have engaged demographers, R. Lesthaeghe, G. Kaufmann, D. Meekers, and J. Surkyn describe some common characteristics and discernible patterns of nuptial change. Their original analysis superimposes ethnographic data from the Human

Relations Area Files on to World Fertility Survey data to sort out some of the socio-economic correlates of variables that have drawn sustained fertility attention in Africa: long birth intervals, polygyny, early female marriage, and wide age gaps between spouses. The study shows that polygyny and long post-partum abstinence continue to be shaped by indigenous forms of production and lineage organization as well as by Christian and Islamic penetrations; polygyny and abstinence appear to be shaped much less, surprisingly, by factors such as female education. The authors turn to a key theme in this volume: the anomalous persistence of polygyny in the face of growing economic, legal, and religious pressures. They document an important demographic basis for this persistence: age gaps between spouses have remained wide, despite marginal rises in age at first marriage for both men and women. The authors point out as well that in many of the places where polygyny appears to be declining, it remains under the guise of 'outside wifeship'.

Age at First Marriage

To some extent the tenuous connection between marriage and childbearing in many African countries can be attributed to measurement problems. Women who report being married, for example, are more likely to be coded by surveyors as being older than those who report being unmarried (Lesthaeghe *et al.*, 1989: 244), making connections between age at marriage and age at first birth extremely problematic. Another problem is that it is not clear which stage of the process should be taken as the start of the union. Many large-scale analyses, since they are concerned with periods of exposure to risk of pregnancy, use cohabitation as their principal criterion. Several chapters draw on contextual ethnographic information to produce more fine-grained answers to the question of when, in the conjugal process, we can categorize a person as married.

One reason why so many births in Africa are reported as 'premarital' is the fact that the beginning of marriage is frequently difficult to date, forcing many surveys to use as a proxy the date when cohabitation began. Étienne van de Walle and Dominique Meekers use some innovative questions developed by the organizers of the Côte d'Ivoire fertility survey of 1980–1 to determine whether this emphasis is justified. They ask whether some groups reorder the commonly understood order of events in the marital process—cohabitation, initiation of sexual relations, and ritual celebration of marriage. They find that several major ethnic groups in Côte d'Ivoire deviate from the expected sequence: these groups recognize unions—and thus their offspring—as legitimate well before cohabitation begins. Although no one magical question will solve all the problems of measuring African nuptiality, van de Walle and Meekers argue that unions among these groups should be dated back to the time when they received the requisite social sanction: perhaps a betrothal celebration or a ritual exchange of drinks and kola nuts. The fact that the time of celebration coincides closely with the time when most couples began sexual relations lends further credence to these findings.

Just as events in the marriage process must be situated in a cultural context, they also derive character from the economic context. The next two chapters, those by Isiugo-Abanihe and Enel *et al.*, examine the effects of economic changes on the age at first marriage among two very different groups.

Bridewealth is often viewed as a distinctive feature of marriage practices in Africa. But bridewealth, like other 'traditions', changes with the times. Sometimes it decreases or even disappears altogether, especially among educated urbanites. Uche Isiugo-Abanihe, however, shows that the opposite has occurred among the highly educated Igbo-speaking peoples of Nigeria: bridewealth costs have escalated far beyond easy reach for most men. Bridewealth nowadays is cast as a form of reimbursement for the costs of education that the woman's family incurs. Hence, bridewealth demands, often set according to the woman's educational level, have risen sharply alongside women's educational achievements over the past few decades. This inflation in bridewealth, coupled with a shift in responsibility from the family to individual men to provide bridewealth, has delayed marriages, increased ages at first marriage, and produced high proportions of never-married individuals. In effect, escalating costs of marriage have made Igbo marriage veer sharply from the standard African pattern of being early and universal. These trends probably delay entry into parenthood, since virtually all intra-ethnic childbearing takes place within marriage. But while prestigious intra-ethnic and bridewealth marriages are delayed, informal unions or polygynous unions and even, possibly, unions with partners from different ethnic groups, appear to be increasing.

Age at first marriage—and, indeed, marriage in general—is also dramatically affected by economic changes in south-western Senegal where Catherine Enel, Gilles Pison, and Monique Lefebvre studied, but these authors describe patterns for the Joola people that seem to differ radically from those among the Ibo. By contrast with the highly educated Igbo, the rural Joola handle the economic costs of marrying by massive seasonal labour migration. Much like the Igbo case, however, responsibilities for providing marriage costs have shifted sharply: from the families to the partners themselves. To meet these costs, young men often migrate to work as palm wine collectors, and most unmarried young women leave each year to work as household servants in large cities to earn their marriage costs. As in the Igbo case, the results are an attenuation of the marriage process and an increase in the age at first marriage (that is, the stage at which a union is recognized by village elders) of about 5 years for both men and women since the beginning of the century; it has now reached 24 years for women and 30 years for men. By contrast with the Igbo case, however, women's age at first birth appears to have stayed about the same—a finding that parallels several other chapters in this volume. The result is an increasing number of premarital births.

Legitimacy

Escalating costs of formal marriage to prospective partners are among the most striking changes occurring in some parts of Africa, particularly in contemporary urban areas. Another is shifts in the status of various kinds of union and the children resulting from them. Examining analyses of different legal codes concerning polygyny in francophone African countries under indigenous, colonial, and post-colonial regimes, Ngondo a Pitshandenge shows that colonial laws created a double system of civil versus customary marriage. Yet attempts by post-independence francophone legislators to draw up appropriate marriage laws that build on the past as well as accommodate present realities are often, despite their good intentions, contradictory and incomplete. Polygyny and the status of children born within customary versus civil marriages are the most salient manifestations of these legal conundrums. Despite efforts to create one consistent and equitable legal system, polygyny has resisted the restrictions imposed by colonial and post-colonial legislation alike.

Even in urban areas, polygyny involves so many people in one way or another, and covert resistance to abolishing it remains strong. As for legal efforts to bolster protection for children, Ngondo a Pitshandenge points out that new legal codes have been written, requiring men to accord equal recognition to their children by all their conjugal partners who were married either under civil law or custom. These codes, however, bear an ironic contradiction: complying with them would accord full legal status to all the children's mothers, in what is now a monogamous legal structure. He argues that in any event, legal measures involving marriage have had less effect on regulating numbers of children than on determining their social standing and their prospects for inheritance.

The kind of esteem a union enjoys is no small issue. How a union is defined can determine whether it is recognized as legitimate. But it may be less useful to categorize legitimacy as an 'all-or-nothing' state than as a continuum of statuses ranging from privileged to marginalized. Such an interpretation is suggested in Marc Pilon's chapter. Examining nuptiality among the Moba-Gurma in northern Togo through a 1985 retrospective survey, Pilon attempts to incorporate local cultural terms for union types to explain the dynamics of partnerships. He finds that according to local understandings, men have different ways of acquiring wives: exchange, reimbursement of women, promise, brideservice, levirate, elopement, and bridewealth. The study documents two reasons why it is important to distinguish among union types. First, the procedure by which a couple marries influences which woman, among a polygynous man's wives, will be the first or 'senior' wife. A woman married through brideservice, for example, is more likely to be recognized as the first wife than a woman married through abduction. Second, how a marriage begins affects its ending: the likelihood of it ending, how it ends, and who initiates the dissolution (the woman, the man, or the woman's family).

Like Pilon, Anastasia Gage and Caroline Bledsoe find 'legitimacy' a problematic concept. Their study examines reported ages at marriage, which they believe may reflect strategic construal as much as the realities of conjugal life. In their analysis of the influence of education on the timing of marriage and the birth of the first child in Freetown, they show that education has remarkably little influence on the age at first birth for these younger women; many become mothers before marrying. Yet education does appear to have a powerful positive effect on the reported age of first marriage. One explanation to this puzzle may be that in reporting their marital status, educated women appear to emphasize civil or church ceremonies, even though they may have already undergone customary marriage ceremonies: events that uneducated women, by contrast, are willing to acknowledge as legitimating their unions. These findings imply that educated women may prefer to define their births as premarital, rather than to admit being married through customary rites. Changes in education also appear to have a strong effect on polygyny—or at least on the official reporting of polygyny. Relating their findings to increasing social stratification in wider society, the authors argue that female education has made polygyny a highly brittle institution. It counteracts the capacity of polygyny to accommodate excess women in marriage by legitimating the visible presence of only one 'civilized' wife. While women with little education become increasingly marginalized, women who progress beyond primary school are increasingly unwilling to accommodate themselves to a polygynous household. These women, because they demand exclusive recognition as monogamous wives, generally enjoy more advantages than other women attached to the same men.

Multiple Unions

Once we turn explicitly to the issue of multiple conjugal partners, it becomes quite clear why a union's legitimacy is a vital question: legitimacy can be the basis for considerable differentiation among various women and their children who claim affiliation with one man. One of the most important cases of differentiation within conjugal relationships is that of 'outside marriage': a long-term union contracted outside church or statutory codes. Kristin Mann's historical study of élite settlers in nineteenth- and early twentieth-century Lagos focuses on conflicts between élite men and women over outside unions in a polygynous Yoruba environment. It shows that outside marriages gave non-élite women access to resources and opportunities for upward mobility otherwise closed to them.

Echoing themes of instrumentality in conjugal relations that appear in other chapters in the volume, Mann shows that actors seek to manœuvre around these crucial points of legal, social, and religious tension. She shows that the demands of living in a rapidly developing colonial and commercial capital made outside marriage an attractive option for a number of élite men and non-élite women, allowing them to circumvent the religious and legal prohibition against polygyny, and Christian marriage. Élite women, by contrast, had little to gain from their husbands' outside marriages, but much to lose if they raised strident public objections. Mann documents the efforts of the church to protect the precarious advantages of 'inside' wives and their children against competition from outside wives. Although the church has claimed some success in enforcing monogamy, Mann shows that many Christian adherents are being absorbed *de facto* into the wider customary polygynous milieu. Indeed, the church's insistence on monogamy receded, as an increasingly educated and sophisticated local populace began to demand equal inheritance rights for all a man's children. The result is that while Nigeria makes no inheritance provisions for 'outside' children born of statutory unions, children born within customary marriages do have such rights.

In a chapter complementary to Mann's, Wambui Wa Karanja explains outside wifeship among the upper classes in Lagos and Ibadan as a way for an élite man to achieve the African ideal of polygyny under a form acceptable to his social group. Focusing on the perspectives of outside wives, she shows that while an outside wife reaps some of the economic and social benefits of marriage, she usually lives apart from her 'husband'. This flexibility of residential patterns, combined with individuals' efforts to negotiate the definition of unions, echoes key themes in other essays. Turning to the possible impact of outside marriage arrangements on fertility, Karanja notes that in her study sample, outside wives seemed to bear fewer children than legal wives, but she cautions that it is unclear whether this pattern stemmed from these women's advanced levels of education, late entry into union, or low initial fertility. She draws primary attention to a surprising fertility strategy that seems to be emerging: because these unions are less stable than legal ones, outside wives seek to keep their options open for future relationships; therefore, most outside wives prefer to suppress the numbers of children they bear for any one man.

Even more so than bridewealth, polygyny is an institution that is associated with marriage in Africa. Yet polygyny, like bridewealth, is not an institution graven in cultural stone; it is changing alongside other aspects of conjugal life. Thérèse Locoh's material on urban Lomé, Togo, however, shows that polygyny is not necessarily changing in expected ways. Focusing on residence patterns, Locoh argues that although rates of polygyny have not declined appreciably among urban dwellers, urban polygyny presents a much more complicated picture than the popular image of African marriage that depicts the entire conjugal unit living together and sharing communal resources. Polygynous wives living in the city are much more likely than their rural counterparts to live apart from their husbands and co-wives, and to separate sharply their own economic resources from those of their spouses.

Locoh's analysis reveals that these patterns do not reflect urban disarray; rather, individuals appear to be using various residence options to circumvent the constraints imposed by monogamous ideals and to regain some of the flexibility that rural polygyny offered. Locoh reminds us that despite the advantages it offers women in some instances, monogamy decreases the opportunities for social advancement that multiple unions offer. A man who commits himself to monogamy effectively divests himself not only of previous spouses but also of their families, who might, under polygynous rules, be retained as important political and financial supporters. Under these conditions, contracting non-co-residential polygynous unions gives men more liberty to accumulate concurrent spouses. The main locus of Locoh's originality, however, lies in her observation that non-co-residential unions not only give women conjugal freedoms that neither conventional monogamy or polygyny would allow, these unions even allow women to pursue social mobility. Although few women have realistic recourse to the full justice of modern law in divorce cases, a woman can distance herself gradually from a soured marriage by moving elsewhere and beginning to pursue alternative relationships. Locoh implies that there are benefits to be gained by moving away even in marriages that remain cordial. A woman can allow her husband to pursue additional relationships that may benefit him, and hence her as well. The irony in both cases—polygyny and divorce—is that both men and women are seeking room to manœuvre through conjugal links in contexts in which mobility through marriage is in theory precluded.

As these papers suggest, multiple-partner unions are not restricted to formal polygyny. Even men who are legally monogamous can be supporting more than one woman at a time, and a woman can be pursuing relationships with more than one man. Starting from a similar observation, that women can have several simultaneous relationships, Jane Guyer pulls together a number of conceptual strands regarding multiple-partner unions and the utility of children stemming from different unions. Using material from a Yoruba-speaking area of Nigeria, her analysis demonstrates very forcefully that changes in marriage stem from changes in the utility of children. Guyer points out that women need contributions from men towards their day-to-day expenses; without them, economic difficulties can undermine women's security. Seeking to buffer the effects of these uncertainties, women increasingly rely on childbearing as a means of creating networks that can offer support. A woman has far greater leverage to press her demands on a man, whether or not they call their relationship a marriage, if she has a child by him. Indeed, some women find it advantageous to bear children by different men in order to gain access to several resource networks. Although a woman cannot marry polygamously, therefore, she can have several concurrent fathers to her children, making marriage largely incidental to her reproductive career. Anthropology's past focus on marriage, Guyer contends, has obscured the fact that a woman may be managing ties with several men at once, through what she calls 'polyandrous motherhood'. Such observations call for a shift in analytical perspective from the traditional lineal interpretations of fertility, in which most benefits from children are derived decades after their birth, to an explicit focus on women's short-term lateral strategies, using children to make demands on a wide range of men and their kin.

Guyer's analysis also highlights shifts in the changing status of children, and their implications for women's lateral fertility strategies—a fact that is particularly significant in the light of Ngondo a Pitshandenge's observation that children's social status has drawn more attention in Africa than their numbers. Guyer shows that although African cultural repertoires have seldom included the idea of the unknown, absent father that historians of Europe have described in such detail, present economic circumstances appear to be making support attenuate to recognition alone, recognition may attenuate still further into selectivity, and selectivity into neglect, at least for some off-spring. Under these conditions, women's lateral strategies may have limited pay-off.

Sexuality and Reproduction in Problematic Unions

While several chapters described above confront the issue of fertility directly, three additional ones examine factors that may affect sexuality and reproduction in the context of changing nuptiality patterns.

Bringing an epidemiological perspective to bear on the issue of nuptiality, Michel Caraël's study implies that conjugal patterns may affect the spread of sexually transmitted diseases (STDs—gonorrhoea, syphilis, etc.). Highlighting the importance of transient sexual partnerships for the diffusion of STDs and HIV infection, the study uses survey data from Rwanda to identify two significant conjugal patterns that are associated with higher rates of HIV infection: the late age at marriage for men and the strong Catholic emphasis on female sexual virtue, both before and within marriage. Having limited sexual outlets, unmarried men tend to patronize prostitutes who, because they have many sexual partners, are both primary victims and transmitters of AIDS. Their clients who become infected are likely to transmit the disease to their wives when they marry. This pattern tends to concentrate the epidemic in a small number of carriers initially, but spread it quickly.

Caraël contends that the greater the imbalance in sexual freedom between men and women, the more rapid the spread of seroprevalence, which has risen to about 30 per cent in the adult population in Rwanda, versus moderate and more stable rates elsewhere in central Africa. In Kinshasa and Dar es Salaam, for example, where men appear to have more sexual outlets, seroprevalence in the population is much lower: around 5 or 6 per cent. He concludes by stressing that urban situations themselves do not necessarily intensify transmission of STDs. Inequalities in male and female sexual behaviour that may be conducive to the spread of AIDS, particularly the double standard that allows men to accumulate sexual partners while women are expected to be faithful to one partner, may be intensified in urban areas. However, urban contexts which allow women to adopt a variety of forms of union may make re-entry into regular unions safer from an epidemiological perspective.

John Caldwell and Pat Caldwell argue that many African societies have long known about methods for abortion, but have utilized them infrequently until now, reserving them to prevent unwanted pregnancies in exceptional circumstances, such as when a girl became pregnant before her initiation or a married woman during her husband's absence. Though illegal in most African countries, the practice of abortion seems to be expanding today. The Caldwells' 1973 survey in Ibadan and their review of more recent studies suggest that abortion is largely an urban phenomenon. Those women most likely to resort to it fall into three major groups: schoolgirls who have not finished their studies, unmarried women, and married women whose husbands were absent when they became pregnant. The authors speculate as well that growing antinatalist messages will almost certainly increase not only the uptake of contraception but the abortion rate as well; both trends will probably be instruments of a future African fertility decline.

Using data from the innovative Côte d'Ivoire fertility survey, Dominique Meekers examines the conjugal correlates of child mortality. Although the persistence of unilineal kinship in most African societies might suggest that marriage is not necessary to legitimize a child, he finds that how a union is defined socially does seem to affect child mortality. Arguing that conjugal relationships contain discrete components, some of which are more important for child welfare than others, he shows that the parents' marital status may be less important for the child's survival chances than whether the mother and child are living with the father, or even with another man. The Côte d'Ivoire survey distinguished some of the key events of conjugal unions: the date of first cohabitation, the marriage ceremony, and first sexual relations with the first husband. The results indicate that children born before the woman begins cohabiting with a man may be 'legitimate' in some legal and social sense, including paternal recognition, yet they suffer higher mortality risks than children born after a more stable union is in place. Hence, marriage *per se* appears less important to the child's welfare than the presence of a stable male partner for the mother. The fact that even children who may have been fathered by a different man have lower mortality levels than those fathered by the woman's first husband suggests that the crucial factor with respect to the survival chances of the child is not biological fatherhood, but whether the mother is living with a man who, we assume, provides vital economic support for her and her children.

Discussion

The papers presented here describe some fascinating perspectives on the changing dynamics of marriage in Africa. The authors were in close agreement that the ambiguity and flux in nuptiality mean that the age at first birth as well as overall fertility rates may be at best tenuously related to reported ages at marriage. Hence, factors such as age at marriage or time spent in union may offer bleak prospects for identifying significant loci of fertility change. Most of the studies are also in agreement that although the number of children women bear has changed little, the reasons why people want children and the conditions under which wives are acquired are undergoing vast changes.

Just as the reasons why people want children may be changing, so are the values placed on different kinds of union and on children. Ranking among wives was clearly present in the past; a woman who came from a local 'landowning' family was more likely to emerge as a senior wife than one who came from a 'latecomer' family (see Kopytoff, 1987, for a general discussion of landowner-versus-latecomer status). But new criteria for ranking are arising based on access to education, development resources, international trade, and national political connections. Claude Meillassoux, a discussant at the seminar, drew forceful attention to these sweeping changes in nuptiality and to their economic underpinnings. He stressed that these criteria become encoded into the marriage system, making conjugal hierarchies more entrenched: élite women are more likely to become 'legitimate' or 'inside' wives, while the status of other women deteriorates.

'Legitimacy' is not important simply for prestige; it becomes the basis for disparity in resource allocations among the various women and their children who claim attachment to the same man. Since nuptiality is in greater flux, children increasingly comprise women's best claims on male resources. But whether a woman becomes a wife or remains an informal partner may spell the difference for herself and her children between full rights to the man's name and property, or none at all. Pressures to avoid connotations of non-legitimate ties may also induce women to terminate pregnancies by non-legitimate partners.

The growing resource gulfs between legitimate and non-legitimate unions bring us back to the question of polygyny which, though it appears to be declining, is in fact the key to the changes occurring in nuptiality and reproduction. The fact that new forms of polygyny are emerging throws into sharp relief the legitimacy status of different kinds of partners and children. While informal unions can be strategies for individuals to initiate potentially advantageous conjugal links, they can also comprise idioms for marginalizing some women and giving them and their children less recognition and support.

Meillassoux underscored these conclusions, arguing that escalating costs are deepening the disparities between 'inside' and 'outside' children, and forcing some women to market sex with ever more loosely attached partners in order to survive. As for children, some—usually those from already advantaged backgrounds—can move up the social and economic scale through education and prestigious marriages. Their upward mobility, however, masks an equally potent tendency for other children to slide down the scale. Certain children who become defined as 'illegitimate' or 'outside' children are more likely to be sent to caretakers at the social margins; they also run greater risks of becoming malnourished, and of producing 'outside' children themselves at early ages (see also Harrell-Bond, 1975: 134). Hence, parents continue to want children, but they—and men in particular—become more selective about investing resources in them and even about claiming them. Similarly, men now tend to select individual wives to retain and support more fully.

The fact that there may be little change in actual fertility levels but more careful targeting of monetary investments in children suggests the factors directly affecting the numbers of children may be less important at this juncture in African demographic history than the changing definitions of unions and reproduction. This is but one of the areas where collaboration between anthropology and demography can be most fruitful.

Notes

1. Committee members consisted of Gilles Pison (Chair), Peter Aaby, Lourdes Arizpe, Caroline Bledsoe, Monica Das Gupta, Tim Dyson, and Valerie Hull.
2. Evidence especially from the Demographic and Health Surveys shows that significant declines, of about one child per woman, have occurred in less than a decade in Kenya, Zimbabwe, and Botswana (Cohen, 1993: 41; Lesthaeghe and Jolly, 1992). Whether these declines will continue or whether other countries will follow suit is unclear.

References

Adegbola, O. (1987), 'A Comparative Analysis of Mortality of Children of Informal and Formal Unions', Paper presented at the IUSSP Seminar on Mortality and Society in Sub-Saharan Africa, Yaoundé, Cameroon.

Ariès, P. (1980), 'Two Successive Motivations for the Declining Birth Rate in the West', *Population and Development Review*, 64: 645–50.

Bongaarts, J. (1983), 'The Proximate Determinants of Natural Marital Fertility', in Rodolfo Bulatao and Ronald Lee (eds.), *Determinants of Fertility in Developing Countries* (New York: Academic Press), 103–38.

Borgerhoff Mulder, M. (1989), 'Marital Status and Reproductive Performance in Kipsigis Women: Re-evaluating the Polygyny–Fertility Hypothesis', *Population Studies*, 43: 285–304.

Burnham, P. (1987), 'Changing Themes in the Analysis of African Marriage', in D. Parkin and D. Nyamwaya (eds.) *Transformations of African Marriage* (Manchester: Manchester University Press), 37–54.

Caldwell, J. C., Hill, A. G., and Hull, V. J. (eds.) (1988), *Micro-Approaches to Demographic Research* (London: Kegan Paul International).

Clignet, R. (1987), 'On dit que la polygamie est morte: vive la polygamie!', D. Parkin and D. Nyamwaya (eds.), *Transformations of African Marriage* (Manchester: Manchester University Press), 199–209.

Cohen, B. (1993), 'Fertility Levels, Differentials, and Trends', in K. A. Foote, K. H. Hill, and L. G. Martin (eds.), *Demographic Change in Sub-Saharan Africa*, Panel on the Population Dynamics of Sub-Saharan Africa, Committee on Population, National Research Council (Washington, DC: National Academy Press), 8–67.

Comaroff, J. L., and Roberts, S. (1977), 'Marriage and Extra-Marital Sexuality: The Dialectics of Legal Change among the Kgatla', *Journal of African Law*, 211: 97–123.

—— (1981), *Rules and Processes: The Cultural Logic of Dispute in an African Context* (Chicago: University of Chicago).

Dinan, C. (1983), 'Sugar Daddies and Gold-Diggers: The White-Collar Single Women in Accra', in C. Oppong (ed.), *Female and Male in West Africa* (London: George Allen & Unwin), 344–66.

Garenne, M., and van de Walle, É. (1989), 'Polygyny and Fertility among the Sereer of Senegal', *Population Studies*, 43: 267–83.

Geertz, C. (1973), 'Thick Description: Toward an Interpretive Theory of Culture', in *The Interpretation of Cultures* (New York: Basic Books), 3–30.

Goldman, N., and Pebley, A. (1989), 'The Demography of Polygyny in Sub-Saharan Africa', in R. Lesthaeghe (ed.), *Reproduction and Social Organization in Sub-Saharan Africa* (Berkeley, Calif.: University of California Press), 212–37.

Goode, W. J. (1970), *World Revolution and Family Patterns* (New York: The Free Press).

Gyepi-Garbrah, B. (1985), *Adolescent Fertility in Sub-Sahara Africa: An Overview* (Boston and Nairobi: The Pathfinder Fund).

Harrell-Bond, B. (1975), *Modern Marriages in Sierra Leone* (The Hague: Mouton Publishers).

Henry, L. (1961), 'Some Data on Natural Fertility', *Eugenics Quarterly*, 82: 81–91.

Karanja, W. W. (1987), ' "Outside Wives" and "Inside Wives" in Nigeria: A Study of Changing Perceptions in Marriage', in D. Parkin and D. Nyamwaya (eds.), *Transformations of African Marriage* (Manchester: Manchester University Press), 247–61.

Kopytoff, I. (1987), Introduction, 'The Internal African Frontier: The Making of African Political Culture', in Igor Kopytoff (ed.), *The African Frontier: The Reproduction of Traditional African Societies* (Indiana: Indiana University Press), 3–84.

Lacombe, B. (1983), *Le Deuxième bureau: secteur informel de la nuptialité en milieu urbain congolais* (Paris: Stateco), 35.

Laslett, P. (1980), 'Introduction', in P. Laslett (ed.), *Bastardy and its Comparative History* (Cambridge: Edward Arnold), 1–68.

Lesthaeghe, R. (ed.) (1989), *African Reproduction and Social Organization* (Berkeley, Calif.: University of California Press).

—— and Jolly, C. (1992), 'The Start of the Sub-Saharan Fertility Transition: Some Answers and Many Questions', IPD-Working Paper, 1992–4, Brussels.

—— Kaufmann, G., and Meekers, D. (1989), 'The Nuptiality Regimes in Sub-Saharan Africa', in R. Lesthaeghe (ed.), *African Reproduction and Social Organization* (Berkeley, Calif.: University of California Press), 238–337.

Page, H. J., and Lesthaeghe, R. (eds.) (1981), *Child-Spacing in Tropical Africa: Traditions and Change* (London: Academic Press).

Parkin, D. (1988), 'Anthropological Perspectives on Changes in African Marriage', Paper presented at the IUSSP seminar on Nupiality in Sub-Saharan Africa: Current Changes and their Potential Impacts on Fertility, Paris.

—— and Nyamwaya, D. (eds.) (1987), *Transformations of African Marriage* (Manchester: Manchester University Press).

Pison, G. (1986), 'La Démographie de la polygamie', *Population*, 1: 93–122.
Radcliffe-Brown, A. R., and Forde, D. (eds.) (1950), *African Systems of Kinship and Marriage* (London: Oxford University Press).
van de Walle, É. (1968), 'Characteristics of African Demographic Data', in W. Brass (ed.), *The Demography of Tropical Africa* (Princeton, NJ: Princeton University Press), 12–87.
—— (ed.) (forthcoming), *The Cultural Roots of African Fertility* (Berkeley, Calif.: University of California Press).
—— and Kekovole, J. (1984), 'The Recent Evolution of African Marriage and Polygyny', Paper presented at the meetings of the Population Association of America, Minneapolis, Minn.
Wrigley, E. A. (1969), *Population and History* (New York: McGraw-Hill).

Part I

Demographic Overview

1 Post-Partum Abstinence, Polygyny, and Age at Marriage: A Macro-Level Analysis of Sub-Saharan Societies

RON LESTHAEGHE, GEORGIA KAUFMANN,
DOMINIQUE MEEKERS, JOHAN SURKYN

Introduction

Classic frameworks that seek to explain fertility, such as those concerned with the value of children or with household-level decision-making, commonly centre around individuals. This paper, by contrast, compares long-term patterns of social organization and exchange as institutional factors that have historically shaped reproduction. While intrasocietal micro-variations are by no means irrelevant to access and control over resources, properties of collectivities need to be understood before micro-level information can be compared across societies.

Since the survival of any society hinges on (1) its technological adaptation to a given environment, (2) its demographic balance between births, deaths, and migration streams, and (3) the economic principles governing its production and distribution, a comparative societal analysis needs to start from the links between production and reproduction. This is in fact a classic starting point: Thomas Malthus examined population dynamics in the context of moral economies of peasant societies 200 years ago; his notions of preventive and positive checks still remain relevant. Although most of his applications dealt with the Western-European system of the eighteenth century, it is now evident that all populations, from hunters and gatherers to more technologically advanced societies, regulate reproductive processes through specific cultural and economic arrangements.

The links between the organizing principles of a society and the specific parameters of the starting, spacing, and stopping patterns of fertility bear obvious importance for demographic variations at a particular point in time. They are of even greater value for understanding the various paths followed in the course of social change. Since changes in the spheres of political and economic organization, social stratification, division of labour, gender relations, cultural differentiation and demographic regulation are not likely to proceed in some uniform, harmoniously synchronized fashion, a clearer view of the historical links between 'production and reproduction' is essential to understand the new strategies adopted by both individuals and groups in resolving ensuing tensions (see also McNicoll, 1975: 89). Such an institutional approach calls for both (1) a better specification of what are sometimes called 'cultural entities', and (2) a comparative methodology which focuses on the holistic nature of societal traits (cf. Ragin, 1987).

Demography usually deals with ethnic, linguistic, or geographical aggregates. The problem typically arises when a simple nominal variable that identifies such aggregates appears to account for a large part of the variance of the independent variable, leading the investigator to conclude that 'cultural factors' are of greater importance than the predictors that have been used so far. To be sure, the forces at work may well be 'cultural', but there may also be other factors underlying the pattern. Hence, the observation that ethnic or regional patterning persists is of little use if no attempt is made to identify new variables that produce the ethnic or regional contrasts.

The preoccupation with micro-level data in demography during the last three decades has shifted attention away from aggregate properties (see also Ryder, 1980: 200 ff.). This leaves us ill-equipped to explain contextual effects when comparing micro-level analysis results across societies. We also need to formulate and test theoretical explanations about mechanisms that operate at the societal level. In this paper, the units are sub-Saharan ethnic groups, but no nominal variable that denotes ethnicity is used in the statistical analysis. Instead, ethnicity is translated into a configuration of organizational characteristics, such as patterns of lineage organization, the productive value of women in the various types of economies, etc. Because the configurations could have comprised other traits, our choice needs a theoretical underpinning.

In this paper, we shall make use of both regression and Boolean analysis. For regression analysis, the various traits that are used to give specific meaning to 'ethnicity' are treated as separate variables which enter classic multivariate regression analyses for individual impact assessment. In this form of analysis, large sample sizes are required for significance testing. In 'case-oriented' research, by contrast, we are interested in outcomes produced by *combinations* or *conjunctions* of causal factors. This logic is akin to Charles Ragin's 'multiple conjunctural analysis' which uses Boolean analysis of truth tables. In such an analysis the characteristics of a case are not broken down into separate variables but remain a single combination. Hence, we are primarily interested in interactions, but use the holistic character of the Boolean approach (combinatorial logic) to determine causes that are either:

1. necessary and sufficient (one cause alone produces the outcome);
2. sufficient but not necessary (a cause is capable of producing the outcome, but is not the only cause with this capacity);
3. necessary but not sufficient (a cause is capable of producing an outcome, but only in combination with other causes);
4. neither necessary nor sufficient, but non-redundant (a cause appears only in a subset of combinations of conditions that produce an outcome).

Finally, the outcome variables need specification. Indigenous sub-Saharan reproductive regimes are characterized by three main traits. First, *long birth intervals*, maintained primarily by long post-partum abstinence, have historically played a central role in reproduction. Second, the proportions of persons in *polygynous unions* are much higher than in any other continent. Third, *ages at first marriage* of both men and women are influenced by the formal demographic features such as those imposed by polygyny (see Goldman and Pebley, 1989). As a result, there is, contrary to the stereotype of universally early marriage, a considerable degree of *variation in ages at first marriage* for both sexes and in *husband–wife age differentials.*

The present paper attempts to assess the impact of societal configurations of traits in social organization (the translation of Murdock's ethnicity) upon the post-partum abstinence period and characteristics of the nuptiality regimes in sub-Saharan Africa. To this end we are using Murdock's files, otherwise known as Murdock's Ethnographic Atlas or the Human Relations Area Files.

The Theoretical Foundations of an Explanatory Framework

The major theories that seek to explain African patterns of nuptiality and the high incidence of polygyny draw attention to (1) the productive value of women in traditional economies, (2) the patterns of exchange of women between lineages and the ensuing rights *in uxorem* and *in genetricem*, and (3) the meaning of male gerontocracy for both young women *and* young men. Hence, several factors related to women's positions are of central importance.

The Productive Value of Women and Lineage Organization

Whilst anthropologists documented the existence and practice of polygyny in early research (cf. Radcliffe-Brown and Forde, 1950), they made little attempt to explain its occurrence. Although structural-functionalist anthropology drew elegant, holistic pictures of discrete societies ('case-oriented' research), it tended to avoid extensive comparative approaches. It was actually an economist, Ester Boserup (1970; see also 1985), who first attempted a major comparative overview of the link between polygyny and the productive value of women.

Boserup's thesis is straightforward: farming systems can be divided into 'male' and 'female' systems. In areas of low population density, such as sub-Saharan Africa, Boserup noticed that shifting cultivation using the hoe and female labour is widely practised. Women's high economic value, she reasoned, would be reflected in high levels of polygyny, fast remarriage for widows or widow-inheritance, and the payment of bridewealth (i.e. a marriage payment from the husband's lineage to the wife's lineage). Bridewealth negotiations in their turn produce a type of exchange with transactions spread over time, making marriage in sub-Saharan Africa more a 'process' than a well-demarcated event. Extensive child-circulation and fosterage also seem to be corollaries of Boserup's 'female system'. The alternative is a system where men do the agricultural work with more advanced technology (plough, irrigation), and women's reduced contribution is reflected in monogamy, less remarriage, and a dowry (i.e. the woman takes away a certain amount of wealth from her own father or kin). Goody (1976) put Boserup's thesis to the test and found that, in general, it held well when run against Murdock's worldwide ethnographic data bank (see Murdock, 1967*a*; 1975).

Goody's own thesis is similar to Boserup's and goes one step further. For Goody, the transmission of property becomes the key factor in the social and economic system. In low-technology hoe agriculture, land is corporately owned and its transmission is no grave issue for individuals or households since all lineage members have access to it. Marriage payments in such a system are less concerned with the transmission of property than with the exogamous exchange of both the productive and reproductive value of women between lineages. In systems with more intensive agriculture and more individualized forms of property, caste and social classes are established at birth and the system evolves away from the principle of circulation of resources towards the concentration of material wealth through monogamy. A father who wishes to protect his daughter's status with a dowry, but keep the family property intact, may solve the problem via endogamy. In-marriage implies strong paternal control with respect to partner choice and avoidance of 'mésalliances' or premarital pregnancies. In many instances that are *not* characterized by neolocal marital residence, this leads to marrying women off at young ages, the socialization of these young brides in the parental households of their spouses, and lowered female status.

Both Boserup and Goody divide the world essentially into two types of society: those with diverging devolution of property (i.e. women alienate parental capital via dowry and/or inheritance), monogamy, plough agriculture, complex polity and social stratification, and caste or class endogamy, and those societies with unilineal descent groups, bridewealth, polygyny, low-technology shifting agriculture, and a high female productive value.

These associations stand the test against the Murdock files (Goody, 1976), but when checked against a worldwide sample of societies, the strength of the associations for any pair of these organizational variables (as measured through phi-coefficients) is commonly of the order of 0.20 to 0.40. Stronger associations with a strength of 0.50 or more are absent. It should also be borne in mind that worldwide samples are strongly affected by the Eurasian versus sub-Saharan African typological watershed. Most Eurasian societies fall in the 'concentration' type, whilst the overwhelming majority of sub-Saharan societies fall in the 'circulation' category. As such, the associations measure a strong Eurasia-versus-Africa contrast, while Pacific and American populations complicate the matter.

Our goal of detecting the discriminating power of the Boserup–Goody variables *within* a sub-Saharan setting is rendered much more difficult by increased homogeneity. The exchange of bridewealth, for instance, constitutes by far the most common form of marriage prestation (84 per cent of the entries in our latest Yale Human Relations Files-based NUPFILE 3 sample for sub-Saharan Africa have substantial bridewealth), and widow-inheritance is almost equally widespread (78 per cent in NUPFILE 3). Goody's diverging devolution of property through dowries is absent, but it occurs in about a fifth of cases through inheritance rights for women. Tolerance and/or preference for endogamous forms of cousin marriage is more common in these societies, but the combination with inheritance for women is only found in 17 per cent of the cases of NUPFILE 3. Of the twenty-nine societies with both diverging devolution of property and endogamy, twenty-one are thoroughly Islamized, which explains the origin of this non-African combination. None of the remaining eight societies are patrilineal and hence are exceptional in more than one respect.

As Comaroff (1980) has pointed out, there is little utility in applying the Boserup–Goody distinction when so much of sub-Saharan Africa falls into a single category. Moreover, Goody (1973) himself, in an earlier critique of Boserup, stressed that female involvement in agricultural production itself would not suffice to explain marriage and polygyny patterns. As Comaroff underscores, the problem with assigning 'value' to women according to their productive potential begs the question of what kind of female and male activities are regarded as valuable. There may indeed be a crucial distinction between the actual economic contribution of women (the more formal definition used by Boserup) with respect to the production of means of subsistence, and its socially perceived value.

The elements to be added to our explanatory framework pertain precisely to these issues. Criticizing Boserup, Goody pointed out that in African societies where the two sexes are equally involved in farming or where men do more agricultural work, women often have other productive activities (i.e. other than housekeeping and associated tasks). More specifically, the Murdock files do not contain a variable that indicates any female involvement in trade, whereas it is obvious, especially from the West African experience, that female trading activity substantially increases both female economic value and status. Consequently, we enlarge our data-base with a 'women in trade' variable which has been constructed from survey results and an elaborate scanning of ethnographic references. Furthermore, in many African societies which rely to some extent on agriculture, animal husbandry is also practised; here the socially perceived value of female agricultural productivity is often lower relative to the socially inflated value ascribed to the exclusively male involvement in cattle-raising and -trading. This feature is typical of eastern- and southern-African populations, where cattle raising is common and where women rarely engage in trading activities. Hence, the presence of pastoralism or cattle-keeping is also introduced as a distinct variable in subsequent analysis.

Bearing these considerations in mind we attempt to operationalize the 'productive value of women in traditional economies'. First, societies with a strong involvement in pastoral activities (with more than 35 per cent reliance on animal husbandry as defined in the Murdock files; twenty-seven cases) are assumed to attach the lowest productive value to women's work in housekeeping or agriculture. At the next echelon, we have grouped societies that are more dependent on agriculture, where women do a substantial amount of such work, but where they are not customarily engaged in trading. Many such societies also have women's organizations, but these are generally restricted to mutual aid societies and agricultural work parties. The political power of these associations is limited. The third group, finally, is made up of populations with agriculture as a dominant sector and extensive engagement of women in trade. In many of these latter societies women gain political power through institutions such as:

- hereditary political positions (queen mothers, female regents in societies with large chiefdoms or state formation);
- eligibility for political office;
- multi-functional female organizations, sometimes culminating in the complete unfolding of the 'dual sex stratification system' with parallel male and female organizations at the various echelons;
- female secret societies and women's social organizations (title-taking societies);
- associations of market women with their own elected officers (market queens);
- forms of group defence via women's strikes, charivarying of men, lodging of complaints via women's organizations and their officers.

The distinction with respect to women in trade is more than a mere economic one: it is also a close proxy of the presence of institutionalized forms of political power and influence. Notably, this category distinguishes, although not perfectly, West-African societies from the others.

The hypotheses to be verified are that a *higher productive value* of women and/or *enhanced female economic autonomy and political power* are associated with:

1. *More polygyny*, since this institution is a form of appropriation of female productivity by a male gerontocracy (Boserup, 1970; Pison, 1984; and especially Clignet, 1970).

2. *Later female and especially later male ages at first marriage*, since women's lineage mates are more likely to consent to the transfer of a woman only after lengthy negotiations and payment of substantial bridewealth. The

formal demographic implication of a high incidence of polygyny is, furthermore, a substantial husband–wife age difference and further retardation of first marriages for men (Saucier, 1972).

3. *Greater incidence of long post-partum abstinence*, and consequently *more marked child spacing*; in a natural fertility population, a higher productive value of women would call for a balance in time allocation between female productive and reproductive activities, respectively. Furthermore, since polygyny facilitates post-partum abstinence (Talbot, 1926; Carr-Saunders, 1936; Herskovits, 1938; Dorjahn, 1959; Murdock, 1967b; Saucier, 1972), there is also an indirect effect on the productive value on women of abstinence via enhanced polygyny.

4. *More restricted transfers of both productive and reproductive rights in women*. This emerges in a variety of forms: the maintenance of ties between married women and their lineage mates, the return of women at the occasion of a first pregnancy or birth to their own kin, greater choice for widows when remarrying, absence of 'true levirate' (i.e. children born to widows belong to the deceased husband), easier divorce, etc. This implies that a woman's own kin keep a closer eye on her interests which, at least in patrilineal societies, facilitates the observance of prolonged post-partum abstinence.

5. Finally, societies with institutionalized forms of female solidarity and self-defence are more likely to protect women against male transgression of, among other things, long post-partum abstinence.

The next factor in our design is lineage organization. This factor operates through two major mechanisms. First, matrilineal societies have patterns of marital residence that are matrilocal or uxorilocal. This reduces the incidence of polygyny, because it is difficult for a man to bring a new wife home to his first wife's kin unless the two women are related. The correlates of less polygyny are, of course, also expected to emerge in matrilineal societies: later age at marriage for women, smaller husband–wife age differentials, and presumably greater pressure to shorten the period of postpartum abstinence. Further, a man would have greater difficulty in finding illicit partners in an environment that is monitored by his affines. Polygyny, if it occurs at all in matrilineal societies, is often associated with a change in marital residence later in life when a man might be able to set up neolocal or even patrilocal residence.

The second mechanism that moulds demographic characteristics in matrilineal societies is that the bond between women and their own kin remains strong, since the rights in children are not transferred upon a woman's marriage. Such better protection of women reinforces the tendency to preclude very early marriage and renders divorce more easy. Bridewealth payments tend to be smaller in matrilineal societies, a logical correlate of the restricted transfer of rights in women. This factor also permits easier divorce, since bridewealth may have to be returned in the event of divorce. In fact, when checked against the Murdock entries, the odds for matrilineal societies to fall in the category of societies with no or small bridewealth are 0.4 against 0.1 for patrilineal or bilateral societies (N = 11 and 51 respectively) (odds ratio = 4). Again using Murdock's classification, the odds for easy divorce are 3.3 in matrilineal societies (N = 17), 2.0 in societies with bilateral descent (N = 8) and only 1.4 in societies with patrilineal descent (N = 49). The odds ratio for easy divorce is more than 2 in favour of matrilineal societies compared to patrilineal ones.

To sum up, matrilineal organization, which, incidentally, is rarely found among African pastoralist societies, may reduce polygyny and the husband–wife age differential, limit the period of post-partum abstinence, preclude very early marriage for women, and facilitate divorce.

Islamic and Christian Penetrations and Female Literacy

The third factor we introduce is alterations in social organization produced by Islamic and Christian penetrations. Since these are historically grafted on to a prior mode of production and lineage organization, we have inserted them at a subordinate location in the hierarchical unfolding of causal combinations. Before discussing their impact upon reproduction, a number of clarifications are in order. First, a further distinction has been made for Islamic societies between those that have diverging devolution of major movable goods (cattle) or real property through an inheritance for women in conjunction with preferential parallel cousin marriage of the father's-brother's-daughter type, and those that have not.

These major deviations from general sub-Saharan rules of no diverging devolution through women and strict lineage exogamy tend to occur in populations that are either Hamitic or have been Islamized prior to the nineteenth century. Many of them have also been Islamic conquerors. Goody (1976) takes parallel cousin marriage of the FaBrDa type as a strong indication of Arabization or profound structural Islamization. Typically these populations also had a caste system consisting of free persons and slaves, and have on occasion introduced the custom of seclusion for women among their élites. Geographically, Islamic groups with diverging devolution and both caste and lineage endogamy live at the northern edge of the Sahel, from Mauritania to Lake Chad and further towards the Red Sea or along the northern Pacific coast (e.g. Zanzibar). Many of them are pastoralists (e.g. Delim, Zenaga, Borroro Fulani, Tuareg, Chuwa Arabs, Baggara [or 'black'] Arabs, Ogaden Somali), their neighbours (Wolof, sedentary Fulani, Soninke, Hausa, Kanuri, Kanembu, Hasania), or mixtures with Arab traders (Bajun, Hadimu). Other Islamized populations in sub-Saharan Africa, by contrast, are more typically farming populations that have undergone conquest by Islamic pastoralists, or populations that acted as intermediaries in the slave trade. Their conversion to Islam was more recent and often carried the connotation of self-protection. Consequently, Islamization has not altered their typical sub-Saharan traits such as lineage exogamy, lack of diverging devolution of property, and absence of caste formation. A few have even maintained matrilineal characteristics. Hence, the distinction that we make within the cluster of Islamized populations centres less on religious criteria than on the pervasiveness of patterns of social organization which are presumably brought in from outside via religion.

The effects of Christian penetration are, if anything, even more complex. In general, we distinguish between mainstream Christian religions (Catholicism or European-type Protestantism) and African churches. The category of 'Christian' used here refers to populations that predominantly belong to mainstream Christianity. African churches, by contrast, have strong syncretic, prophetic, and faith-healing elements and are grouped into the category of 'traditional religion' together with populations that have maintained more indigenous religious systems. Many African churches have also reacted strongly—and publicly—to Christian restrictions on polygyny. Virtually all ethnic groups have maintained a strong ancestral component, but this fact does not affect the main distinctions we establish here. More important is the strong degree of fragmentation of all these churches, even within areas occupied by a single ethnic group, and our resulting difficulty in producing a plausible categorization of the dominant religion. Nevertheless, it is worthwhile scanning the ethnic-specific literature on the history of Christianization for further clues, and to tabulate survey data with rough classifications by religious group.

On the whole, survey data are useful in assessing the dominance of Islam and Catholicism, but are less precise on Protestantism, and extremely diffuse with respect to the distinction between syncretic churches and indigenous African religions. Ethnographic descriptions, on the other hand, give more detailed information on the various churches and sects, but seldom indicate frequencies of membership. As a result, many cases are borderline ones and the classification adopted here is far from airtight.

The expectations with respect to the effect of religion on the components of the demographic regime can be summarized as follows:

1. *Traditional religion* in its indigenous or syncretic form, because of its greater association with African institutions, will be associated with a *higher incidence of long post-partum abstinence* and a *higher incidence of polygyny*, *long periods of lactation*, and (therefore) the *most intact overall pattern of child-spacing*, a feature that was borne out by WFS data for a subsample of predominantly West-African ethnic groups (Lesthaeghe *et al.*, 1989).

2. *Christianity* should be associated with a *reduction in formal polygyny*, whereas *Islam* is especially likely to be connected with a *reduction* in *post-*

partum abstinence (cf. the Islamic prescription of 40 days' abstinence only) (see Schoenmaeckers *et al*., 1981).

3. The ages at first marriage for the two sexes and the husband–wife age differential are likely to be influenced not only by differential polygyny, but also by the very strong impact of the two major religious penetrations on literacy levels and female literacy in particular. Obviously, Christianity is associated with higher-than-average literacy and Islamization with lower levels.

Some of the religion-connected gaps have been reduced by the post-independence drive towards more universal primary education, but the distribution of female literacy levels by religion was still very distinct in the 1970s and early 1980s.

Limited diversity

The problem of limited diversity implies that any analysis based on interactions of causal factors requires a *complete* unfolding of *all* possible combinations. More often than not, fewer combinations are found in reality than are theoretically possible, generally for good reasons.

In our application, using societies listed in the Human Relations Area Files, limited diversity emerges very clearly. We should have 3 (productive value of women categories) × 2 (lineage organization) × 4 (religion) × 2 (literacy) = 48 logical combinations, whereas we have only 20 combinations in the societies represented to work with. Two substantive reasons have emerged:

1. There are no matrilineal pastoral societies in the sample. Because lower female productive value is associated with a full transfer of rights in women, the non-transfer of reproductive rights, typical of matrilineal societies, is less likely to occur in pastoral societies than in others.

2. The very strong connection between Islamization and low female literacy leaves insufficient room for an adequate assessment of the literacy effect in Islamic societies. Moreover, Muslim women with advanced schooling are very atypical in all respects. As a result of this strong selectivity, regression coefficients of length of education tend to be magnified in such societies, especially when compared to societies with a more even distribution of women across schooling categories.

To sum up, the related problems of limited diversity and selection seriously limit our capacity for comparison. The prior conditioning factors in a structure that produce limited diversity and differential selectivity are of paramount importance, and it is advisable to limit comparisons within groups that are homogeneous with respect to the higher-order conditioning factors, rather than to attempt comparisons across such conditioning contexts.

The NUPFILE-3 Data Base

The NUPFILES are based on a linkage between ethnographic information concerning features of social organization and demographic information for ethnic groups or regions inhabited by such groups. The nature of the data sources has implications for (1) the time reference of the variables and (2) the structure of the file.

The ethnographic information stems essentially from George Murdock's compilation, called the 'Ethnographic Atlas' (also known as the Yale Human Relations Area Files), and further addenda based on relatively early ethnographic studies, typically dated prior to 1970. This information is essentially historical since it tends to describe societies before they were affected by major structural alterations associated with rising literacy, urban growth, market-oriented capitalist production, social class formation, privatization of land, etc. By contrast, the female literacy variable and the nuptiality measures (including polygyny) are dated from the last available census or survey: typically between 1960 to 1982. The difference in time reference between the conditioning factors and most dependent variables implies that we can only try to discern whether there has been any lasting effect of prior forms of social organization upon the nuptiality system; besides proxies such as female literacy, we simply lack the proper tools to assess the impact of more recent forms of structural reorganization. Even published census materials seldom allow us to make important distinctions (e.g. wage sector, subsistence sector, entrepreneurial self-employment, size of informal urban sector by sex and ethnic group or region) that could comprise relevant indicators.

Post-partum abstinence, though it should be measured through recent survey information, is only available for a subsample of populations (see Eelens and Donné, 1985, for the World Fertility Survey results of abstinence by ethnic group). Here we analyse the older Murdock information and the addenda produced by Schoenmaeckers *et al.* (1981). The WFS-subsample is analysed elsewhere (Lesthaeghe and Eelens, 1985; 1989). A comparison between Murdock's results and the 1976–82 measurements in the WFS-subsample shows that the original east–west divide and the Islamization effect towards shorter post-partum abstinence are still clearly noticeable. In fact, the characteristics of indigenous social organization in the WFS-subsample are much stronger determinants of the *recent* length of abstinence than are elements of socio-economic development. In other words, most societies have gradually moved towards shorter abstinence, but the patterns on Murdock's original historical map have not been significantly reordered or restructured. The same does not hold for breast-feeding durations and modern contraception, two variables with presumably little historical variation (i.e. breast-feeding was universally long and contraception was universally absent). For breast-feeding duration and contraception, indigenous aspects of social organization associated with ethnicity have a weak discriminating effect, whereas measures of socio-economic development (literacy, urbanization, employment sectors) explain most of the variance within and between ethnic clusters (see Lesthaeghe and Eelens, 1989; 81 ff.).

In terms of its structure, NUPFILES is based on a nested design, and the unit of analysis is an ethnic cluster comprised of a variable number of ethnicities. The clusters correspond, with one exception, to Murdock's aggregation (their location is shown in Lesthaeghe, Kaufmann, and Meekers, 1989: 294–5). The reason for using this larger unit is that clusters produce maximal contrast with respect to patterns of organization. Internally, a cluster is made up of ethnic groups that are linguistically, culturally, and organizationally similar. If these smaller units were maintained as units of analysis, we would have weighted the outcome in favour of ethnic clusters that contribute more ethnic groups. Instead, we have weighted the latter by a fraction equal to $1/n$, where n is the number of ethnic groups in the cluster. The total weighted sample size is 68 clusters, but the file contains information for 170 ethnic groups.

In several censuses and all WFS-surveys, the ethnographic information is matched by demographic information by ethnic group. But in many censuses and a few other surveys, the demographic information is only available by region. As a rule, we use such regional information for those areas that are ethnically homogeneous. When an ethnic group is dominant in adjacent regions, we link the ethnographic information to all available regional data, and reduce the ethnic weight from $1/n$ to $1/n*r$, with r being the number of representative regions. The nested design also solves the problem of analysing groups that share a region, provided that they belong to the same ethnic cluster. In most cases, the ethnographic data for cluster members that share a region are identical with respect to the independent variables. For other traits of social organization in NUPFILE-3 and at higher levels of detail, this is not necessarily so. The instance of two ethnic groups belonging to different ethnic clusters but sharing a region remains without solution; if no representative region can be identified for each, they are eliminated from the file.

Finally, NUPFILE-3 differs from earlier versions as a result of additions (e.g. religion, women in trade, variables pertaining to transfers of rights in women) and upgrading (more recent or better matching demographic information). Data from the Demographic and Health Surveys (DHS) conducted in the late 1980s are not yet incorporated.

Results

The results for each of the dependent variables take the form of simple percentages or means, reported at each step of the branching process that describes the unfolding of causal configurations. Limited diversity produces an incomplete branching process and restricts comparisons. The results are displayed in Tables 1.1 to 1.5.

TABLE 1.1. *Incidence of the long post-partum taboo (1 yr. +) by characteristics of social organization*

	Economy and productive value of women	Lineage organization	Islamic and Christian penetrations		Female literacy (< or ≥ 50%)	
% with long post-partum abstinence taboo $\overline{X} = 55$ (WN = 54)	Pastoralist and hunter/gatherer 23 (10)	Pat./bilat.	Islam + DD	0 (1)	Low	
			Islam	7 (4)	High	
			Xian	39 (3)	High	
			Trad. r.	42 (2)	High	
	Agricultural, no women in trade 41 (24)	Mat. 22 (7)	Islam + DD	0 (1)	High	
			Islam	0 (1)	Low	
			Xian	50 (1)	High	
			Trad. r.	62 (1)	High	
		Pat./bilat. 49 (17)	Islam + DD	57 (2)	Low	
			Islam	20 (1)	High	
			Xian	33 (7)	Low	63 (2)
					High	23 (6)
			Trad. r.	70 (7)	Low	88 (4)
					High	41 (3)
	Agricultural, women in trade 85 (20)	Mat. 25 (1)	Xian		High	
		Pat./bilat. 88 (19)	Islam + DD	57 (2)	Low	
			Islam	100 (3)	Low	
			Xian	84 (8)	Low	100 (2)
					High	80 (7)
			Trad. r.	100 (5)	Low	

Bilat. = bilateral. DD = Diverging devolution of property through women. Mat. = matrilineal. Pat. = patrilineal. Trad. r. = traditional religion.

Notes: 67 ethnic clusters (weighted sample of 170 ethnic groups).

In Table 1.1, the incidence of long post-partum abstinence rises dramatically as one moves from pastoral societies (23 per cent) to agricultural ones with women customarily in trade (85 per cent). Especially the presence of women in trade contributes to this increment.

At the next step, kinship organization, limited diversity emerges and no effect of matrilineal organization can be established for pastoralists. Among agricultural societies, the differences between patrilineal/bilateral and matrilineal societies are substantial and consistent with our hypothesis: long post-partum abstinence occurs much more frequently in patrilineal societies than in matrilineal ones.

TABLE 1.2. *Women aged 15 and over in polygynous unions by characteristics of social organization*

	Economy and productive value of women	Lineage organization	Islamic and Christian penetrations		Female literacy (< or ≥ 50%)	
% Women 15+ in polygyn. unions $\overline{X}$ = 55 (WN = 54)	Pastoralist and hunter/gatherer 27 (13)	Pat./bilat.	Islam + DD	13 (2)	Low	
			Islam	23 (5)	Low	
			Xian	30 (3)	High	
			Trad. r.	40 (2)	High	
	Agricultural, no women in trade 33 (31)	Mat. 27 (10)	Islam + DD	23 (1)	High	
			Islam	27 (3)	Low	
			Xian	27 (2)	High	
			Trad. r.	28 (3)	High	
		Pat./bilat. 36 (22)	Islam + DD	40 (2)	Low	
			Islam	42 (1)	High	
			Xian	31 (9)	Low	37 (3)
					High	31 (6)
			Trad. r.	39 (9)	Low	39 (6)
					High	39 (4)
	Agricultural, women in trade 44 (24)	Mat. 39 (1)	Xian		High	
		Pat./bilat. 44 (23)	Islam + DD	44 (3)	Low	
			Islam	49 (6)	Low	
			Xian	38 (9)	Low	39 (2)
					High	38 (7)
			Trad. r.	49 (5)	Low	

Notes: As Table 1.1.

When we come to religion, the effects on reproductive variables become more complicated and regularities less obvious. In all cases, however, indigenous religion is associated with the highest incidence of long post-partum abstinence, and Christianity tends to come second. As hypothesized, Islam (with or without diverging devolution of property through women) has a strong negative effect on long abstinence, but only if societies are pastoral or matrilineal and women do not trade. For patrilineal agricultural societies, the effects of Islam and diverging devolution are no longer obvious when compared to those of Christianity.

Finally, female literacy has a strong negative effect on long abstinence in patrilineal, non-Muslim societies (limited diversity because of the lack of

TABLE 1.3. *% Women 15–19 still single by characteristics of social organization*

	Economy and productive value of women	Lineage organization	Islamic and Christian penetrations		Female literacy (< or ≥ 50%)	
% single among women 15–19 $\overline{X}$ = 55 (WN = 65)	Pastoralist and hunter/gatherer 54 (11)	Pat./bilat.	Islam + DD	72 (2)	Low	
			Islam	39 (4)	Low	
			Xian	69 (3)	High	
			Trad. r.	49 (2)	High	
	Agricultural, no women in trade 56 (30)	Mat. 58 (10)	Islam + DD	65 (1)	High	
			Islam	53 (3)	Low	
			Xian	69 (2)	High	
			Trad. r.	53 (3)	High	
		Pat./bilat. 55 (21)	Islam + DD	27 (2)	Low	
			Islam	47 (1)	High	
			Xian	63 (9)	Low	45 (2)
					High	71 (6)
			Trad. r.	54 (9)	Low	43 (5)
					High	70 (4)
	Agricultural, women in trade 54 (24)	Mat. 66 (1)	Xian		High	
		Pat./bilat. 53 (22)	Islam + DD	27 (3)	Low	
			Islam	49 (6)	Low	
			Xian	68 (8)	Low	25 (2)
					High	78 (7)
			Trad. r.	50 (5)	Low	

Notes: As for Table 1.1.

high-literacy Muslim societies), but the difference according to literacy is substantially reduced if women are also active in trade. In the latter instance, virtually all societies fall in the long-abstinence category.

The analysis for the other dependent variables is analogous, so we proceed with the summary given in Table 1.6.

The main findings are:

1. A higher productive value of women in indigenous African economies is very strongly associated with a higher incidence of both polygyny and long post-partum abstinence. This is entirely in line with Boserup's prediction, and holds just as firmly within sub-Saharan Africa as across the Eurasian–African cultural watershed.

TABLE 1.4. *Male v. female singulate mean age at first marriage by characteristics of social organization*

	Economy and productive value of women	Lineage organ- ization	Islamic and Christian penetrations		Female literacy (< or ≥ 50%)
Difference between m. and f. sing. mean age at 1st marriage $\overline{X}$ = 6.6 yrs (WN = 61)	Pastoralist hunter/ gatherer 6.5 (9)	Pat./ bilat.	Islam + DD	7.4 (2)	Low
			Islam	6.6 (4)	Low
			Xian	6.3 (2)	High
			Trad. r.	5.4 (1)	High
	Agricultural, no women in trade 6.0 (30)	Mat. 5.9 (10)	Islam + DD	6.6 (1)	High
			Islam	6.4 (3)	Low
			Xian	6.3 (2)	High
			Trad. r.	5.1 (3)	High
		Pat./ bilat. 6.1 (21)	Islam + DD	7.4 (2)	Low
			Islam	7.2 (1)	High
			Xian	5.5 (9)	Low 5.7 (3)
					High 5.4 (6)
			Trad. r.	6.1 (9)	Low 6.2 (5)
					High 5.9 (4)
	Agricultural, women in trade 7.4 (21)	Mat. 7.1 (1)	Xian		High
		Pat./ bilat. 7.4 (20)	Islam + DD	7.7 (3)	Low
			Islam	7.6 (6)	Low
			Xian	7.2 (8)	Low 7.9 (2)
					High 7.0 (6)
			Trad. r.	7.3 (4)	Low

Notes: As for Table 1.1.

2. The formal demography of polygyny dictates a large age gap between husband and wives and consequently a large difference in their respective ages at first marriage. The effect of productive value of women on polygyny should therefore be associated with sex differences in proportions single (SMAM-difference: singulate mean age at marriage). The results support these speculations, but polygyny is unlikely to be the only factor involved. The larger SMAM-difference only occurs in agricultural societies where women are also engaged in trade. Furthermore, the larger SMAM-difference in societies with the highest productive value of women is not the outcome of earlier marriage for women, but of later marriage for men. As both Clignet and Saucier stressed, the burden of high polygyny in such gerontocratic

TABLE 1.5. *Men 20–4 still single by characteristics of social organization*

	Economy and productive value of women	Lineage organization	Islamic and Christian penetrations		Female literacy (< or > 50%)	
% single among men 20–4 $\overline{X}$ = 68 (WN = 61)	Pastoralist and hunter/gatherer 67 (9)	Pat./ bilat.	Islam + DD	83 (2)	Low	
			Islam	61 (4)	Low	
			Xian	73 (2)	High	
			Trad. r.	61 (1)	High	
	Agricultural, no women in trade 65 (30)	Mat. 65 (10)	Islam + DD	74 (1)	High	
			Islam	66 (3)	Low	
			Xian	73 (2)	High	
			Trad. r.	56 (3)	High	
		Pat./ bilat. 65 (21)	Islam + DD	61 (2)	Low	
			Islam	70 (1)	High	
			Xian	65 (9)	Low	58 (2)
					High	68 (6)
			Trad. r.	65 (9)	Low	60 (5)
					High	71 (4)
	Agricultural, women in trade 74 (21)	Mat. 77 (1)	Xian		High	
		Pat./ bilat. 74 (20)	Islam + DD	63 (3)	Low	
			Islam	73 (6)	Low	
			Xian	79 (8)	Low	64 (2)
					High	83 (6)
			Trad. r.	71 (4)	Low	

Notes: As for Table 1.1.

societies falls on younger men, since elders restrict the access to the highly valued productive and reproductive capacities of women.

3. The correlates of matrilineal kinship are straightforward in *all* agricultural societies: it is associated with a lower incidence of polygyny and with shorter periods of post-partum abstinence, and furthermore with later ages at marriage for women. Since there is no apparent association with male ages at marriage, matrilineal kinship organization is also related to smaller SMAM differences. Hence, the special forms of marital residence and the retention of productive rights in women are noticeably associated with reproductive regimes. They would help explain the anomalies—later age at marriage for women, smaller husband–wife age differences—observed in Ghana and the Ivory Coast (presence of the Akan cluster) and along the well-known

TABLE 1.6. *Effect of variables on post-partum taboo, polygyny, SMAM, and % of single women, 15–19*

	Effect of greater female productive value	Effect of matrilineal kinship organization (if non-pastoral)	Effect of religious infiltrations (Islam, Christianity, against traditional religion)	Effect of higher female literacy (if non-Muslim)
Long post-partum taboo	Incidence of long taboo *strongly* increases with female productivity value	Matrilineal organization clearly reduces incidence of long taboo	(1) Long taboo most prevalent if *traditional religion* (2) Islam reduces incidence of long taboo *if pastoral or matrilineal*	Strong negative effect of female literacy on incidence of long taboo
Polygyny	Polygyny *strongly* increases with higher productive value of women	Matrilineal organization reduces polygyny	(1) Islam reduces polygyny if pastoral (2) Diverging devolution reduces polygyny in Islamic societies (3) Christianity reduces polyg. if agriculture dominant and patrilineal organization (4) Polygyny strongest if trad. religion	Tiny negative effect of female literacy

SMAM* difference	Largest SMAM-difference *if* women in agriculture + trade	Matrilineal organization reduces SMAM-difference	(1) Islam produces largest differences (2) Diverging devolution further increases SMAM-differences (3) Christianity generally smallest difference	Small negative effect of female literacy on SMAM-difference
% single women 15–19	No effect	Matrilineal organization increases % still single (later marriage)	(1) Islam has lowest % single (earliest marriage) (2) Diverging devolution lowers % single further if agricultural + patrilineal organization (3) Christianity generally highest % single	Strong positive effect of female literacy on higher % single (later marriage)

* Singulate mean age at marriage.

matrilineal belt of central Africa that runs from the Atlantic coast (Bas Zaire, northern Angola) via Kasai, Zambia, and Malawi to the Pacific coast (the Yao-Makonde cluster of southern Tanzania and northern Mozambique).

4. Both Christian and Islamic penetration appear to suppress the incidence of polygyny and long abstinence (for the latter finding, see also Schoenmaeckers *et al.*, 1981). Conversely, the survival of traditional religion and syncretization seem to have protected polygyny and post-partum abstinence.

5. Islamization and Christianization, however, appear to have had strongly opposite effects on female ages at first marriage and on the SMAM-difference (i.e. age at first marriage differences between the sexes), while indigenous religion holds the middle ground. Islam is associated with early marriage for women and produces the largest SMAM-difference, irrespective of mode of production or kinship system. Moreover, the presence of diverging devolution of property through women enhances the SMAM-difference even more and seems to produce even earlier marriage for girls in Islamic patrilineal agricultural societies. This trait was predicted by Goody: Islamized societies tend to enhance control over women by keeping them married during most of the reproductive age span. (See also Lesthaeghe, Kaufmann, and Meekers, 1989.) There are two caveats, however.

6. The first caveat involves the existence of several Muslim pastoralist groups with diverging devolution, caste endogamy, and preferential parallel cousin marriage (Goody's indicators), which have exceptionally late ages at marriage for women. This is coupled with large SMAM-differences, since ages at marriage for men are much later still. The late age at marriage for women is consistent with the low incidence of polygyny, but otherwise the whole pattern is very different from that of agricultural societies in sub-Saharan Africa that are Muslim and patrilineal.

7. The second caveat with respect to Goody's hypothesis concerns the impact of religion via female literacy. Societies with predominantly Christian adherents, as indicated, generally have the highest ages at marriage for women and smaller SMAM-differences than the others. However, if we compare only low-literacy Christian societies with other low-literacy groups, much of the presumed Christianity-effect on female age at marriage disappears. Instead, the association seems to occur indirectly via the historical connection between Christianity and the promotion of schooling for women on the one hand, and negatively on the other, through the association between Islam and low literacy. (Goody's theory could be justified if we interpret the substantial lag with respect to female education other than Koranic as a manifestation of strong institutional control over women in Muslim societies.)

8. Up to a point, the incidence of polygyny and long post-partum abstinence are clustered. They are also associated in similar ways with indigenous modes of production, kinship organization, and religious penetrations. The association breaks down, however, when the effect of female literacy is introduced. Female literacy produces no variation in polygyny, which is consistent

with the observation that, at least until the early 1980s, polygyny did not seem to be disappearing from the African scene. But female literacy is associated with a lower incidence of long post-partum abstinence, a finding that is consistent with observations at the micro-level.

Finally, as we have shown elsewhere, contrary to W. Goode's prediction of 1963 and other theories based on presumed 'Westernization' or 'nuclearization' of the African family, polygyny has not declined in the time-frame with which we are dealing, with the possible exception of urban polygyny which is being transformed into the practice of 'outside wives' (van de Walle and Kekovole, 1984; Clignet, 1987; Lesthaeghe, Kaufmann, and Meekers, 1989). Indeed, the ethnic and geographical patterning of polygyny has essentially remained intact. The same holds for post-partum abstinence: its decline seems universal, but the original ethnic and regional distribution has not altered in the process. Consequently, the historical weight of very long-standing modes of production and the roles they assign to women are still of major importance in accounting for the present ethnic differentiation in polygyny and post-partum abstinence. Moreover, since post-partum abstinence is significantly associated with child-spacing over and beyond the effect of lactation (see Lesthaeghe *et al.*, 1989), recent regional changes in this 'abstinence bonus' constitute a significant component of differential fertility at present. Since life expectancies vary little between eastern- and western-African societies, weaker post-partum abstinence followed by an earlier reduction of its duration in eastern Africa is partly responsible for record population growth rates of 3.5 per cent or more in that part of the continent.

The Boolean Approach

In contrast to the variable-by-variable approach, the Boolean approach derives a logically minimal equation describing the different combinations of conditions associated with a certain outcome. The final reduced equation is an explicit statement of multiple conjunctural causation. A saturated regression model and a truth-table have in common that all combinations of the predictors (preferably but not necessarily dichotomized) are analysed, but the dependent variable in regression analysis maintains its original format, whereas it is converted to a Success–Failure (or True–False) dichotomy in the Boolean approach. This *S–F* contrast implies that a set of contradictions may have to be resolved from the start. If a similar combination of predictors produces *S* in one case and *F* in the other, classic regression will retain this information and proceed with the proportions of all cases that produce *S* (or *F*) or with the odds for *S* (or *F*). In the Boolean design, the original proportions or averages are immediately converted into *S* or *F*, depending on an arbitrarily chosen cutting point. Our analysis has calculated the average outcome values for each causal combination, and although such subsamples contain contradictions, the outcome *S* has been assigned to combination averages that are greater than the population mean. Conversely, the outcome *F* stands for aggregated results lower than or equal to the grand mean. In this simplification, within-combination heterogeneity has been disregarded. One of the consequences is that Boolean analysis has no use for sample sizes which, in the regression approach, are of prime importance for the simplification of saturated models (i.e. elimination of certain second or higher-order interactions). In the Boolean system, simplification occurs through (1) minimization and (2) the use of prime implicants. The fundamental rule for minimization is as follows:

> If two Boolean expressions differ in only *one* causal condition and produce the *same* outcome (either *S* or *F*), then that causal condition is irrelevant and can be removed to create a simpler, combined expression.

For instance, if 4 conditions (coded as present or absent) are retained for analysis, 16 different combinations of them can be established (from *ABCD* or all present to *abcd* or all absent). For each of these the presence or absence of an outcome is noted (*Y* or *y*). This forms a truth-table which can be manipulated via Boolean minimization to eliminate conditions that yield contradictory results (i.e. if $ABCD \rightarrow Y$ and also $ABCd \rightarrow Y$, then $ABC \rightarrow Y$) and via the search for prime implicants to eliminate tautologies. The result is the most parsimonious expression of the conjuncture of conditions that yields either *Y* or *y*. The purpose of the search for prime implicants is to achieve maximal parsimony. The example in Table 1.7 will illustrate the procedure.

The outcome studied in Table 1.7 is whether the average proportion of women 15–19 still single for each of the combinations of causal factors falls above or below the grand mean. The presence of the causal factor is indicated by upper-case letters (e.g. *A* for agriculture as dominant mode of production) and its absence by lower-case letters. The averages presented along the branching process of Table 1.3 are recombined to produce the aggregated outcome for each of the observed combinations. The design of Table 1.7 links the presence of later marriage for women (*S*) to the combined presence or absence of agriculture (*A–a*), women customarily in trade (*T–t*), matrilineal kinship organization (*M–m*), and above average female literacy (*L–l*). Normally, we should expect the 16 combinations that can be derived from 4 dichotomized predictors. Limited diversity, however, restricts our set to only 9. The primitive Boolean expressions (i.e. the lines of Table 1.7) are minimized for each set of the outcomes (*F* and *S* separately). For instance lines 1 and 2 in Table 1.7 differ in only one condition (*M* versus *m*), so that they are combined in a shorter expression without that condition.

The structure of the 5 available primitive expressions leading to *S* in Table 1.7 is such that minimization produces $S = L$. Higher female literacy is a necessary and sufficient condition for later-than-average female marriage (but the statement assumes homogeneity of proportions single among all the

TABLE 1.7. *Truth-table for the prediction of above-average proportions of single women, 15–19*

	Agricultural *A a*	Women in trade *T t*	Matrilineal organization *M m*	Higher female literacy *L l*	Above $\overline{X}$ single *S F*
1	*A*	*T*	*M*	*L*	*S*
2	*A*	*T*	*m*	*L*	*S*
3	*A*	*t*	*M*	*L*	*S*
4	*A*	*t*	*m*	*L*	*S*
5	*a*	*t*	*m*	*L*	*S*
6	*A*	*T*	*m*	*l*	*F*
7	*A*	*t*	*M*	*l*	*F*
8	*A*	*t*	*m*	*l*	*F*
9	*a*	*t*	*m*	*l*	*F*

Notes:
Boolean minimization S
lines 1 & 2 *ATL* } *AL* }
3 & 4 *AtL* } } *AL* }
1 & 3 *AML* } *AL* } } *L*
2 & 4 *AmL* } }
5 *atmL* (but *a* implies *tm*) → *aL* }

$S = L$

Boolean minimization F
lines 6 & 8 *Aml* }
7 & 8 *Atl* } *ml* }
9 *atml* (but *a* implies *tm*) → *aml* } } *tl*
→ *atl* }

$F = l(m + t)$

Source: Table 3.

ethnic clusters with identical causal combinations). This is a much stronger statement than the one made in Table 1.6 where effects of variables other than female literacy also prove to be relevant. The Boolean procedure is geared to *eliminating variables* whereas regression is geared to *describing all variable-specific effects*.

If a complete truth-table is available, the expression for F is the logical equivalent of the expression for S (here this would be $F = l$), and the use of De Morgan's law permits a direct conversion. However, such logical consistency does not hold in Table 1.7 since we do not derive $F = l$, but $F = l(m + t)$. The condition is still necessary but no longer sufficient. The presence of the extra condition, i.e. $(m + t)$, stems from limited diversity, and the expressions that would have eliminated $(m + t)$ are not present in the truth-table. The common practice to restore logical equivalence consists of artificially assigning outcomes to the missing combinations. But if one has good substantive reasons for explaining limited diversity such a procedure is illogical. Alternatively, one simply accepts that the universe of combinations is more limited. In such circumstances we prefer to report both the expressions for *S* and *F*, although they may not be logically equivalent. Table 1.8 presents an overview of these results. From these we shall select the most striking ones.

The *F*-expressions for the incidence of polygyny, difference in male and female ages at first marriage (DIFSMAM), and percentages of single men aged 20–4, all contain the same necessary but insufficient condition: lower-than-average polygyny, smaller DIFSMAM and earlier marriage for men all require the absence of the high productive value of women in trade (*t*). These characteristics are therefore not likely to be encountered in West-African populations where women do engage in trading and have strong traditional women's organizations. However, the absence of women in trade (*t*) is not enough. To get at lower than the average incidence of polygyny, the expression is

$$F = t(M + r)$$

which means that the absence of women in trade has to be combined with (AND) matrilineal kinship organizations (*M*) OR the absence of traditional/syncretic religions (*r*). Conversely all causal combinations which do not contain *tM* OR *tr* will have an incidence of polygyny that is higher than 36 per cent. This fact is easily verified in Table 1.8. Hence, the Boolean outcome highlights very clearly the fact that the three polygyny-reducing factors in the specific combinations *tM* OR *tr* are behaving according to our theoretical expectations. Low polygyny can therefore be expected to be found in the central-African matrilineal belt (*tM*) and in many eastern-, southern-, and central-African populations (women not customarily in trade), including pastoral ones (*t* implies *a*), which have also been Islamized or Christianized to a significant extent (*tr*).

The outcome for DIFSMAM is highly analogous to the result for polygyny. Smaller DIFSMAM-values are again found in pastoral societies and agricultural societies without trading women that are also matrilineal (*tM*) or non-Muslim (*ti*). This excludes again most of West Africa (presence of *T* and/or *I*), but locates smaller values of DIFSMAM along the central-African matrilineal strip (*tM*) and in eastern-, central-, and southern-African ethnic clusters that have not been Islamized (*ti*). This description fits again the detailed geographical distribution of DIFSMAM-values that we have shown on maps presented elsewhere (Lesthaeghe, Kaufmann, and Meekers, 1986; 1989).

The expression for earlier male first marriage restricts again its occurrence to pastoralist and agricultural societies without trading women (*t*):

$$F = t(AR + Al + aRL)$$

However, this condition needs to be combined with traditional religion OR low female literacy in agricultural societies ($A(R + l)$), or with traditional religion AND higher female literacy in pastoral societies (*aRL*). Like lower polygyny and shorter-than-average DIFSMAM, earlier marriage for men is

TABLE 1.8. *Boolean minimization of truth-tables for societal characteristics of 67 sub-Saharan ethnic clusters*

Dependent variable	Variables used in analysis						Boolean expressions ($S = > \overline{X}$; $F = < \overline{X}$)
	Agriculture *A* or *a*	Women in trade *T* or *t*	Matrilineal organization *M* or *m*	Traditional religion *R* or *r*	Islam *I* or *i*	Female literacy *L* or *l*	
% women 15–19 still illiterate; $\overline{X} = 55$	X	X	X			X	$F = l(m + t)$ $S = L$
% women in polygynous union $\overline{X} = 36$	X	X	X	X			$F = t(M + r)$ $S = T(r + m) + Rm$
DIFSMAM $\overline{X} = 6.6$ yrs.	X	X	X		X		$F = t(M + i)$ $S = T(m + i) + Im$
% men 20–4 still single; $\overline{X} = 68$	X	X		X		X	$F = t(AR + Al + aRl)$ $S = T(l + r) + rL(A + t) + arl$
% with long post-partum taboo $\overline{X} = 68$	X	X	X	X			$F = a + Ar(M + t)$ $S = A(Tm + Rm + Rt)$

not a typical West-African trait, but is predominantly found elsewhere given the presence of traditional religion or low female literacy, i.e. in eastern-, southern-, and central-African agrarian societies, which by virtue of the retention of traditional religion or low female literacy belong to the least developed category ($A(R + l)$). This finding can be linked to the general upward trend in male ages at marriage (cf. Lesthaeghe, Kaufmann, and Meekers, 1986; 1989). Presumably, fast population growth during the last twenty-five years and higher literacy have been factors associated with this upward trend, especially in areas where young men cannot be absorbed by the economy. Apart from West Africa, where male ages at marriage are customarily higher than elsewhere, the link between socio-economic change and later male marriage implies that the areas with early marriage are shrinking, and that the ones which are left over are inhabited by populations that have not been integrated in a wider national economy to the same extent (cf. conditions $R + l$). Finally, the few pastoralist ethnic groups that correspond with the condition *aRL* are located in the East African Rift Valley (Gogo, Ha of Tanzania, and Maasai of Kenya). The condition *aRL* sets them apart from other pastoralists where male ages at marriage are commonly much later.

The absence of long post-partum abstinence is indicated by the most complex Boolean expression

$$F = a + Ar(M + t)$$

which means that long abstinence is absent in (1) pastoralist societies (condition *a*) and (2) in agricultural societies that have *not* retained traditional religion (*Ar*) AND which are either matrilineal (*M*) OR do not have trading women (*t*). This excludes again much of West Africa which is known for long abstinence. However, the expression identifies several of the West-African exceptions as the matrilineal societies (e.g. the Akan, since they satisfy the condition *ArM*) and the West-African pastoralists (they satisfy *a*). The strong erosion of long abstinence is furthermore predicted for many societies in eastern and southern Africa (they satisfy *Art* or *a*).

Another model for the absence of long abstinence involving female literacy rather than agriculture yields:

$$F = L(Mr + tm) + tr$$

which highlights the erosive capacity of female literacy in matrilineal societies without traditional religion (*LMr*) and in the much larger group of patrilineal societies without women in trade (*Ltm*). The absence of long abstinence is also predicted in societies without women in trade (including pastoralists) but which have undergone Islamization or Christianization (*tr*). In this large category of societies, the erosion of long abstinence is independent of female literacy. Typical for this outcome is again the fact that the patrilineal societies with trading women are excluded from the *F*-expression and hence that rising female literacy in West Africa did not reduce abstinence below the duration of one year as easily as in the other parts of the continent. However, micro-studies in West Africa do identify urban subgroups of women with more than full primary education with shorter abstinence in this region (e.g. in Lagos).

A number of caveats are needed before concluding this section. First and foremost, the logical expressions presented here all depend on our arbitrarily chosen cutting point. Any other form of dichotomization (e.g. a cutting point for *F* or *S* set of the first or third quartile rather than the mean) would yield different expressions; but these would also have a different meaning, given a different criterion for dichotomization. The above analysis should therefore be taken as an illustration of the use of a uniform procedure of data reduction. Whereas we lose a substantial amount of information in the process, this is the price to pay for outlining the contours of the forest, despite the multitude of trees. We are well aware that many anthropologists with an eye for detail would not follow us in such a macro-level venture. But we take the view that the social organization characteristics of a particular case can be analysed more adequately if this case is compared to the overall pattern. Finally, the analyses performed so far are based on a sample of African societies, and this sample is by no means a random one. For many areas or ethnic groups, the required information is simply missing; we are only presenting the structures of overall relationships as they are emerging from the data available at present. This is obviously an invitation to revisit the subject once a more complete data set becomes available.

Conclusion

The aim of this paper has been to translate 'ethnicity' into a set of social organization characteristics, and to use these to explain patterns in ethnic variations with respect to demographic components. Despite classification problems and exceptions, the results indicate that there is considerably more order, even at the macro-level, than had been believed previously.

Virtually all predictors used here pertain to the position of women in African societies: they are related to the productive value assigned to women in indigenous economies, to the forms of exchange of rights in women between corporate lineages and to changes in these aspects produced by non-African influences. The concern with women *per se* was not part of our original plan and we did not go through any of the checklists of 'indicators of female status' (cf. Ware, 1981; Oppenheim-Mason, 1985; Safilios-Rothchild, 1986). The fact that precisely these social organization traits have emerged which appear to be structuring gender relations reflects the likely strength of such relations at the societal level.

The analytic features of the regression approach predominantly highlight the *variable-by-variable* effects. These can be straightforward (e.g. the effects of matrilineal kinship organization) or conditional on factors specified at a higher level in our hierarchical unfolding of patterns. The Boolean approach stresses the holistic nature of *configurations* in predicting the presence or absence of demographic traits, and proves particularly useful in accounting for geographical patterning. It is significant that the different perspectives adopted by these two approaches result in highly concordant conclusions.

As in our earlier work based on a smaller WFS-sample, we find that polygyny and long post-partum abstinence are strongly patterned by indigenous modes of production, lineage organization, and the Christian or Islamic penetrations, and less by aspects of recent socio-economic change (e.g. female literacy). Mean ages at marriage, however, show the opposite effects. Recent socio-economic changes associated with various forms of departure from the subsistence economy have increased ages at marriage for *both* men and women, so that the age-differences between the sexes have not been affected to the same degree. The effect of female literacy on DIFSMAM, for instance, is small even in Christianized societies. The maintenance of such age gaps in ages at marriage for the two sexes has undoubtedly facilitated the survival of polygyny. Moreover, wherever polygyny is receding from high levels, it seems to be replaced by the newer practice of keeping 'outside wives' or 'deuxièmes bureaux'. This stresses once more the fact that the Western pattern of conjugal marriage in nuclear households is not likely to be the immediate outcome.

How far have we come? In 1963, W. Goode, who was essentially looking at the same type of data as used here, seemed overwhelmed by the unstructured amount of diversity: 'The changes in African family systems are difficult to interpret adequately because comparable data do not exist for most societies, and the sheer diversity and number of tribes offer too many details to permit simple conclusions' (p. 200). But, one page further on, he offers a strong, but still conditional statement: 'If the new African nations follow the path of many other emerging nations, the next decade will witness an accentuated move away from *tribal family patterns*, and toward a *conjugal system*' (p. 202, our emphasis). Others have presented the same 'Westernization' hypotheses, though rarely in the conditional form used by Goode.

In sum, our systematic comparison of ethnic groups, involving the operationalization of ethnicity and a much broader demographic data base, does not substantiate expectations of any emerging signs of nuclear families or monogamy. Even factors such as the precarious economic situation of the 1980s, which hamper further increases in female schooling—a major correlate of rising female ages at first marriage—will have effects on marriage that will be difficult to predict. Although African nuptiality and family systems have undergone changes since 1963, these patterns are evolving along directions that have little in common with Western-style marriage. The study of sub-Saharan nuptiality systems, therefore, is likely to remain a challenging enterprise, not only because of numerous problems with definitions and measurement, but particularly because the changes we are witnessing are not following the predicted road towards a Western model. Although sub-Saharan

Africa is still firmly located at the other side of Boserup's or Goody's divide, retaining elements of long-standing patterns of surprising resilience, its populations are currently producing their *own* versions of new systems that have not yet fully crystallized.

Acknowledgements

This paper draws extensively on the data contained in the NUPFILES 2 and NUPFILES 3 data bank prepared by G. Kaufmann, D. Meekers, and R. Lesthaeghe. The extension of the NUPFILES has been possible thanks to project funding provided by the Rockefeller Foundation Grant RF 85050–221. The authors would also like to thank O. Frank, J. Caldwell, E. van de Walle, C. Bledsoe, John Massad, and an unknown reviewer for their comments and suggestions.

References

Boserup, E. (1970), *Women's Role in Economic Development* (New York: St Martin's).

—— (1985), 'Economic and Demographic Interrelationships in Sub-Saharan Africa', *Population and Development Review*, 113: 383–98.

Carr-Saunders, A. M. (1936), *World Population* (Oxford: Clarendon Press).

Clignet, R. (1970), *Many Wives, Many Powers: Authority and Power in Polygynous Families* (Evanston, Ill.: Northwestern University Press).

—— (1987), 'La Polygamie est morte, vive la polygamie', in D. Parkin, and D. Nyamwaya (eds.), *Transformations of African Marriage* (Manchester: Manchester University Press), 199–209.

Comaroff, J. L. (1980), Introduction, in J. L. Comaroff (ed.), *The Meaning of Marriage Payments* (New York: Academic Press), 1–47.

Dorjahn, V. (1959), 'The Factor of Polygyny in African Demography', in W. R. Bascom and M. J. Herskovits (eds.), *Continuity and Change in African Cultures* (Chicago: Phoenix Books, University of Chicago Press).

Eelens, F., and Donné, E. (1985), *The Proximate Determinants of Fertility in Sub-Saharan Africa: A Factbook Based on the Results of the World Fertility Survey*, IPD Working Paper 85–3, Vrije Universiteit, Brussels.

Goldman, N., and Pebley, A. (1989), 'The Demography of Polygyny in Sub-Saharan Africa', in R. J. Lesthaeghe (ed.), *Reproduction and Social Organization in Sub-Saharan Africa* (Berkeley, Calif.: University of California Press), 212–31.

Goode, W. (1963), *World Revolution and Family Patterns* (Glencoe, Ill.: The Free Press).

Goody, J. (1973), 'Polygyny, Economy and the Role of Women', in Goody, J. (ed.), *The Character of Kinship* (Cambridge: Cambridge University Press), 175–90.

—— (1976), *Production and Reproduction: A Comparative Study of the Domestic Domain* (Cambridge: Cambridge University Press).

Herskovits, M. J. (1938), *Dahomey* (New York: Augustin).

Lesthaeghe, R., and Eelens, F. (1985), *Social Organization and Reproductive Regimes: Lessons from Sub-Saharan Africa and Historical Western Europe*, Working Papers, 85–1, Vrije Universiteit, Brussels.

—— (1989), 'The Components of Sub-Saharan Demographic Regimes and their

Social and Cultural Determinants', in R. J. Lesthaeghe (ed.), *Reproduction and Social Organization in Sub-Saharan Africa* (Berkeley, Calif.: University of California Press), 60–116.

Lesthaeghe, R., Kaufmann, G., and Meekers, D. (1986), *Data Files NUPFILE1 and NUPFILE2*, IPD Working Papers, 86–4, Vrije Universiteit, Brussels.

—— (1989), 'The Nuptiality Regimes of Sub-Saharan Africa', in R. J. Lesthaeghe (ed.), *Reproduction and Social Organization and Reproduction in Sub-Saharan Africa* (Berkeley, Calif.: University of California Press), 238–337.

Lesthaeghe, R., Vanderhoeft, C., Gaisie, S., and Delaine, G. (1989), 'Regional Variation in Components of Child-Spacing: Measurements and Social Correlates', in R. J. Lesthaeghe (ed.), *Reproduction and Social Organization in Sub-Saharan Africa* (Berkeley Calif.: University of California Press), 122–61.

McNicoll, G. (1975), 'Community-Level Population Policy: An Exploration', *Population and Development Review*, 1: 1–21.

Murdock, G. P. (1967*a*), 'Ethnographic Atlas: A Summary', *Ethnology* 62: 109–234.

—— (1967*b*), 'Postpartum Sex Taboo', *Paideuma*, 13: 143–7.

—— (1975), *Outline of World Cultures*, 5th edn. (New Haven, Conn.: Human Relations Area Files).

Oppenheim-Mason, K. (1985), *The Status of Women: A Review of its Relationship to Fertility and Mortality* (New York: The Rockefeller Foundation), 1–86.

Pison, G. (1986), 'La Démographie de la polygamie', *Population*, 411: 93–122.

Radcliffe-Brown, A. R., and Forde, D. (1950), *African Systems of Kinship and Marriage* (London: Oxford University Press).

Ragin, C. (1987), *The Comparative Method: Moving Beyond Qualitative and Quantitative Strategies* (Berkeley, Calif.: University of California Press).

Ryder, N. (1980), 'Where Do Babies Come From?', in H. Blalock, Jr. (ed.), *Sociological Theory and Research* (New York: The Free Press), 189–202.

Safilios-Rothchild, C. (1985), *The Status of Women and Fertility in the Third World in the 1970–80 Decade*, Working Papers of the Center for Policy Studies, 118 (New York: The Population Council).

—— (1986), *Socio-Economic Indicators of Women's Status in Developing Countries, 1970–1980* (New York: The Population Council).

Saucier, J. F. (1972), 'Correlates of the Long Postpartum Taboo: A Cross-Cultural Study', *Current Anthropology*, 132: 238–49.

Schoenmaeckers, R., Shah, I. H., Lesthaeghe, R., and Tambashe, O. (1981), 'The Child-Spacing Tradition and the Postpartum Taboo in Tropical Africa: Anthropological Evidence', in H. J. Page and R. J. Lesthaeghe (eds.), *Child-Spacing in Tropical Africa: Traditions and Change* (London: Academic Press), 25–71.

Talbot, P. A. (1926), *The Peoples of Southern Nigeria* (London: Oxford University Press).

van de Walle, E., and Kekovole, J. (1984), 'The Recent Evolution of African Marriage and Polygyny', Paper presented at the Meeting of the Population Association of America in Minneapolis, University of Pennsylvania Population Studies Center, Philadelphia.

Ware, H. (1981), *Women, Demography and Development*, Development Studies Center, 3 (Canberra: Australian National University).

Part II

Age at First Marriage

2 Marriage Drinks and Kola Nuts

ÉTIENNE VAN DE WALLE AND DOMINIQUE MEEKERS

> . . . for if the parents agree to [the match], and eat a few *kolla-nuts*, which are presented by the suitor as an earnest of the bargain, the young lady must either have the man of their choice, or continue unmarried, for she cannot afterwards be given to another.
>
> (Mungo Park, 1799, about the Manding of present-day Mali)

Introduction

Demographic surveys in sub-Saharan Africa encounter special problems in defining marriage and measuring changes in nuptiality. These problems stem from patterns that are widespread in the region. In contrast with Western marriage which tends to be more unambiguously initiated by a ceremony or by signing a document, African marriage is often a process with a varying number of potential stages, from the beginning of courting to the customarily sanctioned wedding, through periods where sexual relations are allowed, and periods of full cohabitation. Payment of a bridewealth may also occur in stages. If all goes right, subsequent stages confirm the legitimacy of earlier ones, so that in retrospect it is not always clear when the decisive step was taken, when the clinching payment of bridewealth was made; but if the couple drift apart, the marriage, retrospectively, has never taken place.

Additional problems of definition arise because there is a wide diversity of forms of union, not only by ethnic groups but also within the same group by socio-economic status. The importance of a single event or ceremony which would clearly separate the time before marriage from the onset of married life varies from one person to another. Moreover, customs are rapidly evolving under the forces of modernization and the erosion of indigenous social controls. Education, urban living, and the opportunities they afford young women, and the growing importance of mutual consent in unions have all played roles. To complicate the picture, new imported forms of civil or religious marriage (which clearly involve a ceremony) are often preferred by the élites as more prestigious; sometimes they stand alone, and sometimes they legitimate an earlier customary marriage. Finally, to the extent that indigenous rites and ceremonies are used to establish marriage, their importance may have diminished in recent times (Verdon, 1983: 166).

A priori reasoning about trends in the age of unions in Africa can lead to contradictory hypotheses. On the one hand, increased education, particularly female education, has often led in other international contexts to a higher age at marriage. On the other hand, a streamlining of marriage arrangements and the emancipation of young people from the control of lineages may have led to a younger age at union in many urban contexts (see Robertson, 1984, e.g. p. 182). The age at first intercourse and first cohabitation may fall most rapidly among the educated, who have more opportunities to meet members of the opposite sex without family supervision.

Ultimately, questions about changes in the age at marriage and duration of union must be addressed statistically. For comparing and measuring such changes, unambiguous dates of marriage would be ideal. Moreover, the length of exposure to the risk of childbearing, which serves as denominator for computing fertility rates, could be easily measured if the definition of marriage rested on accurate dates. Several problems impede these tasks. Demographic surveys have elicited retrospective reports of the timing of events: an extremely dubious procedure when several events may have been crucial to the establishment of a union. Most statisticians are trained to expect that marriage will be clearly defined by a single event; our concepts are poorly adapted to the processual nature of African marriage.

Problems in measuring change in the age at marriage are intensified when we attempt to compare data from successive censuses or surveys. There is no guarantee that constant definitions of marriage are maintained over time. Although the accepted rule for defining marital status in censuses is to take the people's word for it, the perceptions of respondents are unlikely to remain constant when rapid social change is occurring. The World Fertility Survey introduced an even more liberal rule. Because analysing fertility was the underlying goal of asking about marriage, the Survey abandoned the focus on marriage as such to concentrate on stable unions in order to identify 'exposure to risk' of conceiving. As such, cohabitation was generally considered the event that determined the existence of a union. Using this indicator might allow us to establish trends in the age at which these particular events took place (assuming we could solve problems raised by difficulties of recall). Such trends, however, would not necessarily reflect parallel trends in the age at which some other event (e.g. age at first intercourse with the partner in that union) had occurred.

Various census and survey takers have tried to identify stages in the marriage process. The WFS itself, despite its insistence on international comparability, yielded to the temptation in one notable instance, the Côte d'Ivoire Survey of 1980–1. Besides asking questions on time of first intercourse and first cohabitation, the authors of the questionnaire attempted to identify a ceremony that would define the precise moment when people were married. This is relatively easy (leaving aside the exact meaning of that moment) when the couple have contracted a civil or religious marriage. Yet very few women had contracted such unions. The very innovative step taken by the Ivoirian statisticians, therefore, was to ask one further question to the many women who reported having contracted a customary union: 'In what month and year did your husband give the drink or the kola for the marriage?' The question seems to have elicited a response from every woman to whom it was asked; it did not include women who reported themselves in a civil marriage (unless this marriage had been preceded by a customary marriage) or in a free union. It is, of course, impossible to reconstruct the exact phrasing of the question. But if we can rely on the tabulated results, the husband had either given marriage drinks or kola nuts for every customarily married woman in the survey.

This paper picks up this thread of questioning that the Ivoirian questionnaire designers may have had in mind when they attempted to discern specific events that comprise a union or to document the prevalence of different kinds of unions. The analysis reveals significant differences among ethnic groups, even within the same country, in the order in which conjugal events (cohabitation, celebration, and consummation) and childbearing occur. Whether one's ultimate interest in nuptiality lies in the traditional demographic question of 'exposure to the risk of childbearing' or to more sociologically oriented issues of conjugal legitimacy, the resulting variances are too striking to be ignored.

Drinks and Kola Nuts

The ritual gift of drinks or kola nuts has been a long-standing component of marriage among many populations of the West African coast. We know of no systematic ethnography of the custom, but drinks and kola are mentioned in many accounts of the marriage process, and not only among populations of Côte d'Ivoire. For example, among the Ashanti of Ghana,

> The decisive formality for the establishment of a legal marriage is the giving of the *tiri nsa* (lit. head wine), which Ashanti describe as a thanking gift (*aseda*). Most commonly this consists of two bottles of gin or an agreed equivalent in cash. It is handed over on behalf of the husband by the head of his lineage to the head of the bride's lineage through her legal guardian in the presence of the representatives of both groups . . . The payment of *tiri nsa* gives the husband exclusive sexual rights over his wife and the legal paternity of all children born to her while the marriage lasts . . . *Tiri nsa* may be paid before or at any time after the couple begin to cohabit (Fortes, 1950: 279–80).

Writing about a related Akan-speaking group some twenty-five years later, Bleek (1976: 92) also makes reference to the *tiri nsa* which 'gives the man exclusive domestic and sexual rights over the girl'. Oppong (1981: 30) refers to it for Akan civil servants' marriages in Accra: 'A marriage contract is sealed by the giving and acceptance of drinks.' However, Bleek also describes an earlier stage in the marriage process, the onset of cohabitation, which is marked when the boy's father sends the girl's father a bottle of schnapps (called the 'knocking fee').

Conjugal customs involving drinks covers a wide variety of alcoholic beverages, though the specific practices and the drinks that are used constantly change. As early as the seventeenth century, Bosman (1705: 198) mentions palm wine and brandy as part of 'the Expence of the Wedding-Day' in the Gold Coast. Afzelius (1967: 126) describes a custom of the Susoo people of Sierra Leone in 1795, called 'drawing wine on a wife'; this consists in sending 'a present commonly a Baberoon [from the French *biberon*] to the Father whose wife is big [pregnant], saying in case the child should be a girl I wish to have it for wife'. Among the Ga of Accra in the late nineteenth and early twentieth century, the negotiations for a high status marriage were opened when the man sent, and the girl's parents accepted, a 'family drink' (*weku daa*); since World War II, the most common form of marriage is *afayinodaa*, 'offering a drink', which signals the beginning of cohabitation and indicates that the man is assuming paternity for the children to be born (Robertson, 1984: 180–1). Today, various European liquors such as schnapps, brandy, rum, or gin have taken on a quasi-ritual function in the marriage customs of various groups.

In non-Islamic Africa, drinks of various sorts appear to be associated with all the various stages of marriages: betrothal, the granting of sexual rights, the start of cohabitation, or the final step in the union. As a token of betrothal, they signify that a young man lays claim to a young woman, and that the latter's family has accepted the claim; thus, the 'knocking fee' in Ga marriage, and the 'Put Stop' [to the proposals of other suitors] among the Creoles of Sierra Leone (Harrell-Bond, 1975: 187). In certain regions, this may occur at a very young age of the future bride, or even before her birth. Meek (1937: 267–9) describes a custom of betrothal in childhood among the Ibo of Mboo, involving libations and drinks of palm wine. The gift of liquor may also initiate a period of sexual intimacy or even cohabitation, during which children born will be legitimate offspring of the man even though the marriage is not complete. And finally, it may be part of the ceremony of customary marriage.

Not only do the marriage drinks appear at various stages, they also play various roles and are used in different fashions. In some instances, they are poured on the ground as a libation; in Abutia Ewe marriage in Togo, 'the "wedding" essentially consisted of a "blessing ceremony" during which elders from both sides would pray with palm wine' (Verdon, 1983: 164). Among the Igbo of Nigeria, drinking in common seals the union of bride and groom:

> Before the father takes the bridewealth, he gives his daughter a cup of palm liquor and asks her to show her husband to the audience by giving him the liquor. The shy girl walks with faltering steps to her husband, sips the liquor, and as she gives it to him, tells her shouting audience: 'This is my husband. Father may take the bridewealth' (Uchendu, 1965: 51–2).

Among the Anlo Ewe, partaking of the drinks by the bride's relatives signifies their approval of the union, and 'a man may continue to give marriage drinks throughout his whole life, depending on when and where he meets affines who demand their drinks' (Nukunya, 1969: 96).

Kola nuts often play similar roles, particularly among Muslim populations where alcoholic beverages are not approved. When eaten in combination with drinks, they often represent unity and good faith. Among the Creoles of Sierra Leone, bottles of brandy are used both for pouring libations to the ancestors and entertaining guests, while kola nuts, symbolizing unity and long life, are among the gifts brought by the groom's family to 'The Gage' or engagement ceremony (Harrell-Bond, 1975: 189–90). Their role in Manding marriage was mentioned by Mungo Park (1799: 266), and it remains symbolically important today among Sahelian populations.

The question asked in the Côte d'Ivoire Survey ('In what month and year did your husband give the drink or the kola for the marriage?') seems to represent an attempt to use a widely recognized form of customary ceremony as a marker to indicate the date of celebration of marriage, and to distinguish unions that the kin groups recognize as legitimate from those that are not. However, the survey responses on marriage drinks and kola nuts may reflect more than one stage in the marriage process, rather than a ceremony identifying the exact time of marriage, and their meaning may vary somewhat among the ethnic groups and the social classes of Côte d'Ivoire.

The Côte D'Ivoire Survey

The Côte d'Ivoire Fertility Survey was conducted between August 1980 and March 1981 (Ministère de l'Economie et des Finances, 1984). It contains information on 5,764 women aged 15 to 49 years, of 'African nationality'. The flow of nuptiality questions is shown schematically in Fig. 2.1. Women were asked: 'Are you married?' or 'Have you ever been married?' Further probes elicited answers about other kinds of unions: 'Do you live with someone?' 'Have you ever lived with someone?' Since cohabitation ('living with someone') was the mark of a union in WFS methodology, the questions that served to establish the age at first union read as follows: 'In what month and what year did you and your husband (friend) start living together for the first time?' 'How old were you at that time?'

Once it had been determined that a first union had taken place, a distinction was made between civil marriages, customary marriages, and free unions. Women in a civil marriage were asked when their marriage was celebrated; women in a customary marriage were asked when their husband had given the marriage drink or the kola nuts. It was assumed that women who reported that their first union was a free union had not gone through a ceremony of any kind. The next set of questions concerned the start of

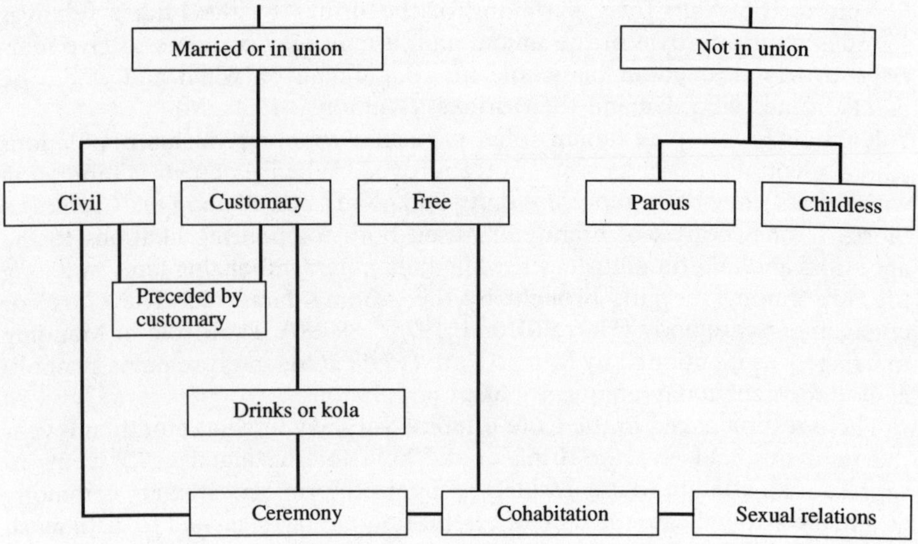

Source: Côte d'Ivoire Fertility Survey Questionnaire, 1980–1.

FIG. 2.1. Flow of questions used to derive the date of the ceremony, cohabitation, and consummation of the first union

cohabitation before marriage and of sexual relations before cohabitation. Thus the survey allows recognition of three stages in the union: (1) the onset of sexual relations with the partner in that union—what we call *consummation* for brevity, even though it can occur before the union's social recognition; (2) *cohabitation*; and (3) *celebration*, i.e. a marriage ceremony, defined as either a civil marriage or the giving of drinks or kola. There is no necessary normative sequence in these three stages. Moreover, there could be a number of women in the survey with truncated or incomplete sequences who were interviewed before they reached the stage of cohabitation that the survey acknowledged as a union.

When computing age at union, or other indicators involving the timing of the union (e.g. the interval between first birth and first union), the published results of the survey conformed to other World Fertility Surveys in using cohabitation as the decisive criterion for the existence of a union. But the ceremony, the drinking of wine, and the eating of kolas, often precede cohabitation. The Survey Report characterizes the intervening period as a period of *fiançailles*, i.e. betrothal. In other instances, there is a period of cohabitation *before* the sharing of drinks or kola. This was interpreted as a stage of free union, which could still lead to a regular (i.e. customary or legal) marriage. Finally, the ceremony can coincide with the beginning of cohabitation, though

Not in union		In union					
No child	With child	Free union	Customary union		Civil union		
			(1)	(2)	(1)	(2)	(3)
	Sex. rel.	[Sex. rel.]	[Sex. rel.]	[Sex. rel.]	[Sex. rel.]	[Sex. rel.]	[Sex. rel.]
		Cohabit	Cohabit	–	Cohabit	–	–
			Ceremony	Ceremony	Ceremony	Ceremony	–
				Cohabit	–	Cohabit	–
					Civ. mar.	Civ. mar.	Civ. mar.
							[Cohabit]

FIG. 2.2. Sequence of components in different types of union

the onset of sexual relations may or may not precede cohabitation. We have indicated the various possible sequences in Figure 2.2.

It is tempting to interpret each of these sequences in terms of the customs of various ethnic groups, although it is likely that new forms have been introduced by urbanization and modern living conditions, and that religion and education have played major roles in these changes. There are certainly populations where child marriage used to result in lengthy periods of betrothal, i.e. a formal, ritualized agreement preceding cohabitation. Among the Senoufo of Côte d'Ivoire, for example, Holas (1957) reported an average period of *fiançailles* of five years in the 1950s. For another Voltaic group, the Koulango of Nassian, the most common form of marriage recently still involved a lengthy betrothal during which the man had to work for his future father-in-law; there was no bridewealth, but the wedding ceremony was marked by great solemnity (Boutillier, 1971). Marriage as such was indissoluble, and the Koulango husband had a right to all the children born to his wife, either before marriage or after separation, extending to the time after his death. But most women spent a substantial part of their adult life in 'para-conjugal' unions with another partner than their husbands, either before marriage or after separation or widowhood.

One would expect cohabitation before marriage (to the same man) to be widespread. Among some Ivoirian groups (such as the Guéré; see Schwartz, 1971) bridewealth, which is high, is paid in instalments over a lengthy period. This may lead to elopements ending in ritualized marriage, and to temporary forms of unions that are confirmed and transformed into customary or civil marriages after further payments of bridewealth. Finally, there are marriages without cohabitation, or where cohabitation is a stage that intervenes only after several children have been born. Among the Baoulé, a matrilineal group of Akan origin, 'marriage is often initiated by a pregnancy':

TABLE 2.1. *Types of union, by current age of the woman* (%)

Age	Type of Union				No. of Cases
	Civil	Customary	Free	Single	
15–19	1.1	36.4	18.5	44.0	1,321
20–4	4.2	59.4	26.0	10.4	1,255
25–9	5.5	70.7	19.3	4.5	962
30–4	6.8	73.8	17.5	1.9	747
35–9	6.8	78.0	14.5	0.7	587
40–4	4.9	83.0	12.0	0.2	493
45–9	3.3	84.7	11.8	0.3	399
15–49	4.3	63.5	18.7	13.4	5,764

Source: Côte d'Ivoire Fertility Survey, 1980–1.

The marriage process is very slow. It is inaugurated by gifts of palm wine not only to the father of the fiancée, but also to her uncles . . . for a very long time the girl continues to reside with her parents, sometimes until her third or fourth pregnancy. It can also happen that she never joins the domicile of her husband or that she divorces before that (Étienne and Étienne, 1971: 175).

This description does not mention a ceremony at the time when the woman joins the home of her husband, even though data from the *Enquête Ivoirienne*, as we shall see below, suggest that a gift of drinks or kola takes place at the time. Guerry (1970: 47) mentions 'a symbolic dowry' received by the Baoulé woman when she joins her husband, and it may be that drinks are part of that dowry.

The Côte d'Ivoire survey attempted to identify both the *fiançailles* and the onset of cohabitation. It also attempted to trace the stable union to its beginning, i.e. the first sexual relations with the future husband or stable partner; and, but only when a birth had resulted, to identify a period of relations (with a man other than the first partner) prior to that birth.[1]

The Results

The distribution of types of unions by age of the women is shown in Table 2.1. The decrease in the proportions single and in free unions with age, and the increase in the proportion of customary marriages, reflect the progressive and processual nature of marriage. Civil marriage has only minor importance in the survey.

Because the survey truncates marriage histories for many women who will still go on to the next stage, the ages at celebration of the first union and at cohabitation, and the intervals between start of union and the first birth

TABLE 2.2. *Never-married parous women*

	Never-married parous mothers as % of			Absolute No.
	Never-married women	Parous women	All women	
Ethnic Group				
Baoulé	37.1	10.9	8.0	79
Other Akan	31.4	6.6	5.3	37
Krou	28.6	5.5	4.4	34
Mandé Nord	23.8	3.3	2.7	19
Mandé Sud	26.2	3.8	3.0	17
Voltaique	24.6	3.2	2.7	17
Other African	32.7	3.0	2.5	35
Religion				
Christian	34.8	8.9	7.0	122
Islamic	17.2	1.6	1.3	26
Traditional/none	33.3	5.4	4.5	90
Literacy				
Can read	26.8	14.5	9.5	96
Cannot read	34.2	3.5	3.0	142
TOTAL	30.9	5.1	4.1	238

Source: Côte d'Ivoire Fertility Survey, 1980–1.

presented in later tables are underestimates. On the other hand, many a woman who is recorded as 'never married' at the time of the survey may already have started on the route that will eventually lead to fully recognized marriage. If this involves the same partner, the age at consummation of the union may be retrospectively re-evaluated downwards. To further complicate matters, maternity does not only occur in union. Some women who begin sexual relations with a man, and even conceive, may abandon the track that would have led to marriage. In Table 2.2, we show the percentage of women who had never been in a stable union at the time of the survey, yet had a child.

The sequencing of the various stages of union reported retrospectively clearly differs by ethnic group. Table 2.3 presents a classification of such sequences from the survey, and Table 2.4 shows their distribution by ethnic group, religion, and literacy. Fewer than half of the women in the survey followed pattern A where the three stages—ceremony, cohabitation, and the onset of sexual relations—occurred in the same month. There is a great deal of variability among women classified by ethnic group, religion, or literacy. The Baoulé are distinguished by the frequency of patterns D, F, G, H, and I where consummation and/or celebration precede cohabitation. From a

TABLE 2.3. *Marriage sequences observed in Côte d'Ivoire*

Marriage sequence	First event	Second event	Third event
A	Cohabitation, ceremony, and consummation		
B	Cohabitation and consummation	Ceremony	
C	Ceremony	Cohabitation and consummation	
D	Consummation and ceremony	Cohabitation	
E	Ceremony	Consummation	Cohabitation
F	Consummation	Cohabitation	Ceremony
G	Consummation	Ceremony	Cohabitation
H	Consummation	Ceremony and cohabitation	
I	Consummation	Cohabitation	
J	Cohabitation and consummation		

Source: Côte d'Ivoire Fertility Survey, 1980–1.

TABLE 2.4. *Marriage sequences* (%)

	Marriage sequence*									
	A	B	C	D	E	F	G	H	I	J
Ethnic Group										
Baoulé	12.4	5.8	2.4	4.9	1.1	8.6	8.1	10.4	21.8	24.6
Other Akan	26.6	10.6	3.5	2.3	1.2	10.7	2.8	6.9	11.3	24.1
Krou	40.3	10.8	5.9	.6	.2	8.3	.8	6.4	10.2	16.6
Mandé Nord	67.6	3.3	14.2	1.4	2.1	1.9	.7	3.8	1.5	3.6
Mandé Sud	49.2	10.4	10.0	1.3	.4	5.2	.8	7.7	5.2	9.8
Voltaique	58.6	4.5	6.3	.4	.4	2.9	.5	3.6	7.7	15.3
Other African	61.7	3.0	17.8	.7	.6	2.1	.6	3.4	2.8	7.3
Religion										
Christian	33.2	7.7	4.9	1.9	.7	8.6	2.2	6.8	11.9	22.0
Islamic	66.3	3.0	16.7	1.1	1.0	2.0	.7	3.3	1.3	4.5
Traditional/none	35.3	8.5	5.9	1.9	.7	6.2	3.4	7.7	13.0	17.3
Literacy										
Can read	19.7	7.9	3.9	2.6	.8	10.6	3.5	7.2	18.1	25.7
Cannot read	50.2	6.0	10.5	1.5	.8	4.5	1.9	5.6	6.9	12.1
TOTAL	46.3	6.3	9.6	1.6	.8	4.5	1.9	5.6	6.9	12.1

* See Table 2.3 for meaning of letters.

Source: Côte d'Ivoire Fertility Survey, 1980–1.

methodological point of view, these findings suggest that there is little justification for surveys to date the beginning of union systematically at the time of the onset of cohabitation if the preferred sequence in a particular group (such as the Baoulé) involves contracting a fully legal marriage several years, and several children, before the partners begin to live together.

As we pointed out in the previous section, the gift of drinks and kola nuts may not always be interpreted as a ceremony that initiates customary marriage. For several ethnic groups, these events occur at an early stage in the marriage process, at the onset of a vaguely defined 'betrothal' (*fiançailles*) period. A distinctive feature among several groups in Table 2.4 is a waiting period between the drinks or kola and the consummation of marriage (patterns C and E). These patterns are followed by fewer than 5 per cent of the Akan, but by more than 15 per cent of the Mandé Nord and the other Africans. In those instances, the situation may correspond to a period of courting. But there are obviously other instances where the word betrothal is probably inappropriate, because the ceremony serves to normalize sexual relations, the bearing of legitimate children, and perhaps cohabitation, even if the partnership stops short of a fully recognized, irreversible customary marriage.

A common justification for including nuptiality questions in fertility surveys is that being married (or in union) provides a socially recognized context for legitimate reproduction, and a denominator for the computation of marital fertility rates. Hence it should be especially interesting to look at the interval between marriage, defined by a ceremony (or between stable union, defined in the WFS by cohabitation), and the first birth. Table 2.5 gives the proportion of unions where the first birth preceded the marriage ceremony or cohabitation. The first child, of course, may have been conceived with a partner other than the man with whom the mother had her first union. This should appear in the table as births (or conceptions) before the consummation of the first union.

It is difficult to reconcile the results in Table 2.2 (percentage of women never in union at the time of the survey who had a child) and those in Table 2.5 (percentage of first births before the onset of the first union). One possible explanation of these differences is that 'current status' and retrospective measures of the same phenomenon are often quite different. It is possible either that young women underreport the extent of their 'living with a man' at the time of the survey, or that older women have 'legitimized' their first births retrospectively by assigning them to a time when they were already in union with the man who later became their husband. In other words, some of the single mothers in Table 2.2, if interviewed in a later survey, will report their births retrospectively as legitimate offspring of the man whom they eventually married.

In Table 2.5, the matrilineal Akan and especially the Baoulé follow a distinctive pattern. The data confirm the arguments presented in the literature

TABLE 2.5. *Intervals from various stages of the marriage process to first birth, by ethnic group (%)*

	Length of Interval (Months)			No. of Cases
	Negative	0–7	8+	
A. *Interval from cohabitation to first birth*				
Baoulé	39.0	17.7	43.2	643
Other Akan	18.4	23.0	58.7	479
Krou	15.5	21.3	63.2	517
Mandé Nord	7.3	13.9	78.8	481
Mandé Sud	11.0	18.3	70.8	400
Voltaique	7.4	17.4	75.3	461
Other African	6.2	13.6	80.2	1,009
All Groups	14.9	17.3	67.8	3,990
B. *Interval from ceremony to first birth*				
Baoulé	45.9	13.0	41.3	361
Other Akan	29.4	20.1	50.5	323
Krou	20.5	17.0	62.5	405
Mandé Nord	7.5	12.4	80.1	483
Mandé Sud	11.2	15.7	73.1	357
Voltaique	7.4	14.5	78.1	366
Other African	6.3	10.5	83.2	932
All Groups	15.6	13.9	70.5	3,227
C. *Interval from consummation to first birth*				
Baoulé	17.4	17.0	65.6	642
Other Akan	11.7	18.6	69.7	478
Krou	12.7	17.6	69.7	518
Mandé Nord	2.5	14.3	83.2	481
Mandé Sud	8.0	17.5	74.5	400
Voltaique	4.8	16.5	78.7	461
Other African	4.4	12.4	83.2	1,008
All Groups	8.6	15.8	75.6	3,988

Note: Because free unions do not involve a marriage ceremony, they are excluded from computation of the interval between the ceremony and first birth.

Source: Côte d'Ivoire Fertility Survey, 1980–1.

that Baoulé women commonly bear children before cohabitation. Celebration (giving drinks and kola) among the Akan appears to occur at a later stage in the family building process than cohabitation. The Akan also appear to have a larger proportion of births with a partner other than their future husband (first birth before consummation of the first union).

Determining an Age at First Marriage

A growing literature points to some evidence of a rise in the age at marriage in sub-Saharan Africa in the recent past (see e.g. Lesthaeghe *et al.*, 1989; United Nations, 1990). But almost insuperable problems of definitions arise in attempting to measure the extent of changes in the age at first union. There are two ways of computing the date of marriage or first union. The most practical method is through a question on current marital status, because most people are able to report reliably whether they are 'married' or not at the time of the survey. If women's own definition of their marital status is accepted, the singulate mean age at first marriage can be determined (United Nations, 1983). Although the perception of what 'being married' means may vary over time and across social strata, comparisons remain considerably easier than when the measurement relies on retrospective questions about first marriage. Surveys such as the WFS, of course, rely on the latter type of questions.

The problem is highlighted by the results from the Côte d'Ivoire survey. Even if we can determine that a particular union has been legitimated, the issue of when that union began still remains. To which one of the three initiating stages should it be dated back: consummation, celebration, or cohabitation? Table 2.6 shows significant differences in age at each of these stages.

In general, the first sexual relations (consummation) occur almost at the same age as celebration (ceremony) among most of the ethnic groups of the Côte d'Ivoire; the Akan (especially the Baoulé) constitute the major exception. Among the Baoulé, there is on average a delay of two years between the onset of sexual intercourse and the giving of drinks or kola. It would seem that cohabitation for them starts shortly before the ceremony. Among other groups (Muslims, Northern Mandé, and 'other Africans') celebration is on average sooner than the onset of sexual relations, a pattern that seems to suggest the existence of betrothal or courting customs.

The WFS solution of defining the age at first cohabitation as the age at first union is not appropriate when a socially recognized union, with the potential of bringing forth legitimate children, existed without cohabitation. In such cases, one should date the union back to the time when the union received the requisite sanction from kin and other socially relevant actors; that sanction may or may not have taken the form of a ceremony involving drinks or kola nuts. Still, there is no short cut, no universally valid question (such as a question on drinks or kola) to determine when the critical stage was reached.

TABLE 2.6. *Mean age at various stages of the marriage process*

	Mean Age at		
	Cohabitation	Ceremony	Consummation
Ethnic Group			
Baoulé	18.7	19.2	17.1
Other Akan	17.1	17.8	16.7
Krou	16.5	16.5	16.2
Mandé Nord	16.6	16.1	16.4
Mandé Sud	16.3	16.2	16.1
Voltaique	16.8	16.8	16.6
Other African	16.9	16.3	16.8
Religion			
Christian	17.2	16.8	16.7
Islamic	16.7	16.8	16.5
Traditional/none	17.3	16.8	16.5
Literacy			
Can read	17.1	18.0	16.6
Cannot read	17.0	16.7	16.6
All women	17.0	16.8	16.6

Source: Côte d'Ivoire Fertility Survey, 1980–1.

It is important to point out that although both cohabitation and social sanction may have been preceded by a period of sexual relations, the time of 'consummation' does not appear relevant to date the union. To use it for this purpose would raise intractable issues of truncation. Although many women who were still single at the time of the survey had already started the relationship which would eventually culminate in cohabitation or celebration, this is only known in retrospect. Many other women will abandon the particular partner with whom they were involved at the time of the survey, and one cannot under these circumstances talk about the consummation of a union that will never take place.

Conclusion

The integration of nuptiality variables in fertility surveys is a neglected and difficult topic. Marriage is a central concern of anthropologists, because it gives them clues to the organization of the whole social system. Demographers have generally steered clear of the topic because of its complexity and their concern for comparability between surveys.

Unfortunately, although a birth is always a birth in all international comparisons, there are enormous differences in the definitions of marital status in which that birth may occur. Most World Fertility Surveys (and other surveys of the same kind) slighted the issue of nuptiality, which seemed of secondary importance compared to more programmatically crucial topics such as contraceptive prevalence. Retrospectively, in a continent that is confronting the major emergency of the AIDS epidemic, it may prove a costly mistake to have squandered an opportunity to investigate more fully the social context of early sexuality and of the multiplicity of sexual partners.

There are also good arguments for giving more weight to the study of nuptiality as a proximate determinant of fertility. In many African countries, fertility may decline sooner through a change in the age of marriage than through the control of marital fertility. But at this stage, it is difficult to ascertain the course of a variable, age at first union, which is not consistently defined.

It seems unlikely that a single question will solve all the problems of nuptiality measurement across all populations in one ethnically diverse country, let alone the issue of international comparability. Drinks and kola nuts may serve multiple functions in the process of contracting a union, and various groups in the population may use them in different ways. None the less, these issues underscore the importance of those surveys that have attempted to innovate in the definition of marriage and the measurement of trends. The Côte d'Ivoire Fertility Survey's effort to ascertain whether a marriage ceremony had taken place is well worth evaluating and perhaps emulating.

The existence of such a question in a survey presents a challenge that demographers who typically analyse their results are poorly equipped to meet. The very notion of celebrating a marriage raises many issues normally defined as anthropological turf. What is the purpose of such a celebration? Does it have a legitimizing function? Do the drinks and the kola symbolize components of particular rights in women that are being transferred? Whereas including such tantalizing questions in the survey would imply that these issues were thoroughly examined during the questionnaire design, analysts are largely on their own for interpreting their meaning and significance. We do not have complete access to relevant information on marital practices in the country at the time when the survey was taken, and cannot be certain that we are using the information in the way it was originally intended.

But even short of complete success, innovative questions are instructive. Because the onset of cohabitation will not work everywhere, a question on the celebration of a marriage ceremony will be quite relevant in some situations. The innovative approach followed in Côte d'Ivoire reveals the diversity of customs and behaviour, and underscores a need to adapt survey concepts and procedures to particular populations.

The goal of ensuring representativeness at the national level, which has dominated the field of demography, should be tempered by a respect for cultural diversity, which alone will lead to a fuller understanding of essential mechanisms. We need more studies of demographic variables, including nuptiality, at the regional level. In a more homogeneous cultural context, path-breaking questions such as the one we have examined here may prove invaluable.

Note

1. The questionnaire presented in the First Country Report contains no question on the date of first sexual relations for never married women; but it is clear from the data tapes that such a question was asked of at least some women in the survey (perhaps using a later version of the questionnaire). It would seem that it was added to a later stage of the fieldwork, *after* the main urban areas, Abidjan and Bouaké, the very locations where premarital liaisons are probably most frequent, had been surveyed.

Acknowledgements

We thank the Government of Côte d'Ivoire and the Dynamic Data Base of the International Statistical Institute for permission to use the data tapes from Côte d'Ivoire. The credit for the quality and richness of this data set should, of course, go to the Ministère de l'Économie, des Finances et du Plan, in Abidjan.

References

Afzelius, A. (1967), *Sierra Leone Journal 1795–1796*, ed. A. P. Kup, Studia Ethnographica Upsaliensa, 27.

Bleek, W. (1976), 'Sexual Relationships and Birthcontrol in Ghana: A Case Study of a Rural Town', Ph.D. dissertation, University of Amsterdam.

Bosman, W. (1705), *A New and Accurate Description of the Coast of Guinea* (London: J. Knapton).

Boutillier, J.-L. (1971), 'Histoire et stratégies matrimoniales chez les Koulango de Nassian', *Cahiers ORSTOM, Série Sciences humaines*, 8/2: 153–64.

Étienne, P., and Étienne, M. (1971), ' "À qui mieux mieux" ou le mariage chez les Baoulé', *Cahiers ORSTOM, Série Sciences humaines*, 8/2: 165–86.

Fortes, M. (1950), 'Kinship and Marriage among the Ashanti', in A. R. Radcliffe-Brown, and D. Forde (eds.), *African Systems of Kinship and Marriage* (re-edited by International African Institute, 1987. London: KPI).

Guerry, V. (1970), *La Vie quotidienne dans un village Baoulé* (Abidjan: INADES).

Harrell-Bond, B. E. (1975), *Modern Marriage in Sierra Leone: A Study of the Professional Group* (The Hague: Mouton).

Holas, B. (1957), *Les Sénoufo (y compris les Minianka)*, Institut International Africain (Paris: PUF).

Lesthaeghe, R. J., Kaufmann, G., and Meekers, D. (1989), 'The Nuptiality Regimes in Sub-Saharan Africa', in R. J. Lesthaeghe, (ed.), *Reproduction and Social Organization in Sub-Saharan Africa* (Berkeley, Calif.: University of California Press), 238–337.

Meek, C. K. (1937), *Law and Authority in a Nigerian Tribe: A Study in Indirect Rule* (Amherst, NY: Humanities Press Reprint of the 1949 re-edition).

Ministère de l'Économie et des Finances (1984), *Enquête ivoirienne sur la fécondité, 1980–1981, Rapport principal* (Abidjan).

Nukunya, G. K. (1969), *Kinship and Marriage among the Anlo Ewe* (London: University of London).

Oppong, C. (1981), *Middle Class African Marriage: A Family Study of Ghanaian Senior Civil Servants* (London: George Allen & Unwin).

Park, M. (1799), *Travels in the Interior Districts of Africa in the Years 1795, 1796, and 1797*, facsimile reprint (New York: Arno Press, 1971).

Robertson, C. C. (1984), *Sharing the Same Bowl: A Socioeconomic History of Women and Class in Accra, Ghana* (Bloomington, Ind.: Indiana University Press).

Schwartz, A. (1971), 'Formes de mariage et stratégie sociale dans la société Guéré traditionnelle', *Cahiers ORSTOM, Série Sciences humaines*, 8/2: 221–31.

Uchendu, V. C. (1965), *The Igbo of Southeast Nigeria* (New York: Holt, Rinehart, and Winston).

United Nations (1983), *Manual X: Indirect Techniques for Demographic Estimation* (New York: Department of International Economic and Social Affairs).

—— (1990), *Patterns of First Marriage: Timing and Prevalence* (New York: Department of International Economic and Social Affairs).

Verdon, M. (1983), *The Abutia Ewe of West Africa: A Chiefdom that Never Was* (The Hague: Mouton).

3 Consequences of Bridewealth Changes on Nuptiality Patterns among the Ibo of Nigeria

UCHE C. ISIUGO-ABANIHE

Introduction

Relative to most of the 374 identifiable ethnic groups in Nigeria (Otite, 1990), a major distinctive feature of marriage among the Ibo is high bridewealth. Among the six major ethnic groups, of which the Ibo rank third with about 17 per cent of Nigeria's population (NPB, 1984), the Ibo seem to be alone in maintaining high bridewealth demands, which include large cash payments, incidental expenses, and gifts to the parents of the bride. By contrast, groups such as the Hausa and Fulani in the north have maintained relatively low bridewealth; it may be even lower among the Yoruba in the west, if we discount high material demands such as clothing for the bride. Among the Ibibio and Ijaw, neighbours of the Ibo in the south-east, bridewealth has declined to mere token payments involving modest cash prestations by the groom. High bridewealth or high cost of marriage prestations in Iboland has been a source of concern to governments and individuals for a long time (Balonwu *et al.*, 1955).

Lately, Ibo high bridewealth and its social consequences have become a subject of regular discussion in the Nigerian media, apparently with the aim of raising public consciousness to bring about change. For demographic concerns, high bridewealth is important because it may affect the timing, tempo, and stability of marriages, which are of considerable significance for fertility, especially in a population such as the Ibo, where virtually all child-bearing takes place within marriage.

This paper examines the past patterns of Ibo marriage and bridewealth, their transformation to their current state, and the consequences of the prevailing high bridewealth on nuptiality patterns. The work is based on data and experience gathered from the 1987 Bridewealth, Age at Marriage, and Fertility (BAF) Project, comprising six months of field work in twelve out of forty-four local government areas in Imo and Anambra states, the two exclusively Ibo-speaking states in Nigeria. A random sample of 200 adults of 15 years and over were interviewed in selected towns or villages in each local government area, using a comprehensive questionnaire which elicited information in closed- and open-ended formats. A total of 2,105 people were interviewed, comprising 838 ever-married females, 334 single females, 518 ever-married males, and 415 single males. In-depth interviews were also conducted with 30 key informants such as chiefs, opinion leaders, and woman leaders to provide detailed information and insights with respect to the changing trends in bridewealth and nuptiality in Iboland. The present paper is a summary of the final report of the project.

Traditional Ibo Marriage

Everything concerning marriage is taken seriously in Iboland. Writing on the significance of Ibo marriage, Basden (1938: 213) observed that:

> It looms upon the horizon of every maid and youth as an indispensable function to be fulfilled with as little delay as possible after reaching the age of puberty. The idea of a celibate life finds no favor whatsoever: to the Ibo it is rank foolishness, as well as being utterly contrary to the laws of nature.

To be a woman without a husband is unfortunate and disgraceful. Such a one is ridiculed and associated with evil. A man without a wife is regarded as useless. Monogamy and polygyny exist side by side, the latter being preferred, since a man with more than one wife is considered wealthy.

Rules of exogamy forbid marriage with close consanguinous relatives. Kindred units (*umunna*) are usually exogamous as well; an entire village may constitute an exogamous unit. Endogamy is allowed only among the Osu caste, descendants of cult slaves who in times past were dedicated to the gods (Basden, 1938; Uchendu, 1965).

Ibo marriages were characteristically early. The collective nature of marriage involvement and practices of child betrothal ensured this. The collective aspect of marriage derived from the belief that a woman was being married not only by her husband, but by a wide circle of relatives who encouraged their members to marry and begin bearing children early to ensure the continuation of the lineage. Even today, marriage is not simply an association between the husband and his wife for mutual support; it involves a wide alliance between the kin of the husband and the wife. Therefore, members of the extended family or lineage usually show as much interest as parents in arranging the marriage of the group's members.

Until it was legally abolished in 1956, child marriage was the most common form of marriage in Iboland (Uchendu, 1965), though isolated instances still occur today, especially in the Nsukka area. Girls, sometimes infants, were betrothed to older men or to boys only a few years older than themselves. However, rarely was marriage consummated until the girl reached puberty. Reasons for child betrothal included desires to create alliances through marriage to influential chiefs, priests, or warriors, or even to the mother of many children. Sometimes betrothed girls were used as pawns in lieu of debts.

Another important feature of Ibo marriage, as among many ethnic groups in Nigeria, is its polygynous character. Even though Christian adherents have been fighting for its eradication, polygyny is still common among Ibo Christians more than a century after the entire Iboland was Christianized. Especially in the past, polygyny was held as a worthy institution since the multiplicity of wives enhanced a man's social status and increased his farm workforce and the number of children bearing his name and his lineage.

Polygyny was also compatible with the tradition of long periods of abstinence after the birth of a child. Children were not weaned for two to four years since, in the absence of artificial feeding, a long period of nursing was necessary to enhance their survival. During this time the husband and wife were expected to observe sexual continence. This period was thought to give the mother a rest from work and allow her to gain strength through nourishment, to recuperate fully before her next confinement. Men therefore sought second or higher-order wives to meet both sexual and domestic needs on a regular basis. Given its benefits and the relatively low level of bridewealth in the past, the plurality of wives was a common ambition.

Though the payment of bridewealth was the general mode of contracting marriage, some forms of exchange marriage occasionally obviated this. Meek (1937) described one such variant, where two men could agree to exchange their sisters, daughters, or cousins as brides. Another variant was one in which a father could arrange for his son's marriage by giving his female child in exchange for a young bride for his son. Meek (1937) and also Uchendu (1965) observed an unusual type of exchange: two men agreeing to exchange their wives, with the consent of the women and probably their kin. These patterns were the exception rather than the rule, since the payment of some form of bridewealth was—and remains—the most acceptable means for procuring a wife throughout Iboland.

Indigenous Bridewealth among the Ibo

Ibo marriage involves some recompense given by the husband, or by his father or relatives on his behalf, to the parents or relatives of the wife. This usually takes the form of bridewealth: a transfer of goods, services, or money from the groom and his kin to the bride's people (Goody, 1973; Comaroff, 1980). Whatever the medium of exchange, bridewealth is regarded as a symbolic means of compensating parents for bringing up their daughter and for the loss of her services. Also, bridewealth guarantees the husband custody of all children born by his wife (Meek, 1937; Okafor-Omali, 1965). Further, people argue that bridewealth helps to stabilize a union by dissuading a wife from leaving her husband's home at will or at slight provocations. Since bridewealth should be refunded in full upon divorce, it is to the advantage of the bride's parents and relatives to try to settle disputes and re-establish cordial relations between the couple. In some places, this refund takes place only when the woman remarries; the second husband pays enough bridewealth to cover the costs borne by the first. Yet until the refund occurs, the woman rightfully belongs to her former husband, who still has a claim to her possessions and to the children she bore him.

Monetary bridewealth is a relatively recent phenomenon. Before cash or even goods became the media of exchange, conjugal payments often took the form of brideservice: labour, mainly on the farm, by the prospective husband for his in-laws. This period sometimes spanned several years, depending on the age of the girl when the marriage was initiated. The years were marked with periodic presentations of gifts such as meat, wine, or yams to the girl's parents, who could cancel the engagement if not satisfied with the man's performance. Presents, mainly household items, were often given when the bride finally moved to her new home, after the customary marriage.

Later, bridewealth was reckoned in more tangible goods such as cows, goats, or stacks of yams. When some early forms of money were introduced, bridewealth came to be settled with 'manillas' (kinds of metal ornament) or bars of iron and brass, and later by cowrie shells (Balonwu *et al.*, 1955; Ofonagoro, 1979). Neither the years of service nor the quantity or quality of goods were fixed, and prolonged bargaining often took place before agreement was reached (Basden, 1938). Generally, the stock of wealth that exchanged hands depended on the status of the girl's family, her physical attractiveness and manners, and her resourcefulness and skill in farm work and domestic chores.

In the traditional bridewealth system, a man had ample time to fulfil the requirements, especially if he had betrothed a child. But a prospective bridegroom could try to shorten the period by bringing in a work-team of his relatives and age-grade mates to work for his prospective in-laws, perhaps promising to continue to do so even after the marriage had been consummated. If tangible goods were required, he could expedite the process by asking his father or his kinsmen to help provide them. If necessary, he could promise payment by instalments.

By the early 1900s, the colonial government had started to introduce subsidiary coinage in southern Nigeria to replace cowries as the medium of exchange (Ofonagoro, 1979). Starting with bronze coins, the new exchange system also involved aluminium coins until the more acceptable nickel-bronze coins were introduced in 1908. However, cowries continued to be used as late as 1920, and in many places even during the Second World War. Paper money was introduced during the First World War amidst considerable local prejudice against it. This prejudice stemmed from difficulties in carrying it about, since no one carried wallets or bags, and in storing it: unlike nickel money, people could not bury it or hide it away, leaving it vulnerable to termites and rodents (Ofonagoro, 1979).

Even when cash money came into circulation, some parents still preferred to use cowries to settle bridewealth, just as they used cowries for ceremonies such as funerals, initiation, and title-taking. However, the sharp rise in prices generally, and bridewealth in particular, during the Second World War hastened the demise of the use of cowries in Iboland. When bridewealth rose to £20 or more, it required the services of many porters to convey bags of cowries to the bride's father, who in turn encountered difficulties in storing such quantities (Ofonagoro, 1979).

In this initial period of monetized exchange, cash bridewealth was low, ranging from a few shillings to £10 in most places, and payment could be spread out over some time. Between 1920 and 1945, however, cash bridewealth increased precipitously because of:

1. The increasing importance of Western education, a high demand for educated wives, and parents' desires to recover the cost of a daughter's education through her bridewealth (Balonwu *et al.*, 1955).

2. Increased commercial activities stemming, directly or indirectly, from the Second World War (Ofonagoro, 1979). At the same time, there have been enormous increases in employment opportunities in the flourishing private and public sectors, and in commercial activities stemming from the War, colonial development efforts, and later post-independent economic activities—especially the oil boom of the 1970s.

3. Higher employment opportunities in the flourishing private and public sectors, and the increased availability of money from salaries and wages.

One of the most important reasons why bridewealth surged was that soldiers of Ibo origin who fought in the War on the British side sent large sums of money to their parents to engage wives for them. Some visited home and married quickly, taking advantage of the marriage allowance programme in the Army. When the War ended in 1945, ex-servicemen returned with enormous amounts of money, which gave them access to any compound, no matter how exclusive, in search of wives. Girls were offered to them with enthusiasm, and staggering sums, especially for educated women, were demanded in exchange. It became obvious from then that having a daughter and educating her were a means to wealth. The bridewealth situation has never been the same since.

Contemporary Bridewealth in Iboland

Currently, Ibo bridewealth is at an unprecedented level, one that most people regard as excessive. One of the major reasons is the high premium placed on education. Formerly, bridewealth was fairly standardized, though the social rank, age, and personal qualities of the bride were considered. The costs of her training and education, however, have created sharp differences. Indeed, high bridewealth demands first emerged in areas that pioneered female education in Iboland (Balonwu *et al.*, 1955). Female education is now viewed as an investment for which a befitting reward can be demanded; hence, the considerable haggling that usually takes place between parents and grooms during bridewealth negotiation sessions. Among other things, parents consider the monetary value of the presents that would be given to the bride when she moves to a new home. Because such presents must reflect her education, parents generally make high bridewealth demands to be able to purchase appropriate presents for their daughter after stashing away some part for future use.

Important in the calculus of parents is not simply the money spent on their daughter's education, but also her labour and access to her future earnings. In theory, Ibo patrilineality means that once a woman marries, she and all she has and owns now belong to her husband. Therefore, high bridewealth is a means by which parents try to derive immediate compensation because they know that prospects for future remuneration are slim; they know in particular, moreover, that the man will probably resent his wife giving money to her parents, especially if he feels that their bridewealth demands have already bled him dry. A 28-year-old single man explained, using an Ibo proverb:

> If a man is made to literally buy a girl, as parents now do in the name of bride price, he would not take it kindly seeing his wife spend money on her natal family. But if my father-in-law was considerate during the bridewealth settlements, then I will co-operate by helping them financially, if I have the means to do so. As far as I am concerned, 'If right hand washes left hand, left hand will also wash right hand.' [That is, if the bride's father moderates his bridewealth demands, the groom will reciprocate by rendering help in future.]

Perhaps to forestall hostility, some parents now demand that the groom educate or train another daughter to replace the one whose education he is now finding advantageous. This practice is surprisingly widespread in all the areas studied. On their part, parents see this as a new extended family responsibility that the groom has incurred. The bride may, in fact, take in this sister and foster her while she is getting an education. The bride finds it more fulfilling to take in a sister rather than a more distant relative or even a stranger, as she may be asked to do sooner or later, given the prevalence of child fostering among the Ibo (Isiugo-Abanihe, 1984).

Some people in our intensive interview sessions argued that bridewealth has not increased; rather, the cost of living has. So although the actual money paid may have risen, the value of this amount has fallen relative to the old days. Table 3.1 corroborates this (see also Fig. 3.1). It shows the average nominal bridewealth (the stated price) and real bridewealth (adjusted for inflation) in naira between 1970 and 1987 for two categories of women: those with high education (some secondary education) and their sisters with low education (primary school or lower). The nominal bridewealth was adjusted using the composite consumer price index (CPI) for Nigeria to control for inflation at constant 1980 prices. It is clear from these data that nominal

TABLE 3.1. *Bridewealth in Iboland, 1970–1987: nominal and real, by bride's educational level*

Year	CPI	Low education Nominal	Low education Real	High education Nominal	High education Real
1970	23.8	340	1,429	1,072	4,504
1971	27.7	364	1,314	1,284	4,635
1972	28.4	375	1,320	1,526	5,373
1973	30.0	445	1,483	1,782	5,940
1974	33.8	508	1,503	1,926	5,698
1975	45.1	572	1,268	2,182	4,838
1976	55.1	632	1,147	2,209	4,009
1977	66.9	813	1,215	2,326	3,477
1978	81.4	861	1,058	2,468	3,032
1979	90.9	897	986	2,768	3,045
1980	100.0	933	933	2,854	2,854
1981	120.8	1,015	839	2,935	2,430
1982	130.1	1,104	849	3,476	2,672
1983	160.3	1,241	774	3,561	2,221
1984	224.0	1,301	581	4,763	2,126
1985	236.1	1,380	584	5,217	2,210
1986	248.8	1,507	606	4,534	1,822
1987	274.3	1,740	635	4,741	1,728

Notes: Prices are in naira. Nominal figures are averages derived from survey question: 'How much bridewealth did your husband pay?' The low education categories include women with primary or lower education at marriage, while the high education brides had secondary school education or higher at marriage. The Consumer Price Index (CPI) is for Nigeria as a whole, and derived from the World Bank's African Economic and Financial Statistics.

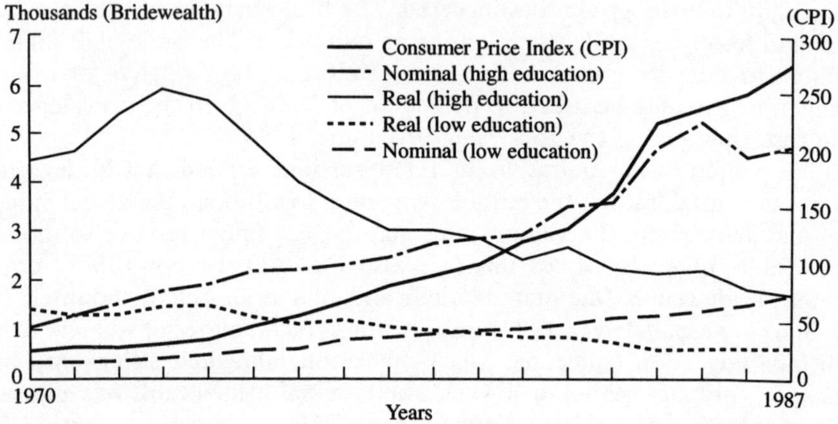

FIG. 3.1. Bridewealth in Iboland, Nigeria, nominal and real, 1970–1987

TABLE 3.2. *Basic salary levels in the Nigerian Federal Public Service* (naira per annum)

Status	Year				
	1965	1975	1979	1983	1988
Permanent secretary	5,880	12,696	13,970	15,080	15,080
Medical officer	2,232	4,368	6,448	6,768	6,930
Engineer	1,944	3,564	4,860	5,598	5,760
Administrative officer	1,440	3,264	3,710	4,374	4,524
Clerical officer	396	1,164	1,290	1,878	1,920
Unskilled labourer	218	720	788	1,572	1,596

Source: Several editions of *Recurrent and Capital Estimates of the Government of the Federal Republic of Nigeria* (Lagos: Federal Government Printers).

bridewealth has been increasing over the years (faster for women with higher education) at a pace closely following the inflation rate.

Yet although the nominal price of bridewealth has soared, especially for the better educated, the real value of money needed to marry has indeed declined over time because of inflation. This decline in real bridewealth started in the mid-1970s, a period corresponding with the post-civil war recovery in Iboland (the core of former Biafra), and the Nigerian oil boom. As Figure 1 shows, real bridewealth levels have declined. Thus, although parents are charging high bridewealth for their daughters, the value of this bridewealth is really less than the value of what they invested over the years in their daughters. Moreover, what they can actually buy in exchange for the bridewealth they receive has declined considerably from 1970.

But although real bridewealth may have declined, so has real income among workers. Table 3.2 shows the salary structure in the Nigerian Federal Service over time. The nominal income shows only modest increases, even for permanent secretaries, the highest-paid civil servants. However, the situation with respect to an administrative officer, the usual starting position for university graduates, demonstrates how high bridewealth might delay marriage, especially because this is the stage when graduates usually marry. Because a university graduate prefers a bride with at least some secondary education, an administrative officer groom earning ₦7,237 in 1975 may have found it possible to pay ₦4,838 for real bridewealth at 1980 prices (see Table 3.1). However, a groom of the same cadre in 1979 or 1983 earning less real income—₦4,081 or ₦2,729—faces greater problems paying for a wife, especially one with some secondary education, given real bridewealth of ₦3,045 or ₦2,221.

Taken together, these data suggest that although the value of bridewealth has actually declined, the financial burden (at least for federal civil servants) that high bridewealth imposes remains high because real income has declined. One consequence of this has been less propensity to save among workers, most of whom are shopping around in the marriage market.

Aside from cash bridewealth, the bridegroom incurs additional costs from incidental expenses. While in the past these were primarily symbolic gestures, they have now assumed new dimensions. With increased prosperity, especially since the oil boom of the 1970s, grooms are expected to buy consumer durables such as televisions, radios, refrigerators, motorcycles, or clothes for their brides' fathers, and trinkets and textiles for mothers-in-law. We even found men who had to buy a car or a piece of land or to build a house for their fathers-in-law as a condition for marrying. In addition, the bride's relations must be given presents and small money gifts, which may amount to a substantial sum, given the usual large number of beneficiaries. With the declining purchasing power of the naira, these incidental expenses represent steep financial outlays, often amounting to much more than the physical cash that is paid as bridewealth. To these can be added the costs of providing drinks to entertain members of the bride's extended family during the often-protracted marriage negotiations and ceremonies. In the past, only kegs of palm wine were required. Now, apart from palm wine, the bridegroom is usually given a long list of 'modern' drinks to buy, including beer, whisky, and schnapps. He must also provide a substantial amount of money for the wedding ceremony. Finally, starting a new conjugal home entails a budget of its own, especially in a culture where a child is usually expected nine months or less after the wedding. Thus, although it would appear from the declining real value of bridewealth that grooms are getting a better deal, declining earnings make these heavy incidental expenses problematic.

Although federal civil servants are used for illustrative purposes, difficulties in paying bridewealth apply virtually across the board to all workers or wage-earners. Further, civil servants (both federal and state as well as workers in private enterprises) are generally regarded as better off than most members of the labour force. If, therefore, they are now hurting because of declining real earnings, in spite of their frequent participation in second jobs or the informal economy, then most others would be much worse off. The case for civil servants or white-collar workers is also critical because there is a certain level below which they probably cannot acquire partners with high education who attract high bridewealth. Not surprisingly, these categories of workers appear likely to marry outside the ethnic group.

It is clear from the foregoing that a young man contemplating marriage is hard pressed to afford the anticipated financial commitments. Among single respondents interviewed in the BAF project, more than 84 per cent stated that high bridewealth is a major concern in their area. Almost 70 per cent agreed that 'Ibo bridewealth is now a form of business', having lost its cultural foundation to a strong materialistic ethic. Asked whether they are worried by the continued increases in bridewealth, given their single status, 95 per cent of males and 80 per cent of females replied they were. Feelings of frustration and resignation were commonly expressed, such as those by a 31-year-old teacher, who lamented the effects of high bridewealth on his prospects for marriage:

> The situation is disturbing, to say the least. How and when can people like me marry with this exorbitant bridewealth being charged? It is an anathema, an evil which has only succeeded in breeding discontent among the growing number of bachelors and spinsters. My father married at 19 years of age. At 31 I have not saved up much money yet. But I am ready to wait it out, since I cannot steal for any man, no matter how virtuous his daughter. My father is not exactly rich either, so I have little choice than to wait and pray.

The problems associated with the large number of spinsters in Iboland are regularly discussed in the recent Nigerian media. Igwe (1984: 8), paraphrasing the comments of a female undergraduate, notes:

> With young men quietly staying away from marriage for a long time because they cannot afford to pay the bride price, some of them growing to as much as forty years without being able to marry, the ultimate losers are the girls who easily slip into the wrong side of age. A man can always get a woman to marry at almost any age, it is not the same for girls. It is difficult for girls to marry after a certain age because people start calling her old maid.

Writing from the Ibo perspective that sees marriage and motherhood as the ultimate measure of femininity, Moebonam (1987) argues that a new wave of women is suffering under the yoke of high bridewealth that has forced them to remain spinsters. His comparison of a woman with a flower was quoted repeatedly during the BAF fieldwork by rural and urban respondents alike:

> A woman's life is like a flower: dazzling in the morning under the refreshing dew of the night and in the evening, pitiably lusterless after the heat of the sun. Incidentally, too, nature plays a lot of tricks with the anatomy of the woman. Her life is so very carefully graduated by nature that each year that passes renders her less able to become a mother (Moebonam, 1987: 8).

Evidence from this study clearly indicates a heavy burden of marriage payments on young men. In the past, the extended kin would usually help, the first son could marry with bridewealth money coming in from his sister's marriage, and men could take wives after a down payment. By contrast, bridewealth is now expected in full in most places before a bride moves in with her husband. Until payment is completed, marriage cannot be said to take place, and a man has no custody over his children. At the same time, the diminishing importance of the collective aspect of marriage and the weakening of the elders' influence in choosing partners tend to place marriage and bridewealth responsibilities squarely on the groom. Of the male respondents, 69 per cent agreed that in their community the individual suitors must now raise bridewealth through personal savings, compared with 20 per cent who reported that relatives still help. As Phillips (1953: 18) observed, even in the early 1950s, the weakening importance of the kinship group in Africa more generally transforms marriage into a relationship between two individuals rather than involving the wider circle of relatives, and leaves the completed transaction more circumscribed:

> The contract is tending to acquire the finality of a 'cash transaction'—and this is indeed no far-fetched analogy, for under modern conditions the bride-price tends to assume a more mercenary aspect than it usually bore in the past. Exorbitant demands are often made, and the resulting dissatisfaction on the part of the younger men raises a new threat to the continuance of the bride-price system . . . there is a growing, and no longer negligible, agitation, not merely for the limitation of current rates, but for the abolition of the institution itself.

These comments are very much applicable to the Ibo who have lived with the problems of high bridewealth for a long time. Following the lead of the *Report of the Committee on Bride Price* (Balonwu *et al.*, 1955), and common sentiments expressed in the media, we asked our respondents whether any social evils were associated with high bridewealth. The aim was to corroborate the observations of the Committee, which noted that:

> [high bride price] is to a large extent responsible for the fall in morality in the Region. . . . The practice of men and women living together is becoming widespread. . . . Many children are today born out of wedlock. Girls are flocking to the towns to practice prostitution merely because as long as their parents demand an exorbitant [bride] price, they know there is little likelihood of their getting married if they remain at home. . . . Men wishing to marry are driven to extremes to raise money. . . . Sometimes men are tempted to steal and misappropriate public funds for the purpose (Balonwu *et al.*, 1955: 6).

In our survey, moral decadence (specifically, high incidence of prostitution, keeping mistresses, promiscuity, etc.) was the most frequent response, mentioned by 30 per cent of men and 37 per cent of women, to the question of the main social evil of high bridewealth. Illicit methods of raising money, such as stealing or embezzlement by men, were identified by 13 per cent of male and 9 per cent of female respondents. Elopement and cohabitation were mentioned, but by a small number of respondents. Completely unexpected was the large number of people (40 per cent of males and 29 per cent of females) who regarded late marriage or remaining single permanently as social evils. That such a large proportion of respondents did so reflects the continued importance attached to marriage in this population and the extent to which it has been affected.

Yet despite considerable agitation and numerous campaigns, high bridewealth has continued to increase. Many communities and Christian groups have recently set up guidelines for reducing bridewealth and incidental expenses to affordable levels. However, such community-level enactments rarely work. They fail for the same reasons that the government bridewealth reduction laws failed for Eastern Nigeria in 1956 and Imo State in 1982: lack of commitment on the part of the authorities and the reluctance of individual bridegrooms to involve in-laws in litigation over bridewealth.

Despite the complaints against high bridewealth, many of the people interviewed argued that trying to legislate bridewealth is unrealistic because it ignores basic Ibo ways of life. They pointed out that for a people who revere their culture, any artificial imposition of uniform bridewealth costs is ludicrous, if not sacrilegious, and constitutes an interference with natural law, if not with adaptive societal evolution. In addition, nearly 70 per cent of the sample believed that high bridewealth was appropriate because it attracted much-needed presents from the bride's parents for her new conjugal home. Since a bridegroom spends considerable money in the marriage transaction, the bride's parents are expected to give enough property or presents to enable the couple to start their new home smoothly. Yet the converse of this argument, commonly heard among older respondents, makes sense as well: since parents want to give worthy gifts to their daughters yet avoid impoverishing themselves under conditions of inflation, they are justified in charging high bridewealth.

Some parents, in fact, declared that the bridal presents they gave were worth more than the cash bridewealth they received. Evidence from this study suggests that Ibo marriage now involves double transmission of property: from the bridegroom to his in-laws (bridewealth), on the one hand, and from parents to daughters (dowry) on the other. However, all or part of what the bride's parents give is supposed to derive from what the groom provides. Also, since the bride's father or kin determine both the bridewealth and the bridal gifts, a young man in search of a wife remains at a disadvantage. Since he must pay the bridewealth and incidental costs before the bridal presents are considered, the initial accumulation of wealth for marriage payment is the real problem.

Consequences of High Bridewealth on the Demography of Marriage

Nuptiality patterns reflect demographic characteristics such as age and sex structure, economic forces (the costs of entering a union and the availability of alternatives to marriage such as education or employment), and sociocultural factors such as the mode of marrying, individual versus collective involvement, cultural norms, and so on. These forces are interrelated, and a change in one is likely to affect the others. Therefore, continuous rise in bridewealth payments among the Ibo is probably affecting many factors which affect young people's abilities to marry. These interrelations make it hard to assess the independent impact of changes in bridewealth. Highly educated women, for example, are likely to be older than the average marriageable woman; they are at a high premium, and less readily affordable. But they also are less amenable to the authority of a husband. Therefore, although high bridewealth demands may outprice some educated women as wives, these delays in marriage have also made women more independent, a factor which itself makes them less apt to marry.

None the less, local people have long suspected that high bridewealth itself leads to marriage postponement. Writing on Ibo bridewealth, Basden (1938: 218) notes:

> Young men find it extremely difficult to raise the ever-increasing prices now demanded, and they must either postpone marriage until well advanced in years, or contract loans. Grave evils arise from the latter course, for a man may be burdened by the incubus of debts incurred in connection with his marriage which will cripple him financially throughout his life. . . . On the other hand, girls will not wait an indefinite period. Faced with almost insurmountable obstacles, the young people are much more readily disposed to cohabit without marriage.

Reporting to the Government on the consequences of high bridewealth in Iboland, the Committee on Bride Price observed that 'both men and women are forced to remain single after the age at which they would normally marry, and the adjustment and adaptation necessary in both partners to a marriage is more difficult and generally less complete' (Balonwu *et al.*, 1955: 6).

In the past, Ibo girls married early, usually between 15 and 19 years of age (Ekanem, 1974). By contrast, the mean age at first marriage among the Yoruba in 1969 was about 20 years (Olusanya, 1969). By 1981/2, however, when Nigeria concluded its World Fertility Survey (WFS), both the mean age at marriage and singulate mean age at marriage (SMAM) for the Ibo were higher than those of the Yoruba, who hitherto had exhibited the highest age at first marriage in Nigeria (Adegbola, 1987). Individuals' perceptions corroborate these findings. In our survey, 57 per cent of the single males and 41 per cent of the single females reported that they could have married if bridewealth in their area had been lower. Indeed, about 70 per cent of males and 56 per cent of females claimed to have experienced an aborted marriage proposal or engagement because of high bridewealth demands.

The results of this study confirm that age at marriage has undergone a rapid increase in Iboland. Respondents who married before the civil war, that is, in 1967 or before, were compared with those who married since the end of the war in 1970. The following mean ages at first marriage were derived: for females, 17 years among those who married before the war and 24 years among those who married after; for males, 26 years before the war and 32 years after. The proportion of never-married, moreover, reflects sharp differences in Ibo nuptial patterns compared to those for other ethnic groups. Nearly 76 per cent of Ibo women aged 15–19 years, and 30 per cent of those aged 20–4 years, were still single at the time of the survey. About one woman in six (17 per cent) and one in 13 (8 per cent) had never been married at 25–9 and 30–4 years respectively. WFS data show that, by contrast, only 18 per cent of Nigeria's Yoruba women are still single at 20–4 years of age and virtually all Hausa/Fulani girls have married. (The corresponding WFS figure for the Ibo is nearly 40 per cent.)

The public is highly aware of this emerging pattern of late marriage and attributes it primarily to increasing bridewealth. More than 86 per cent of the survey respondents stated that age at marriage is currently rising relative to the pre-civil war period. The most common cause they identified was high bridewealth (45 per cent of males, and 34 per cent of females). Economic recession/unemployment and long years of schooling ranked second and third, respectively. When specifically asked, in another question, whether the current pattern of late marriage was related to bridewealth, 79 per cent of men and 76 per cent of women agreed.

It would be easy, as demographic studies commonly do, to attribute the rising age at marriage in Iboland to increasing levels of female education. However, marriage age is rising much less sharply in Yorubaland, an area of comparable, if not higher, education. Moreover, the marriage age is rising among Ibo illiterates and rural dwellers. The mean age at first marriage for rural women married since 1970 is about 21 years, and 26.4 years for urban women, compared with about 16 years and 20 years respectively before the civil war of 1967. Among women with no schooling, primary school education, and secondary school education or above, the mean ages at marriage since 1970 are 19, 20.8, and 25.7 years respectively, compared with about 16 years, 18 years, and 21 years for those marrying before 1967. For educated women, even apart from the time they spend in school, most now seem to spend many additional years before they eventually marry, suggesting that some other factor is enforcing a delay. Affording bridewealth costs, I contend, necessitates postponing marriage because of the number of years it takes a man to accumulate enough wealth to marry.

How do people adapt to these new marital constraints? The high cost of marriage in Iboland has probably intensified high migration rates as young men leave to find work (Isiugo-Abanihe, 1987). These rates of migration, in turn, may have influenced what was found to be an increasing rate of marriage among Ibo men to women outside the ethnic group. It is significant that younger men are making such marriages. There is also a stated willingness on the part of single people to choose their life-partners outside Iboland (53 per cent of single males said they would) if bridewealth continues to be high and unaffordable. The youth, moreover, are now well disposed to a marriage system based on romantic love and individual choice in partner selection. Related to this has been the growing neolocal residence and nuclear families in Iboland. Of our single respondents, 75 per cent would prefer a conjugal residence independent of their parents. Because of these tendencies Ibo men now increasingly choose partners outside their locality and among ethnic groups with negligible bridewealth prestations. They also start their families far away from their natal homes. On their part, women increasingly marry non-Ibo men according to procedures lacking the traditional marriage rites and high bridewealth payments. (Though most parents are usually alarmed to hear that their daughter is in love with a non-Ibo man, they rarely try to boost their bridewealth demands to undermine the marriage.)

Many people believe that polygyny has absorbed the excess of single females in the wake of rising bridewealth costs. However, data from the Nigeria Fertility Survey indicate that the Ibo have the lowest incidence of polygyny among the major Nigerian ethnic groups (although this is undoubtedly related to the predominance of Christianity in Iboland), and a majority of the single respondents said they would prefer not to enter polygynous unions. Moreover, the fact remains that because all women must be paid for, only wealthy men can afford more wives. However, 'outside polygyny' or keeping of mistresses (see Karanja, this volume) was mentioned as one of the social ills of high bridewealth. Conversely, the increasing incidence of outside wives or mistresses is one of the major reasons why Christian denominations oppose high bridewealth.

Still, some evidence suggests, and people certainly believe, that women who become desperate for husbands are likely to marry less desirable men when age starts to catch up with them. For instance, more than two-thirds of our respondents agree that women now end up marrying wealthy men irrespective of their age, education, or marital status, because 'they have enough wealth to marry at any cost'. Also, bridewealth probably tumbles sharply if women fail to marry while young. But since the demand for older wives is low, many of them marry as second or higher-order wives in polygynous households.

While high bridewealth seems to delay entry into marriage or to induce people to marry outside their ethnic group or social status, most people believe that it has a compensating effect, once marriage takes place, in enhancing marital stability, since families find it difficult to refund such large amounts of money. As Balonwu and his associates observed, when bridewealth was still low in Iboland, if a woman 'approached her father with a long tale of maltreatment by her husband, he would in a glaring case set about refunding the [bridewealth], but today when he has accepted a heavy sum which he knows he has no chance of refunding, he is obliged to turn a deaf ear on her entreaties' (Balonwu *et al.*, 1955: 6).

Among our respondents, 39 per cent of males and 45 per cent of females agreed that there would be more divorce if bridewealth was not high. Indeed, only 6 per cent of ever-married men and 5 per cent of ever-married women have been divorced, an extraordinarily low level among African populations. The Nigeria Fertility Survey also shows that the Ibo have the lowest divorce rate in the country.

Conclusion

An examination of high bridewealth in Iboland and its effects on emerging nuptiality patterns and family life reveals that the prestations demanded of suitors have changed considerably over several generations. Constituted in previous times as brideservice for the woman's family, these demands transformed gradually into goods and money. Changes have occurred in amounts as well. The enormous increases in employment opportunities in the private and public sectors in the wake of the Second World War and of colonial and post-colonial economic activities have inflated bridewealth levels. Today, the money needed to marry has soared. Though real bridewealth may have declined, given prevailing prices, young men who want to marry face a heavy financial burden, given declining real incomes, the diminishing importance of collective responsibility in marriage, and the cost of marrying a highly educated woman. To accumulate the requisite wealth, long years of frugal living and work, often involving labour migration, are necessary. These rising bridewealth costs have meant that Ibo marriage has ceased to be early and universal.

What are the likely effects of rising bridewealth amounts on fertility? High bridewealth may stabilize marriages, since the necessity to reimburse such large sums makes divorce problematic. Ordinarily, more stable unions produce higher fertility because of more exposure to intercourse. Yet the escalating costs of marriage are more likely decreasing child-bearing, which still occurs almost exclusively within marriage among the Ibo, by delaying the timing of marriage, shifting child-bearing to the less fecund older years, and perhaps changing norms about family size and preferences.

Whereas the ultimate effects of high bridewealth on fertility are not immediately obvious, high bridewealth does appear to be changing the nature of marriage. Concerning the effects of high bridewealth on women and their marital prospects, the excess of single females left in the wake of high bridewealth is partly absorbed by polygyny, giving an ironic new life to an institution which otherwise might have waned. There is some evidence as well that women tend to marry beneath their status after many years of spinsterhood, or convince their parents to reduce bridewealth demands. Other solutions to high bridewealth problems include marrying partners from other ethnic groups, acquiring 'outside wives' and so-called 'sugar-mummies' (mistresses of younger men), and even elopement. Directly or indirectly, high bridewealth engenders individualism and autonomy, which may increasingly induce Ibo women to remain single, especially if they can have children, comfort, security, and love without marriage. If these effects grow in importance, of course, then the speculation that high bridewealth reduces fertility by acting only on marital fertility would have to be reconsidered.

How does high bridewealth affect the woman's parents? Although grooms complain about higher prices during times of inflation, and romanticize about the more affordable prices of the past, bridewealth is actually less costly in real terms than ever before. The fact that grooms get less help from the family than before might help explain their complaints. Yet it could be that the people who lose the most are the woman's parents, because they put in most of their investments at a time when prices were lower in nominal terms, yet because of inflation they are recouping less in real terms. Hence, many of the increases in marriage ages may stem from delays on the parts of parents. In some cases, there are extensive periods of haggling between parents and grooms: parents struggle to recover some of their investments, although they quickly drop their steep demands as their aging daughters veer dangerously toward permanent spinsterhood. The fact that people report quite a number of aborted marriage proposals would seem to support this.

In any event, Ibo bridewealth comprises a complex story, of which all the parts are not yet clear. The conflicting interests of various parties may help to explain why efforts to check the level of bridewealth throughout Iboland have met with little success. Despite their complaints about the drawbacks of high bridewealth, most Ibo expressed a desire to maintain it. They commonly boast of the 'superiority' of their marriage custom, relative to other ethnic groups, because of its high bridewealth base. To the Ibo, the family, and indeed a woman, is a gem which should be cherished and appreciated. What one gets cheaply is therefore valueless; a woman for whom considerable bridewealth has been paid is admired and respected in the community. Men often boast of how much they have paid for their wives, while wives boast of what their bridewealth cost.

In sum, the evidence indicates considerable transformations occurring in Ibo marriage and nuptiality patterns under high bridewealth burdens. Yet there is no sign that the custom is about to change. On the contrary, bridewealth levels may continue to increase with inflationary tendencies, as past trends suggest; this would continue to test Ibo adaptability. This implies that the observed changes may become even more marked in the future, especially if the prevailing austere economic conditions continue unabated.

Acknowledgements

I am grateful to the Rockefeller Foundation for support for the project (grant RF 86067 Allocation 24); to Caroline Bledsoe, Ifeoma Isiugo-Abanihe, and Chris Udry for helpful comments and suggestions; and to Ibiba George for research assistance.

References

Adegbola, O. (1987), *Regional and Socioeconomic Fertility Differentials in Nigeria, 1981–1982*, IPD Working Paper, 87/6, Vrije Universiteit, Brussels.

Balonwu, M. O., Wogu, A. N., Essessien, D. F. E., Appah, F. T. F., and Cousins, J. G. (1955), Report of the Committee on Bride Price (Enugu: Government Printer).

Basden, G. T. (1938), *Niger Ibos* (London: Frank Cass & Co.).

Comaroff, J. L. (ed.) (1980), *The Meaning of Marriage Payments* (New York: Academic Press).

Ekanem, I. I. (1974), 'Fertility in Eastern Nigeria', in F. O. Okediji (ed.), *Population Dynamics Research in Africa* (Washington, DC: Interdisciplinary Communications Program, Smithsonian Institute), 169–85.

Goody, Jack (1973), 'Bridewealth and Dowry in Africa and Eurasia', in J. Goody and S. J. Tambiah, *Bridewealth and Dowry* (Cambridge: Cambridge University Press).

Igwe, Dimgba (1984), 'Bride Price in Iboland', *Sunday Concord*, 18 Nov.

Isiugo-Abanihe, U. C. (1984), 'Child Fostering and High Fertility Interrelationships in West Africa', *Studies in Third World Societies*, no. 29.

—— (1987), 'High Bridewealth and Age at Marriage in Iboland', in ed. E. van de Walle, *The Culture Roots of African Fertility Regimes*, Proceedings of the Ife Conference, 25 Feb.–1 Mar. 1987.

Meek, C. K. (1937), *Law and Authority in a Nigerian Tribe: A Study in Indirect Rule* (London: Oxford University Press).

Moebonam, T. (1987), 'The Bride Price Trauma', *Weekly Star*, 15 Mar.

National Population Bureau (1984), *The Nigeria Fertility Survey 1981/2*, Principal Report, i. World Fertility Survey, International Statistical Institute.

Ofonagoro, W. I. (1979), *Trade and Imperialism in Southern Nigeria: 1881–1929* (New York: NOK Publishers International).

Okafor-Omali, Dilim (1965), *A Nigerian Villager in Two Worlds* (London: Faber & Faber).

Olusanya, P. O. (1969), 'Rural-Urban Fertility Differentials in Western Nigeria', *Population Studies*, 23/3: 363–71.

Otite, Onigu (1990), *Ethnic Pluralism and Ethnicity in Nigeria* (Ibadan: Shanesor CI Ltd.).

Phillips, A. (ed.) (1953), *Survey of African Marriage and Family Life* (London: Oxford University Press).

Uchendu, V. (1965), *The Ibo of Southeast Nigeria* (New York: Holt, Rinehart, & Winston).

4 Migration and Marriage Change: A Case Study of Mlomp, A Joola Village in Southern Senegal

CATHERINE ENEL, GILLES PISON AND MONIQUE LEFEBVRE

Introduction

Migration is an important demographic phenomenon in developing countries where it represents one of the main forces for social change. It affects marriage particularly strongly. Because of migration we see a florescence of new types of marriage that are linked to the relaxing of social control that takes place when young people live far from their families, especially in cities. Even in rural locales traditional marriage forms, while still dominant, are affected by migration: young girls may postpone marriages or even cancel marriage agreements while waiting for their migrant fiancés to return home, which leads to an increase in age at first marriage.

Although there is clearly a relationship between migration and marriage change, its mechanisms have rarely been described in a detailed way. Small-scale studies allow a more detailed exploration of what one only begins to glimpse from large surveys. The example chosen here is the village of Mlomp, in the Departement of Oussouye, in the Ziguinchor Region, south-western Senegal; we have been studying this village for several years. It has a population of slightly more than 6,000 people, of Joola ethnicity. The great majority are animists, but a substantial proportion are Catholics and a few are Muslims.

The Ziguinchor Region has some of the most important migration flows in all of Senegal. The majority of women who are not yet married work in the city as domestic employees for more than half the year. Most of the men, single or married, are also absent for a large portion of the year. We are looking at the influences that these long-established seasonal migrations have had in the past and continue to have today on marriage change in this population.

The data presented here come from three surveys:

1. a census of the population of the village of Mlomp which we carried out in 1984;
2. an ethnographic survey with open-ended interviews on marriage, also conducted in this village;
3. a comparative survey based on questionnaires administered in January of 1988 to a limited sample consisting of married women who stayed in

the village and women born in the village but who had outmigrated and married in the city.

We will first show the prevalence of seasonal migrations and their principal characteristics through the use of the census of population of Mlomp. We will then describe the most notable features of marriage and will point out the correlation between marriage change over the past fifty years and the development of the seasonal migratory movements. Finally, we will present our methodology and a few results from our comparative survey of forms of marriage adopted by women who have stayed in the village and those who have outmigrated to the city.

The Agricultural Calendar and Seasonal Migration

Agricultural production in the region's villages is dominated by the cultivation of rice that is consumed locally and not sold. In Mlomp, the intensive agricultural technique with transplanting of rice seedlings is still practised (Pelissier, 1966). The agricultural calendar is organized around the yearly alternations of a rainy season, from May to October, and a dry season, from November to April (Fig. 4.1). The tillage and the maintenance work required for the dikes and canals surrounding the rice paddies, tasks which fall on the shoulders of men, stretch from July to September; the transplanting of rice, from August to October; and then the harvesting of crops, from November to February, is done by women.

The less busy times of the year are filled with activities such as the construction or maintenance of houses, the mowing of straw to cover the roofs, the collection of wood for fires, and so on, but these periods are used especially for going away to earn money. From the end of the period of tillage, the single and married men leave the village one by one to go to harvest palm wine in the areas surrounding the cities or in other rural areas of the country. They come back home for the sowing or tilling of the soil the following year, after absences that sometimes exceed six months. Mlomp, of course, is not the only town in the area to experience important seasonal migratory movements. Many studies conducted in the Ziguinchor Region mention the widespread seasonal migrations in this part of Senegal (Pelissier, 1966; Diop, n.d.; De Jonge *et al.*, 1978; Linares, 1984).

After the rice harvest or even at the end of the transplanting, girls and young women in their turn leave to find work in the city as cooks, housekeepers, or child-care workers. They come back to the village six to nine months later to help transplant the rice. The married women do not participate in these migratory movements, but stay in the village for the entire year. If they leave, it is to seek medical attention or to visit family, and their absences do not last for long.

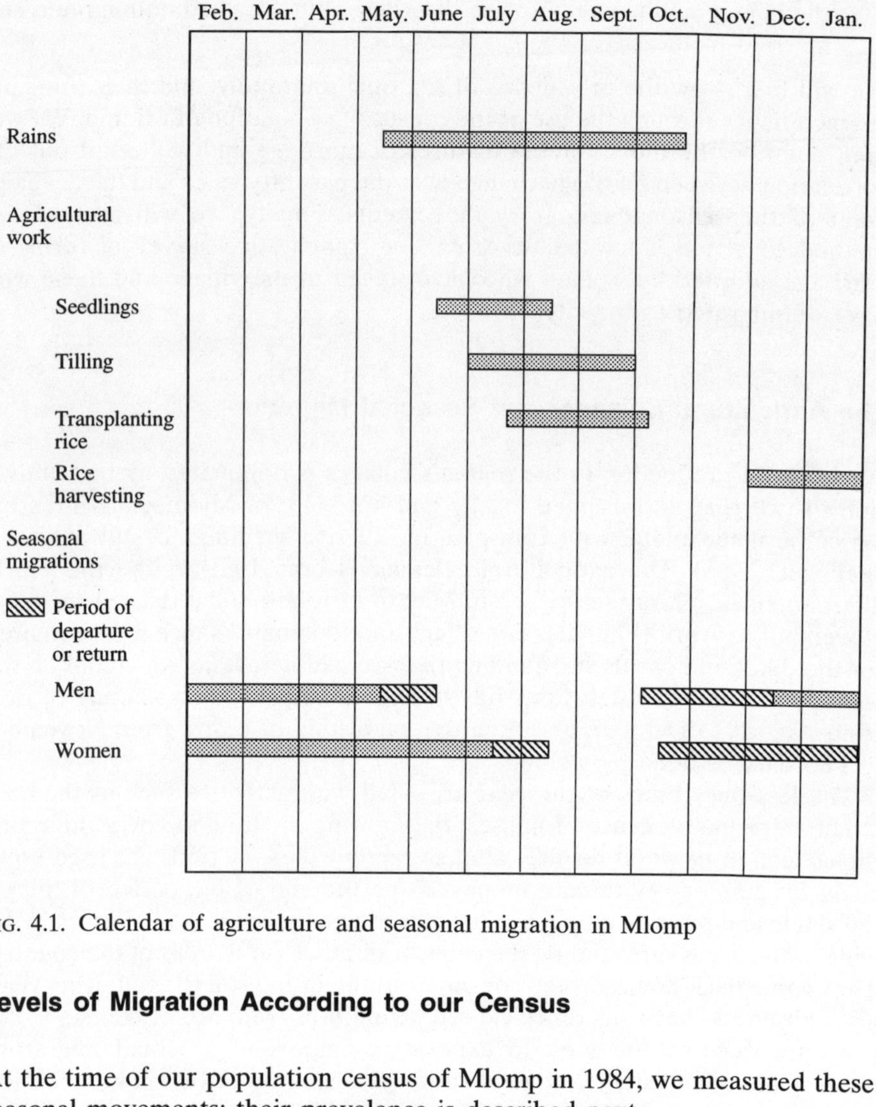

FIG. 4.1. Calendar of agriculture and seasonal migration in Mlomp

Levels of Migration According to our Census

At the time of our population census of Mlomp in 1984, we measured these seasonal movements; their prevalence is described next.

Definitive or Long-Lasting Migration

Age distribution pyramids are a useful tool to describe the natural flow of births and deaths. They also reflect patterns of departures and arrivals affecting different age groups. The age pyramid of Mlomp's population, including seasonal migrants (Fig. 4.2), shows two large irregularities. First, there is a deficit of population in the adult ages, between 25 and 44 years of age for

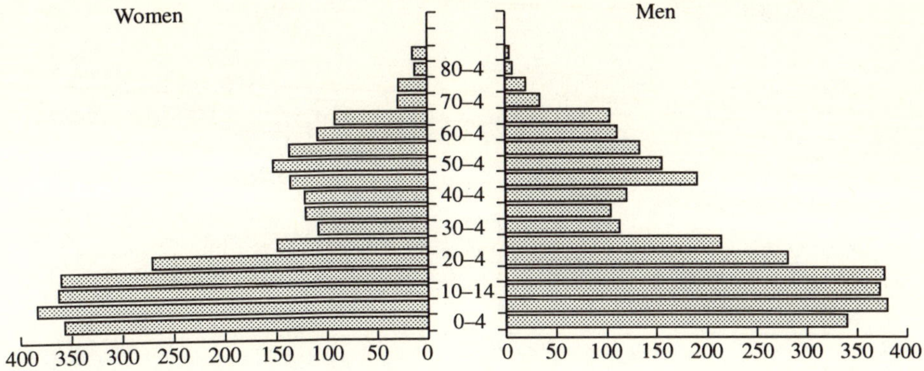

FIG. 4.2. Age pyramid of 1984 Mlomp population

women, and between 30 and 49 years of age for men; and then a second deficit at the base of the pyramid, between 0 and 4 years. These anomalies are too significant to result from errors in the data and age misreporting, though these are typical problems encountered when working with demographic data from Africa.

The first trough is definitely the mark of past outmigration. Long affecting most of the rural areas of Senegal, migratory movements increased in Mlomp among people born about 40 or 50 years ago. Thus we see a lack of people who reached adulthood towards the end of the 1950s and the beginning of the 1960s.

The second deficit at the base of the pyramid, less pronounced than the first, could be the result of a recent decline in the birth rate, resulting from a smaller and smaller proportion of women at fertile ages. In fact, however, it is explained by a relative excess of children over 5 years of age. Many of these children have mothers who live in the city and as a result they have been placed with foster families, in general at the home of one of their grandmothers.

In Mlomp, then, as in many rural areas, migrations are very strongly reflected in the age distribution of the population. The distortions in the age pyramid resulting from long-term migration are exaggerated further by the contribution of seasonal migrations.

Seasonal Migrations

At the time of our census in 1984, we systematically collected information on the migratory movements of each inhabitant over the preceding year, in particular the destination of each move, the dates of departure and return, and the reason for the move. To limit omissions and serious errors in the dates, we registered the presence or absence of each person at certain key

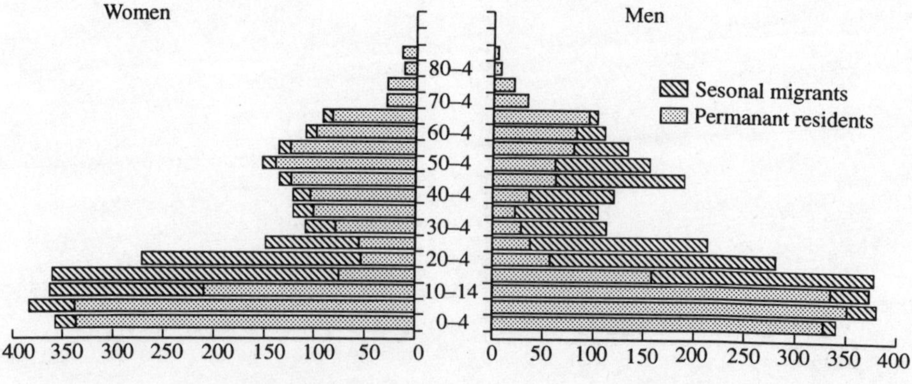

FIG. 4.3. Age pyramid of 1984 Mlomp population: permanent residents and seasonal migrants

moments during the year: the holiday of 'kaamaangen', for example, which marks the end of the harvest in February; or 15 August, which has become a celebration of the reunion of all of the inhabitants of the village, and is thus an important reference point for understanding the events that interest us here. This information on moves permitted us to classify the inhabitants of the village in two categories, seasonal migrants and permanent residents, by whether they had been absent for more than a month during the preceding year.

The age pyramid in Figure 4.3 takes Figure 4.2 and adds the distinction between seasonal migrants and permanent residents. One can see that 38 per cent of the population were absent for part of the year preceding the census, and that seasonal migrations affect both sexes, although in different ways. For males, seasonal migrants are few in number among those under 15: 10 per cent or less. After this age, migrants represent the majority: between 50 and 80 per cent in all age groups up to age 60. Above 60, the proportion declines rapidly. The maximum, 82 per cent migrants, is reached by those in age groups 20 to 40, but the proportion stays very high up to age 60.

For women, seasonal migrations start earlier, at about age 10, and stop earlier as well. After 30, only 10 to 20 per cent of women are seasonal migrants. Between these two ages, migration is very heavy, the maximum of 80 per cent being achieved by women of 15 to 24 years of age.

The distribution of migrants by the reasons for their migration confirms the two dominant causes mentioned already: the harvest of palm wine for men, which employs 60 per cent of male migrants over 20 years of age, and work as 'maids' for young girls, which employs 56 per cent of migrant women aged 10 to 25. These are clearly the second most important jobs after the cultivation of rice. Many children are educated in the village and those who attend school elsewhere make up only a small proportion of migrants, even among those under 20 years old. In contrast, children who have never attended school or those who left school early migrate at very young ages. Boys accompany their fathers or big brothers, and girls accompany their older sisters. The duration of the time spent in migration is on average 7 months for men and 8 months for women.

Earning money is the principal reason for seasonal migration among all the young girls of Mlomp who have chosen to go to work in the city rather than continue their studies. With this money, they acquire a degree of financial autonomy which they use especially to buy new clothes, trinkets, beauty products, and some kitchen utensils to keep for the future when they marry. (We often heard the laments of married women in the village who had suffered from the financial dependence they had experienced once they married.) Certain young women are also pressured by their families to help support their brothers' and sisters' schooling, and to contribute to the daily household finances.

Seasonal movements thus dominate the life of Mlomp and other villages in the area, and have important consequences for their demography. The first of these demographic consequences is the substantial fluctuation in the population present in Mlomp; it varies by about 100 per cent with a 'full' period from August to October, an 'empty' period from February to April, and intermediary periods during which departures and returns are staggered. In the empty period, as our population pyramid in Figure 4.3 shows, the village is populated mainly by children, old people, and adult women; all of the youths and adult men are missing.

Besides these consequences, the migrations have others which are less easy to discern. The separation of husbands and wives for several months of the year is undoubtedly responsible in part for relatively long spacing between births—29 months on average. In spite of the partial abandonment of sexual abstinence during breast-feeding after childbirth, women abstain from sexual relations during 4 months on average and do not use contraception. The long periods spent away from home provide the opportunity for married men to enter other unions, and for unmarried boys or girls to experiment with certain types of union before marrying by Joola custom. We now examine more closely what the development of seasonal migration implies for changes in marriage among the Joola.

Marriage in Mlomp and How it has Changed during the Past Fifty Years

The Stages of First Marriage

Previous ethnographic studies (Thomas, 1959) and our own survey of several older people allow us to reconstitute the main stages of first marriage:

1. The marriage promise is often concluded very early by the parents, when the future spouses may still be young children.
2. The young man ratifies this promise only much later, generally in the year of the marriage, by the symbolic offering of a chicken to the young woman promised to him; he still often knows her only by name.
3. During the high palm wine season (October to June), he must pay a marriage compensation or *bufos* in the form of palm wine or a pig as a sign of respect to the ancestral spirits and their shrines in the neighbourhood and his fiancée's kin.
4. On the occasion of the festival at the end of the harvest, the young couples to be married that year get to know each other on a compulsory retreat. Since the important stages of marriage take place at more or less the same time for everyone, the ties created between couples of the same year are permanent, and all married people, whatever their ages, are able to remember clearly the names of those who were married the same year that they were.
5. At the end of the rainy season and during the dry season, the future husband builds his house, helped by the men of his compound.
6. Finally, the young couple are installed under the same roof in the month of June; at that time, a gift of rice is made from the relatives of the bride, a sort of reimbursement for the wine given at the time of the marriage compensation. It is also at this stage of the marriage that the young bride receives household utensils for her trousseau (mortar, baskets of all sorts, water pots, and so on).

Nowadays, parents rarely intervene in the choice of a spouse, but overall, the stages of marriage are always the same and a new couple do not obtain permission to settle under the same roof in the village if they have not respected these traditions. None the less, new marriage forms are appearing: certain people marry by civil or religious ceremony (basically through Catholic ritual). If the young couple who marry in the church are allowed to live together after this ceremony, it is on the condition that the young man first pay the relatives of his wife a sort of *bufos*, a sum of money whose amount is set by these relatives.

Marriage Compensation, the Trousseau, and Other Expenses

In Mlomp, as in nearly all Africa, marrying is a costly business. Before the beginning of this century, the inhabitants of Mlomp did not customarily harvest the palm wine for the *bufos*; they exchanged their rice in neighbouring villages to obtain it. And it was the parents of the future husband who assumed this expense, since the young man had no rice; rice paddies were distributed only among married men. As for the items for the trousseau that young women would take to their new marital homes, they were produced locally and did not require money to buy; it was the relatives of the girls who provided these things.

The Joola always lacked one important thing, namely cloth. The result was that already, at the beginning of the twentieth century, in order to get what they needed to clothe themselves, men were leaving for the forest to harvest rubber that they would sell in Ziguinchor. With the money that they earned they bought clothes. After the fall in the price of rubber, they began to go and sell the rice or peanuts they cultivated. Then they began to harvest palm wine, an activity which, before that time, had low prestige. The reason was that if a man was obliged to climb palm trees to harvest the sap, he was suspected to be a mediocre cultivator who could not produce sufficient rice for the needs of his family and the requirements of trading.

The wave of seasonal migration touched young women several decades after the men; they went to work, most notably in the port of Ziguinchor, filling the bags of peanuts in the cargo ships leaving for Europe. The possibility of working as maids developed only later, perhaps under the influence of Christian missions that had always employed young girls. These missions thus served as intermediaries for the first colonial administrators who were looking for personal domestic help. Then, with the establishment of a class of officials and people earning salaries, the need for servants increased, but the role of the missions in placing the women became less important. Now the girls look for work on their own, either by going door to door, or by asking relations and friends who have already found work. The fact that the young girls who go to work as maids in Senegal, a country that is largely Islamic, are mostly from areas where the Christian missions are firmly ensconced bespeaks the crucial role that the missions played in stimulating migration in their areas of origin. This is especially important for explaining why the Ziguinchor Region is today one of the main sources of maids for Dakar, Banjul, and for all the other cities of Senegal and The Gambia.

With the increasing monetarization of the economy and the evolution of people's needs in the direction of manufactured goods, most things today must be bought with money. Bit by bit, families have passed on the responsibility for acquiring these things to their daughters' fiancés who will be obliged to assume these expenses from now on. A young man must bear the burden of buying several hundred litres of palm wine, which is sold in the village for 100 Francs CFA per litre in 1988 (1,000 Francs CFA = 20FF = \$US4), a large pig which costs at least 30,000 Francs CFA, and finally, the costs of constructing and furnishing the house, which range from 100,000 to 300,000 Francs CFA, depending on the size of the building and the quality of the materials used, and whether the husband constructs it himself or relies on builders.

As for the future bride, she must buy kitchen utensils, household linens, and clothing, to prepare for the years to come when she will no longer have the financial autonomy that she enjoyed as a single woman. In addition, the amount of rice that is being offered by the girls' parents to her fiancé as counter-reimbursement, which before consisted of a reserve that might serve for several years, now amounts to only a few sheaves of rice in certain cases. It is therefore necessary to have money to buy bags of rice.

The increase in volume and worth of the goods that the young people must furnish or acquire in order to marry, and the migrations that take place in order to collect the necessary money, have lengthened the period separating the ratification of the marriage promise and the end of the payment of the marriage compensation and the start of cohabitation. As we have seen, fifty years ago this period lasted a bit more than a year; today it stretches on for several years. One may well ask if the families actually find it in their own interest to prolong this situation where the marriage date is pushed back further and further, so that single people, especially daughters, contribute for longer periods to the family's finances in the village. We shall see the importance, from a demographic perspective, of these recent changes related to nuptiality.

Smaller Marriage Years

As in nearly every rural area in sub-Saharan Africa, marriages are not registered in Mlomp, except when they are accompanied by a civil or Catholic wedding, a rather recent phenomenon in the village and one far from common. We have none the less been able to estimate the number of marriages concluded each year over the past fifty years, because of the existence of the 'marriage years' mentioned previously.

At the time of the census in 1984, we systematically asked adults who had been married at least once to give us the names of all of the people in their year. These lists included the names of the marriage cohort members still living in the village, of those who were dead, and of those who outmigrated. Since each list was drawn together nearly as many times as there were numbers of people in the marriage group left in the village, we were able to correct for certain errors. We excluded information on marriage groups which had no member still alive or residing in the village. We thus had to cut off the reconstitution of marriage cohorts at 1930, since the risk of omitting entire groups before that year was too great. After this date, however, this risk is very low. To make our analysis as reliable as possible, we focus on the period between 1936 and 1986. The year of marriage of each cohort (in the sense of the year they began living together, which often takes place in the month of August, and not in June as happened earlier) was determined by examining official documents for members of the younger generations who married in the church or in civil ceremonies the same year that they had customary marriages. For the other cohorts, we relied on the reports of those surveyed, using a local historical calendar, or the dates indicated above the doors of certain houses constructed at the time of the first marriage of the household head. In spite of the numerous ways of checking, this reconstitution

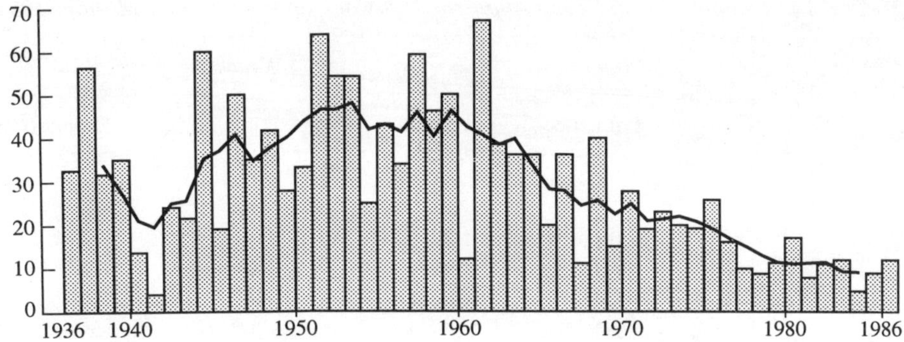

FIG. 4.4. Change in the annual no. of first marriages in Mlomp between 1936 and 1986: observed figures and shifting 5-year means

of the marriages over the past fifty years must still contain a few errors. None the less, it permits what we consider a highly accurate way of examining the important trends in nuptiality during this period.

Figure 4.4 shows the change in the annual number of first marriages we were able to reconstitute that were entered at Mlomp between 1936 and 1986. The figure shows a net increase at the beginning, and peaks starting in the early 1950s; then the trend turns around towards the end of the 1950s. The decrease which follows accelerates during the decade of the 1960s, ending up in a very low overall level today. From 35 marriages a year on average before the Second World War, the figure went to a maximum of about 50 marriages a year in the 1950s, then descended to 20 at the beginning of the 1970s, and to only 10 a year at the start of the 1980s. Note also on the one hand the very few marriages entered during the War, between 1940 and 1943, because of the mobilization of a sizeable group of young people who were of marriageable age at the time, and on the other, the revival of marriages that followed their return in 1944.

The increasing trend in the number of marriages in the 1930s and 1940s is certainly due in part to the increase in the population. There is no reason to think that this would have stopped around 1950 and that the population suddenly began to decline in size. On the contrary, population growth probably increased. There are three possible causes of the stagnation in the number of marriages starting in 1950 and the subsequent drop: the outmigration of a certain proportion of young people before marriage, delay in marrying, and even refusal to marry, at least in the traditional form. We can immediately set aside this last possibility, because it is still impossible for a couple to settle under the same roof in the village without having paid marriage compensation. And, as already stated, even in the city, marriage with the *bufos* has a tendency to be maintained. In contrast, the two other explanations proposed for the drop in the annual number of first marriages seem confirmed: we

TABLE 4.1. *Sample size and proportion single, by sex and age group: Mlomp, 1984*

Age groups (yrs.)	Men		Women	
	Total no.	Proportion single %	Total no.	Proportion single %
0–4	342	100	355	100
5–9	381	100	383	100
10–14	371	100	363	100
15–19	376	100	360	99
20–4	281	99	269	89
25–9	215	92	149	59
30–4	116	66	109	25
35–9	106	34	121	14
40–4	122	14	119	3
45–9	190	7	134	1
50–4	156	2	150	1
55–9	135	2	136	1
60–4	114	1	110	1
65–9	102	2	90	1
70–4	36	3	30	0
75–9	21	0	30	0
80–4	8	0	12	0
85+	4	0	15	0
TOTAL	3,076		2,935	

had suspected the outmigration of young people when we looked at the age pyramids, and we shall examine more closely now the delay in age at first marriage.

The Characteristics of Marriage in Mlomp

The definition of marriage used in our census of the population of Mlomp in 1984 is that of a union recognized by the families of the spouses and by the ethnic group in general, which permits the new couple to live together. When it is a matter of a couple in the village, as noted, the young man must fulfil all the obligations of the marriage compensation. That does not preclude these unions from further sanction by a civil or religious authority. (In the village, one can easily distinguish between informal and formal marriages, in so far as cohabitation is not permitted in the former.)

If we use this definition of marriage, the proportions of single people by age and sex are indicated in Table 4.1 and illustrated in Figures 4.5 and 4.6. The seasonal migrants are included in the population here. One notes a near absence of terminal celibacy among women, a characteristic typical of African

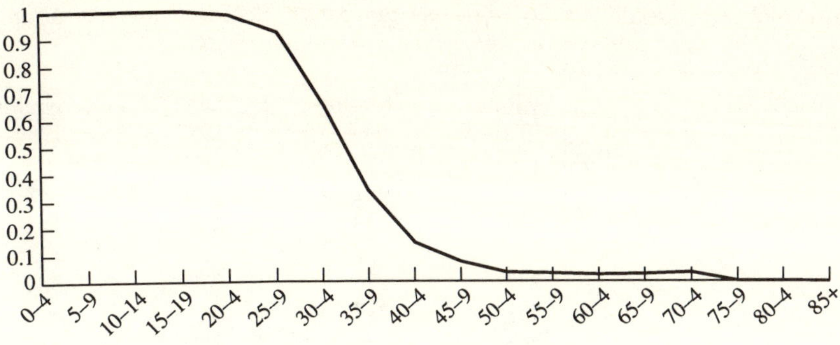

FIG. 4.5. Proportions of single men, by age: Mlomp, 1984

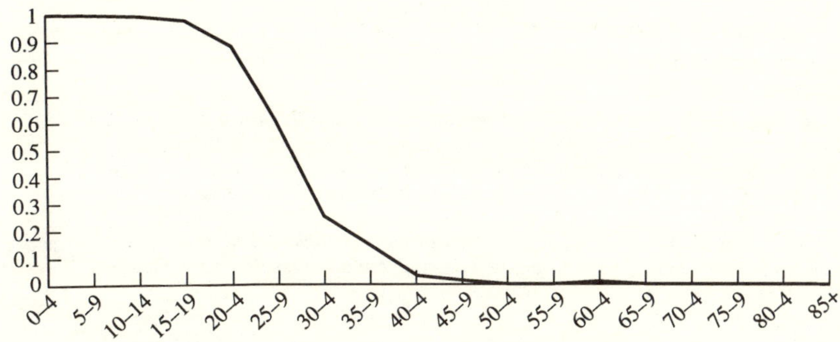

FIG. 4.6. Proportions of single women, by age: Mlomp, 1984

demography. In contrast, and here we add a result not shown in this table, there is a very low rate of polygyny: less than 2 per cent of married women are in polygynous households. This is a surprising fact for Senegal, a country known for its widespread polygyny: in 1986, 46 per cent of married women were in a polygynous union (Ministère de l'Économie et des Finances, Senegal, 1988). Another surprising characteristic is that age at first marriage is very high: 24 years for women and 30 years for men in the period between 1980 and 1986. In all Senegal, women marry much younger: 17 years on average (ibid.).

Although imprecise, the information on the older generations shows that age at first marriage has been fairly late for a long time, in comparison with other populations of Senegal; but it did not reach the high levels it has today. There is therefore, surely, a recent increase in age at first marriage.

As we explained earlier, only unions that are socially recognized in the village are taken into account here. Table 4.1 is missing all the other relationships which bind young people in the village. Young people get together at dances, which are the main source of entertainment during the winter season, and

TABLE 4.2. *Age-specific fertility rates (per thousand). Recent changes in Mlomp and comparison with whole Sénégal and Niakhar*

Age groups (yrs.)	Mlomp 1970–9	Mlomp 1980–90	Senegal 1982–6	Niakhar 1984–90
10–14	5	1	0	3
15–19	35	39	159	161
20–4	167	115	279	323
25–9	249	176	278	334
30–4	281	247	261	309
35–9	268	239	199	259
40–4	157	156	107	143
45–9	35	40	40	31
50–4	12	6	0	0
TOTAL FERTILITY RATE	6,1	5,1	6,6	7,8

Sources:
Mlomp: Lagarde, E. *et al.*, *Niveaux et tendances démographiques à Mlomp (Sénégal)*. Dossier et Recherches de l'INED (forthcoming).
Senega: Ndiaye, S. *et al.* (1988). *Enquête démographique et de santé au Sénégal 1986*. Ministère de l'Economie et des Finances et Institute for Resource Development/ Westinghouse.
Niakhar: Projet Niakhar (1992). *Population et santé à Niakhar. Niveaux et tendances des principaux indicateurs démographiques et épidémiologiques de la zone d'étude 1984–91*. ORSTOM, Dakar, Sénégal, 76 pp.

very often the evenings end in the bedrooms of the boys. The table is also missing the sorts of union that seasonal migrants are involved in, those who are considered unmarried in the village and declare themselves as such in the census. The final types of union the table does not show are the relationships entered by married men during their seasonal work.

Did the development of seasonal migration and the rise in age at first marriage affect fertility in Mlomp? To examine recent fertility changes, age-specific fertility rates for the periods 1970–9 and 1980–90 were derived from birth histories collected in 1984 and combined with the data collected afterwards by the prospective survey (Lagarde *et al.*, 1994). They are presented in Table 4.2 and Figures 4.7 and 4.8. National estimates obtained for Senegal by the DHS survey of 1986 (Senegal, 1988) and recent estimates for Niakhar, another rural area of Senegal (Projet Niakhar, 1992), are also presented for comparison.

Total fertility has decreased in Mlomp from 6.1 children per woman in 1970–9 to 5.1 in 1980–90. The decrease is concentrated in young age groups, especially at 20–4 and 25–9 years: fertility fell by one-third at these ages. By contrast, there has been little or no change among women aged 30 and older. In 1986, women aged 50–69 had delivered 7.1 live children on average, an indication that fertility started to decrease before 1970. Figure 4.7 shows the very large differences in age-specific fertility rates at young ages, before age

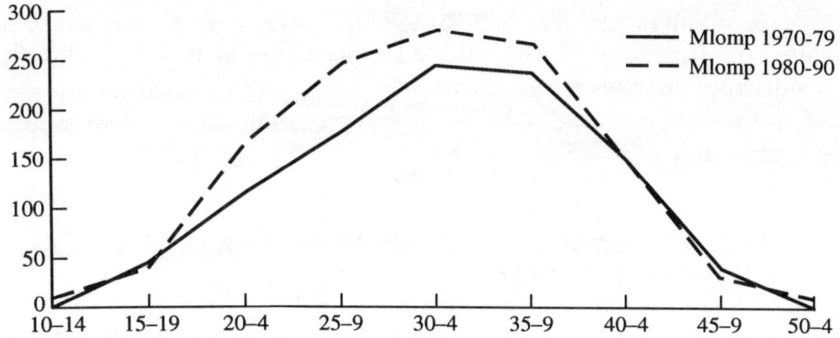

FIG. 4.7. Age-specific fertility rates, Mlomp, 1970–9 and 1980–90

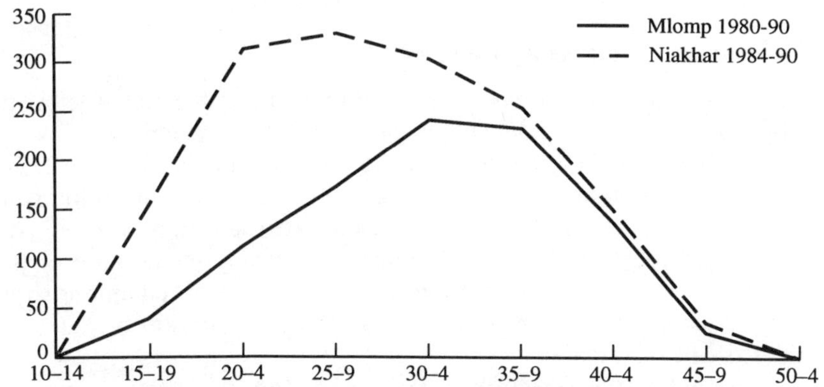

FIG. 4.8. Age-specific fertility rates, comparison between two rural areas of Senegal: Mlomp, 1980–90 and Niakhar, 1984–90

30, in the 1980s in Mlomp and in Niakhar. In contrast, after age 35 the rates are similar.

The drop in fertility in Mlomp before age 30 is associated with a late entry into childbearing: the median age at first birth was 21.6 years during the period 1985–90 (Lagarde *et al.*, 1994). This is 2.6 years higher than the median age at first birth of young Senegalese women in general (19.0 years for women aged 20–9 years in 1986) and 3.6 years higher than for all rural women (18.0 years) (Senegal, 1988). The other factors of fertility, in particular the length of birth intervals and the prevalence of contraception, are not different from other parts of Senegal. The length of birth intervals was 29 months (median) in Mlomp during the period 1985–90 and did not vary according to birth order; the first interval especially was not different from the others. And contraception was at very low levels, as in the whole country.

The postponement of marriage, therefore, is not directly responsible for the postponement of childbearing in Mlomp, since it is celebrated several years after the first birth. Rather, these changes appear to be parallel. Both may result from changes in the status of women, who gained increasing independence in the last decades, in part because of their engagement in regular seasonal migrations.

A Comparison of Women who Remain in the Village with those who have Migrated to the City

Using a small survey on marriage which we conducted in the village and in the city at the same time, we tried to examine what differentiates the marriage experience of women who stay in their villages of origin from that of those who have outmigrated to Dakar.

Methodology: A Comparison of Sisters

In order to compare women who had more or less the same characteristics, aside from their place of residence, we chose to interview pairs of sisters by the same mother, in which one lived in the village, and the other in Dakar. It was necessary to limit the number of participants in this pilot study, although we had decided to look only at sisters who said they were married, since we wanted to examine the differences specifically in the area of marriage. Of course, such a selection criterion implies that informal unions should be out of our scope, but we will see later in the presentation of the results that this was not systematically the case.

To operationalize the selection of pairs of sisters to interview, we used the retrospective fertility survey that accompanied our census in 1984. We researched women's birth histories mentioning the existence of at least two married daughters in which one lived in Mlomp and the other in Dakar. This, of course, could only involve the birth histories of relatively old women. We found twenty-five of them in this situation, thus obtaining twenty-five pairs of sisters who fitted our criteria.

Using the questionnaire designed especially for this study, we first interviewed the sister living in the village, and we asked her the address of her sister in Dakar, hoping to find her at a later date. Certain pairs of sisters had to be eliminated from our study because the sister who was thought to be in Dakar had left the Senegalese capital between 1984—the date of our census—and January 1988—the date of this study. For those who had stayed, the addresses were usually very imprecise. Thanks to the competence of the Joola informant who helped us with the Dakar portion of the study, we were eventually able to relocate and interview sixteen women in the space of one week. A seventeenth woman was interviewed in Mlomp where she was visiting in January of 1988. In all, then, we were able to interview seventeen of the twenty-five pairs of sisters picked out at the start.

The First Results

The first part of this section will provide a brief summary of the socio-economic characteristics of the women interviewed as indicated by our survey. Then we will look at the range of the types of union revealed by the findings.

1. *Socio-economic characteristics of the women interviewed.* With regard to age, no woman interviewed was under 25 years of age, and the majority of the women were between 35 and 55 years old. The fact that they are not very young is largely an artefact of how the sample was constructed. We included only married women and, as we have seen, age at marriage is rather late among the Joola. The sisters who had migrated to Dakar are, on average, slightly younger than the sisters at Mlomp, and a look at birth order confirms that they are more often younger daughters (in eleven pairs out of seventeen) than older (in the six remaining pairs). But one cannot speak about a generational difference since the spread in ages is very limited. The women in the sample have no schooling, with the exception of four of those living in Dakar. In these cases, two had completed primary education and two had attended secondary school.

If the spread in age and schooling are negligible between the women living in rural areas and sisters who left for the city, this is not the case for other factors, such as religious affiliation. In Mlomp, ten women consider themselves animists, while seven consider themselves Catholics. In Dakar, animism has completely vanished as an affiliation: twelve of the women said they were Catholic, one converted from animism to Islam to marry, three converted from Catholicism to Islam to marry; the seventeenth belongs to a protestant sect (she is a Jehovah's Witness). The dominant religion, after animism, is thus, in the two cases together, Catholicism. It seems that among women from Dakar, conversion to Islam is not motivated by religious conviction; conversion appears to be merely one of the conditions that a woman must fulfil when her future husband is Muslim.

As for the occupations that the women are in, those in Mlomp all cultivate rice. As one might expect, this is not at all the case in Dakar where, in contrast, there is a great range of occupations. Three women are retail sellers of palm wine—an occupation that has its origins in Joola culture—and we interviewed them in their little stalls. They sell mostly wine harvested by the men from the village who come to the Dakar area seasonally.

Less than a third of the women in Dakar (five out of seventeen, of whom two were out of work) work as maids, which is surprising when one knows that most of the women came regularly to the city for several years preceding their marriages to find this type of work (see also Diop, n.d.; Myrstad, 1980). There remain six women who declare themselves to be housewives, that is, they have no paid work and their husbands have jobs. Finally, there is a seamstress, a retail seller of fish in the market, and a management secretary. Curiously, the small-scale commerce so common among African women is

TABLE 4.3. *Declared reason for migrating to Dakar, among seventeen women*

Reasons for outmigrating to Dakar	No. of women living in Dakar	No. of these women having current or last husband in Dakar[1]
To sell palm wine	2	2
To work as a maid	11	9[2]
To study	1	0[2]
Divorced from a union in the village	2	1[2]
'Had too many children at once'[3]	1	1
TOTAL	17	13[2]

[1] Two women are widows who have not remarried.
[2] Those who make up the difference knew their future husband from the village.
[3] Had triplets twice and twins once.

not well represented among these women, but it is true that in general the Joola are not business people.

Among the husbands of the women living in Dakar, there is only one who cultivates rice, while the seventeen husbands of the women living in Mlomp all do this for a living. The husband from Dakar who cultivates rice is bigamous and splits his time between his first wife in Mlomp, and his second wife in Dakar. Five of the seventeen rice-cultivating husbands from Mlomp also have another job: four harvest palm wine, one is a mason. In contrast, among the four husbands who are of Joola origin and live in Dakar with wives in our study, we find none of the professional activities pursued in the village: one is a bookkeeper, one is a retired mechanic, one is an instructor at the university, and the profession of the fourth is unknown. The sample of professions becomes even greater when we consider non-Joola husbands: these men are more likely to be artisans (three) and office workers (five).

Finally, with regard to migrations to the city, the seventeen women living in Mlomp had all been to work in the city at one time or another before they married and settled in the village. After their marriages, three divorced and subsequently stayed in the city for several years before remarrying in the village. As for the seventeen sisters in Dakar, Table 4.3 summarizes the reasons why they came to the city before the current union or the last union (since we have two widows who remarried among these seventeen women). We see clearly that work is usually the motive for going to the city (thirteen out of seventeen cases). We also see that nearly all the women met their current partners in Dakar (eleven of these thirteen cases).

2. *The diversity in types of marriage*. The survey among the seventeen women of Mlomp shows that all respected the customary marriage traditions of the

TABLE 4.4. *Type of marriage and place of residence, among seventeen women*

Type of current or last marriage	Women living in	
	Mlomp	Dakar
Customary monogamous endogamous	10	—
Customary monogamous exogamous	—	3
Customary bigamous endogamous	—	1
Customary + civil monogamous endogamous	1	2
Customary + Catholic monogamous endogamous	5	—
Customary + Catholic bigamous endogamous	1	—
Customary + Catholic + civil monogamous endogamous	—	1
Customary + Catholic + civil monogamous exogamous	—	1
Customary + Catholic + monogamous exogamous	—	2
Customary + Muslim monogamous exogamous	—	1
Customary + Muslim + civil monogamous exogamous	—	2
Free union monogamous endogamous without cohabitation	—	1
Free union monogamous exogamous with cohabitation	—	1
Free union bigamous exogamous without cohabitation	—	1
Free union bigomous exogamous with cohabitation	—	1
TOTAL	17	17

Joola: sixteen live in monogamous unions, only one is in a polygynous union. Note, however, that six of these women also had their marriage blessed by a Catholic priest. One woman who was married in the Catholic Church has a husband who is bigamous. Only one woman was married in a civil ceremony in addition to the traditional marriage. These data are summarized in Table 4.4 where they are also compared to the data from the part of the survey done in Dakar.

In looking at this table, one is struck by the diversity of marriage forms in Dakar. One might have expected that the customary norms of monogamous and endogamous marriage (marriage with member of the same ethnic group; exogamy refers to marriage outside the ethnic group) would have a weaker hold in the urban setting. Yet one clearly notes that the obligatory marital compensation of Joola marriage is often respected (in eleven out of seventeen cases) and that, on the other hand, women who entered exogamous marriages (seven cases) are as respectful of this obligation as the others.

We have seen that in the village civil marriage is very rare. In the city, by contrast, six women out of seventeen are married by law, and five of these six cases also involved a religious marriage (Catholic or Muslim).

To conclude, the data shown in this table raise questions about terminology, since one cannot truly speak of marriage defined as a union sanctioned either by an ethnic group's customary law or by a civil or religious authority: four of the seventeen women interviewed do not fulfil these conditions. Furthermore, in two out of four cases, the partners do not cohabit. These women were included in our study because their sisters in Mlomp declared them to be married, although they themselves said they were in informal unions. This implies that informal unions, which we did not see at all in the village, are more easily accepted, even by the people from the village, when they take place in the city.

Discussion

Has our pilot study brought to light any fundamental differences between the women married in Mlomp and their sisters married in Dakar? The women who live in Dakar seem not to marry later than women in the village, but our sample is restricted and non-representative because of the way it was selected. Two-thirds of the women who outmigrated and married in Dakar are younger sisters. One may ask if there is a family strategy at work here with regard to marriage, such that older daughters are expected to marry in Mlomp and leave more liberty for their younger sisters to marry outside the village. If marriage were very expensive for the family of the young woman, this would make it possible to avoid making this payment repeatedly. This is not the case, however, since these marriages require neither sumptuous ceremonies nor heavy expenses for dowry or trousseau. Parents do not get rich from the marriage of their daughter in the village; this daughter, once she has arrived at the home of her husband's family, no longer has any obligation towards her family of origin, and she does not inherit any of her husband's family's wealth. On the contrary, it may be in the interest of the families to see at least some of their daughters marry in the city where they have easier access to money. Very often women married in the city send money home to their parents, while women married in Mlomp do not have the means to assist their parents financially. Furthermore, the women married in the village have no financial autonomy, unless they have a little business of some sort.

The sisters living in the city have an overall level of education that is slightly higher than that of their sisters in the village. This may explain certain differences in their marriage strategies, most notably a certain social ambition to separate themselves from their place of origin.

As for religion, animism seems to distinguish these women since in Dakar they all state that they have left tradition behind them to some extent, but that they have adopted Catholicism. This religious conviction does not pose any obstacle to a marriage in the city with a Muslim, and conversion to Islam is prevalent in Dakar.

Professions also differentiate the sisters from Mlomp and those from Dakar. Among the occupations of urban sisters, nothing remains of the traditional activities, even if certain women return sometimes to the village to help with the transplanting or harvesting of the rice. Only the retail sellers of palm wine keep in regular contact with the activities of people from the village. Some of the women from the city can even boast that they have no need for salaried work, since they have husbands who can support the entire family. In the village, women have very long days of work, a laborious life which guarantees only what they need to subsist. The attractions of a lifestyle that is rather less hard could be an important element in the decision to settle in the city, with easier access to money being a second motive (see also Bledsoe, 1976).

When these women get married in the city, the family exerts pressure to respect certain traditions, and the marriage ideally combines Joola tradition and the demands of the religious affiliation of the partner. One way to avoid all these pressures is to choose an informal union, even if living in such a union means that the partners will not live together.

Finally, one notes a slight tendency in the city to accept polygynous marriages, or at least bigamous ones, even though the rural Joola rarely enter polygynous unions. This could stem from the fact that urban women are more likely to marry men from other ethnic groups and Muslims, both of whom are more likely than the Joola to practise polygyny. These new pressures are significant enough to offset the usual trend among urban populations to reduce numbers of legally married partners.

Conclusion

In this study of changes in first marriage in a village in southern Senegal, and the relationship of these changes to seasonal migrations, we carried out different types of surveys, demographic and ethnographic, in an effort to develop questionnaires that were adapted to specific local conditions. The resulting information permitted us, in spite of the almost total absence of marriage registration, to collect marriage data for up to fifty years ago.

We also experimented with a survey method that consisted of interviewing pairs of sisters born in the village, in which one had stayed there and the other had left to settle in the city, and comparing them with one another. In spite of its flaws and biases, this method seems promising. We had thought that it would be difficult to find people in a big African city when we did not have precise addresses. We were astonished to be able to find, fairly rapidly, most of the sisters we were looking for living in Dakar. Our success encourages us to extend this study to types of migrant other than married (or declared married) women who have outmigrated.

For this first study, which focused particularly on the situation in the village, we were limited to looking at the first socially recognized unions, those in which marriage compensation permitted cohabitation. We omitted the unions that were not recognized, those that were less stable, or those that did not involve cohabitation. We found that age at first marriage, in the sense of age at the time that the couple settles under the same roof, has increased during the last few years, reaching 24 years on average among women married today, and 30 among men, which is at least five years more than was the case fifty or even only thirty years ago. Such high ages are extraordinary for a rural, western-African population, particularly for a country like Senegal. As in many other regions, the increase in the age at first marriage is linked to the increase in the cost of marriage compensation and to its monetarization, both of which push young people to migrate in order to find work and earn money. The fact that the increase in age at marriage has reached such proportions in this village is due to the fact that the migrations are very intense there.

As in many villages in the Ziguinchor Region, and contrary to what happens in most of the other regions of Senegal, migrations affect girls as much as boys. For what has now been several generations, girls have commonly left for the city for part of the year to work as maids, and this happens for several years until they marry. This contact with the city and the money that they earn there are important factors in their emancipation. Removed from family pressures, they may choose the partners that they want from a wide ethnic and professional range. Schooling also plays a role, but it competes directly with seasonal migration or replaces it altogether. In effect, parents have to choose between pushing their daughters in their studies or sending them to be maids. It is not clear that increasing schooling can further raise the age at first marriage since it is already so high; schooling will certainly promote outmigration, however.

In what might be a first stage in a fertility transition, fertility has clearly declined recently in Mlomp with the particularity that there has been no change in childbearing patterns within marriage so far. The decline of one or two children per woman is concentrated before the age of 30 years and corresponds with a rise in age at first marriage; but it is not directly caused by this rise since first births occur several years before marriage. The delays in entry into childbearing and in first marriage appear to reflect changes in the status of women, who gained increasing independence in the last decades partly because of their participation in systematic seasonal migration.

The follow-up to our study will examine the consequences of this tremendous mobility on the current AIDS epidemic. The seasonal movement of young people between the city and the countryside probably results in an increase in the number of sexual partners during this increasingly long period in which the young remain single. Thus migration represents an important risk factor in the spread of the AIDS epidemic in rural areas, particularly in a region where we know that the cities (especially Ziguinchor) have already been touched significantly.

Acknowledgements

This study benefited from the financial support of the Institut National d'Études Demographiques, the Musée National d'Histoire Naturelle, the Centre National de la Récherche Scientifique (UA 49), and the Institut National de la Santé et de la Récherche Medicale (contract number 858028).

References

Bledsoe, C. (1976), 'Women's Marital Strategies among the Kpelle of Liberia', *Journal of Anthropological Research*, 32: 372–89.

De Jonge, K., van der Klei, J., Meilik, H., and Storm, R. (1978), *Les Migrations en Basse Casamance (Sénégal)* (Leiden: Afrika Studiecentrum).

Diop, M. (n.d.), 'Les Femmes migrantes de Basse-Casamance à Dakar: Conditions de vie et de participation au développement des zones de départ', ENEA/Dakar, typescript account.

Lagarde, E., Pison, G., Lefebvre, M., and Enel, C. (1994), 'Niveaux et tendances démographiques, 1984–1992', *Dossiers et Recherches* (forthcoming) (Paris: Institut National d'Études Démographiques).

Linares, O. F. (1984), 'Households among the Diola of Senegal: Should Norms Enter by the Front or the Back Door?', in R. McNetting, R. R. Wilk, and E. J. Arnould (eds.), *Households: Comparative and Historical Studies of the Domestic Group* (Berkeley, Calif.: University of California Press), 407–445.

Ministère de l'Économie et des Finances, Senegal (1988), *Enquête demographique et de santé au Sénégal, 1986*, Dakar.

Myrstad, L. (1980), 'Migration et urbanisation au Sénégal: La Situation des immigrants dans un quartier central de Dakar: Fass-Delorme', Record of Human Geography, Institute of Geography, University of Oslo.

Pelissier, P. (1966), *Les Paysans du Sénégal: Les Civilisations agraires du Cayor à la Casamance* (Saint-Yrieix: Imprimerie Fabrègue).

Projet Niakhar (1992), *Niveaux et tendances des principaux indicateurs démographiques et épidémiologiques de la zone d'étude, 1984–1991* (Dakar: ORSTOM).

Thomas, L. V. (1959), 'Les Diola: Essai d'analyse fonctionnelle sur une population de Basse Casamance', *Mémoire de l'IFAN*, 55, 2 vols. (Dakar: Institut Français d'Afrique Noir).

Part III

The Question of Legitimacy

5 Marriage Law in Sub-Saharan Africa

IMAN NGONDO A PITSHANDENGE

Introduction

Marriage has perpetually drawn the attention of scholars interested in the fertility and, therefore, the future growth of a population. Because marriage has been treated as the traditional base of human reproduction, transformations within marriage systems have been used to help predict changes in fertility levels, especially in contexts where, as in many African societies, contraceptive practices remain non-existent or ineffectual. Yet as this paper will show, African societies do not worry simply about how many children result: they worry more about how to regulate and protect the status and welfare of women and children. Contemporary concerns, therefore, involve the attempt to clarify the status of unions and of the children who stem from them. Problems presented by polygyny have been especially challenging. Depending on the kind of legislation the state passes, a particular woman may be classified as a wife or a mistress. She and her children may have full rights to inherit her husband's property, or she may have none.

Africa legislators have long recognized the complexities and potential conflicts that multiple partnerships can create. Drawing on elements from precolonial practices, colonial influence, and contemporary movements, they have wrestled to draw up appropriate measures for an Africa that at once builds on the past and accommodates present realities. This task is obviously not easy. Although legislators have been well-intentioned in trying to construct a new order by combining what they considered the best of customary ways and new ideas, the laws they create to govern marriage are often contradictory and incomplete.

Still, despite the difficulties these efforts inevitably encounter, legal measures involving marriage have important effects on reproduction. But they appear to have had less effect on regulating numbers of children than on determining their social standing in their families and their prospects for inheritance. This work will examine legal efforts, primarily in francophone Africa, to regulate:

- the definition of marriage and its various ways of celebration;
- types of union: monogamy or polygyny;
- the inheritance rights of widows.

We begin by describing customary African marriage broadly, laying out a few ordering principles as a basis for analysing the changes that occurred on contact with Western society. We focus then on the legal innovations in indigenous civil codes that the colonial powers attempted to implement. Finally, we show how African legislators in the post-colonial era have tried to adopt the most useful elements from the past as well as to accommodate present concerns in constructing legal measures to govern marriage in contemporary Africa.

Customary African Marriage

Let us first look at the main characteristics of African marriage described by Phillips (1953: 9):

- marriage is potentially polygynous;
- procreation is the foundation of marriage (although a marriage without offspring may survive because of polygyny);
- making 'dowry' (i.e. bridewealth) payments and counter-payments on the occasion of ceremonies is the basis for the principle of legitimacy;
- it is potentially possible to dissolve marriage;
- marriage involves two family groups;
- marriage brings together two people generally within the framework of ethnic endogamy and of clan or lineage exogamy, which produces sexual prohibitions between certain members of a family as well as an often-expressed preference for cross-cousin marriages;
- the woman is generally very young at the time of her first marriage;
- the woman does not have the same power/rights as the man;
- marriage establishes mutual rights and responsibilities concerning financial aid to the two partners.

To these characteristics we may add others. For example, marriage is a matter in which almost every adult individual is involved at some point. Further, African marriage is the end-product of a process, a long flow of events that take place over an extended period, in contrast to the more condensed tradition of marriage in the West. This staging of events makes it quite difficult to specify demographic events consisting of civil or religious ceremonies and the start of sexual activity. The ambiguity of such a marriage start complicates the tasks of demographers as well as legislators in contemporary Africa.

To create peaceful relations among kin groups in the community, African societies developed guidelines for establishing marriage: types of marriage, age at marriage, possible partners, the status of married women, the family membership of children, laws concerning residence, etc. Those things which were forbidden were backed up by sanctions. The risks of sterility, bad fortune, sickness, or death to the guilty party or to friends and relations were the most immediate sanctions. Societies also developed a moral system that operated through self-discipline: a feeling of moral dignity that drove one to fear dishonour or shame, humiliation or disapproval.

This social order was made up of practices and norms that varied from one region to another, but most could be interpreted with considerable flexibility. By contrast to the imperative, punitive cast to Western law, a person who broke the law was as likely to be pardoned or asked to participate in a ritual of atonement as to be sanctioned. Moreover, a good deal of latitude was allowed in the interpretation and application of rules. Among the Bakongo (Zaïre, Congo, and Angola), sexual games among young people were allowed without the sexual relations being licit in character (Mertens, 1942, cited in Bernard 1968: 26). Despite severe condemnation of adultery (especially that of women), sexual relations with the younger brothers of the husband or the younger sisters of the wife were not considered adulterous. Moreover, a man might accord hospitality in the form of the right to intimate relations with a member of the household. Hulstaert (1961) indicates that in equatorial Mongo society (Zaïre) there were forms of official and recognized concubinage: a client could obtain from a polygynous man the right to live with one of his wives. The woman, however, remained the legitimate spouse of the husband, and the children of this new relationship, should there be any, belonged to the husband. In our own research in Yaka country, among people known for their puritanism and their sexual mores, we have observed numerous cases of appealing to a potential 'genitor' when a husband proves sterile. In such instances, people do not speak of the violation of spousal rights (Ngondo a Pitshandenge, 1980). In short, the interpretation of customary morality may stem from particular circumstances.

The Regulation of Marriage during the Colonial Era

French and Belgian colonial legislation concerning the rights of individuals and families was dominated, at least initially, by an effort to avoid clashing with local ways of thinking, and to maintain customary institutions in so far as they did not contradict colonial order. This effort created a legal dualism that marked the entire colonial period. On the one hand there was a body of civil law, modern and written, and on the other there was customary law, passed on verbally, spoken in tribunals and by native judges, and conforming to customs that did not contradict the colonial order. Two types of marriage thus came to be recognized officially: civil marriage, celebrated according to written laws, which included indigenous laws that appeared to conform to modern expectations; and customary marriage, celebrated among the family and regulated by the customary laws of both parties (Verdier, 1972: 12).

For the colonies of French-speaking Africa, two major legal actions spurred this change of direction: the Mandel Decree of 15 June 1939 specified the minimum age at marriage, mandated the consent of both spouses, and affirmed the right and the liberty of a widow to remarry. The Jacquinot Decree of 14 September 1951 limited the total amount of bridewealth, required the husband at the inception of the marriage to state whether the marriage would be monogamous or not, and backed the right of a widow to inheritance claims on her dead husband's property. For the Belgian colonies, additional measures were taken:

1. The decree of 9 July 1939 concerning the protection of the preadolescent girl;
2. The legislative ordinance 37/AIMO of 31 January 1947 concerning polyandry;
3. The decree of 5 July 1948 with respect to monogamous indigenous marriage and the way it would be carried out;
4. The decree of 25 July 1948 concerning adultery and bigamy;
5. The decree of 15 July 1949 on the subject of abandonment of families.
6. The decree of 4 April 1950 with respect to polygyny: a measure that preserved the benefits of legitimacy only for customary polygynous marriages concluded before 1 January 1951.

Even more important for changes in the institution of marriage than these specific decrees were the normative assumptions conveyed through Judaeo-Christian, especially Catholic, doctrine:

- the individual character of the marital vow;
- the legitimacy of marriage conferred exclusively by the sacrament;
- the struggle against polygyny;
- the emphasis on procreation as the first goal of sexual relations between spouses;
- the emphasis on marriage as a permanent link; divorce was not allowable even in case of sterility;
- endogamy with regard to religious persuasion;
- the value placed on virginity before marriage;
- a fundamental equality between a husband and a wife.

Because of the historical complicity between the State and the Church,[1] these precepts imbued colonial marriage initiatives with a Christian flavour. Needless to say, many were quite strange to the African way of thinking.

After Independence: Fusing Cultural Authenticity with Western Norms

This background helps to explain what happened subsequently in legislation concerning African marriage. Shortly after national independence, African governments were confronted by enormous problems of social and economic development. Underdevelopment was perceived not only as an economic phenomenon but as a mental state that demanded the re-establishment of customary structures: a kind of mental decolonization.

Various states set themselves the task of reworking the legal code concerning marriage: Mali (2 February 1962), Guinea (1964), Côte d'Ivoire (7 October 1964), Cameroun (7 July 1966), Senegal (12 June 1972), Gabon (29 July 1972), Congo (1972), and Zaïre (1 August 1987). In initiatives that generally sought to fuse written and customary law, each state tried to make marriage a truly modern institution while preserving local values. In Zaïre, the members of the Reform Commission for civil law who were responsible for editing the legal code on the family[2] were invited to take inspiration from 'the concepts, the modes of expression and the thought of traditional law as well as its methods of conciliation' (Pauwels, 1972: 7). Five essential directives, outlined to guide the editors of the Senegalese code on the family, are illustrative:

1. to refer to the spirit of the customary law rather than to the law strictly defined, even when consistently practised;
2. to respect customary values, but to suppress customs that were anachronistic or maladaptive for socio-economic development;
3. to keep constantly to the spirit of the constitutional principles of secularity, the respect of religious liberties, and the equality of citizens before the law;
4. to respect scrupulously that which is tangible for the faithful, while distinguishing that which is religious from that which is called religious by error or exploitation of the term;
5. to adapt the principles in a clear way to conditions of modern-day life.[3]

In general, marriage legislation in sub-Saharan Africa translates the desire to reconcile modernity and mental decolonization with respect for ancestral legislative tradition through:

- the secularization of marriage;
- the freedom of partner choice combined with the need for the consent of the spouses and parents;
- regulating engagement and bridewealth;
- prohibiting or regulating polygyny;
- clarifying the relationship of a child to its parents and recognizing a child born outside marriage;
- regulating the fate of the surviving spouse if the partner dies.

To assess the success as well as the problems of these efforts to harmonize African marriage of the past with alternative visions, we can examine the legislative decisions on several important points.

Celebrating Marriage

One of the most important moves was to eliminate religion as a requirement of valid marriages. Codes of civil law in the region now state that any union that has been concluded only according to religious code is excluded from legal protection. The code of the Côte d'Ivoire goes so far as to make the celebration of a religious marriage itself contingent on the legal proof of its previous civil celebration.[4] Such moves appear to have comprised strategies within the overall quest for 'mental independence' particularly from Christianity.

Another measure involved freedom of partner choice. While most new African legislation insists on the principle of freedom to choose partners (Congo, Gabon, Cameroon, Mali, and Zaïre) and punishes those who force or oppose the marriage of any citizen, it also decrees, in conformity with African traditions, the consent of individuals exercising paternal power over individuals under 21 years of age.[5]

As to how marriage is celebrated, two doctrines are followed: legal exclusivity reserved only for civil marriages (Côte d'Ivoire, Article 18, and Mali, Article 8); and customary marriage celebrated among one's family. Each option presents advantages and disadvantages to which we will return later. Effectively, however, this uneasy tension does not move far beyond the judicial dualisms in colonialism that we described earlier.

Certain bodies of law (Congo, Senegal, Zaïre) regulate engagement. To maintain the independent character of engagements and ensure that they can be broken freely, the Zaïrian code states that legal action cannot be taken against the couple or their parents if the engagement is broken. When an engagement is broken off, presents must be returned and money and bridewealth refunded according to custom, but reparations cannot be charged for the damage caused by the rupture.[6] Once the marriage process gets under way, the engagement gives couples the right to live together, a possibility expressly laid out, for example, in the legal code of the Congo: not an unrealistic proposition, given the frequency of sexual activity outside marriage. This concession conforms to a philosophy of trial marriage to which many African societies adhered in the past. (Needless to say, distinguishing between the end of an engagement and the start of a marriage is extremely difficult for the demographer to discern.)

The principle of giving bridewealth is not questioned; Article 367 in Zaïrian legislation even stipulates bridewealth as a basic condition of marriage; its payment and acceptance indicate the consent of all parties. Yet African legislators are concerned to regulate the amount of bridewealth (Congo, Gabon, Cameroon, Mali, Côte d'Ivoire, Zaïre, etc.). They seek both to try to keep parents from making bridewealth into a purely commercial transaction and to accommodate regional practices. (According to African custom, the value of a bridewealth may be restored should a marriage end. In Malian law, returning the bridewealth is necessary only when a woman is at fault in the divorce.)

African civil law emphasizes 'the stability of the family' and 'the interest of children' to discourage the legal dissolution of marriage. Customary means of reconciliation included appealing to family elders, resorting to 'conventional separation',[7] prohibiting a legal dissolution of a marriage in its first two years, and so on. These more indigenous means are recognized as ways to preserve the unity of the household, while retaining the intimate and individual character of the marriage.

Polygynous Marriage

On the controversial topic of multiple marriages, one of the most sensitive in family law, francophone African countries have diverged widely. There are three main legal currents.

1. *Exclusive monogamy and the concomitant prohibition of bigamy and polygyny* (Côte d'Ivoire, Zaïre). Rigid tendencies in this model can be perceived immediately: an absence of legal protection for polygynous unions, legal grounds for complaint about bigamy by any wronged person or civil servant, and penal sanctions against anyone who helps to formalize the celebration or registration of a bigamous marriage. In the eyes of the law, polygamous marriages are null. In legal terms, the man and the woman are strangers to one another and have no obligation to feed one another, to be faithful to one another, or to assist one another in any way. The rupture of their marriage which, in effect, does not exist, is a licit act.

2. *Optional monogamy*. Some countries have mandated a contractual promise of monogamy made by the husband when the marriage is celebrated (Congo, Mali, Togo—see also Locoh, this volume). That is, the husband must declare at the time of the marriage whether he may take another wife in the future. In Congo and Mali, however, the law allows the husband to go back on his promise and take a second wife.[8] The solution of optional monogamy, however, implies a legal recognition of polygyny.

3. *Regulated polygyny with respect to the number of wives a husband may have at one time*. Such countries are primarily those with strong Muslim influences. In Mali and Senegal, for example, the limit is a maximum of four wives.

These solutions obviously represent a broad spectrum. With the exception of the first strategy, however, they suggest that African legislators are inclined toward solutions that accommodate the situation rather than ignoring reality. How do such accommodations affect key issues such as the conjugal status of women, the legitimacy of children, and the inheritance rights of women?

Informal Unions

Polygyny was viewed with disfavour during the colonial period. Because of this negative perception, the practice is today surrounded with circumlocutions that reflect precautions of propriety and respectability. In Zaïre, Gabon, Congo, Cameroon, and Côte d'Ivoire, one speaks today of the 'deuxième bureau' (Lacombe, 1987). African codes on marriage generally remain silent on the topic of individuals who, in spite of the fact that they live in a stable and durable relationship, live outside the institution of legal marriage.[9]

Childbearing

Desires to claim a child have rarely constituted a problem in societies such as those in sub-Saharan Africa where the number of children is both a symbol of strength and a condition of survival in the community. Whether the child is the fruit of a legitimate or non-legitimate union, it is always clearly linked to a lineage and a clan, either through the father or the mother. However, with the transformation of society and the reduction in the role of the extended family, children have tended increasingly to be associated with the nuclear family.

In emphasizing the spouses' union as the basis of the family, though, legislation on marriage subjects a child born outside marriage to the risk of being without a family, of being abandoned or disowned. With this problem in mind, many African legislative bodies try to facilitate links of parentage. They may offer a man the possibility of declaring paternity before a judge, automatically associating a child with its father's family as a result of the birth process, etc. Some legislation goes to the point of requiring the father to recognize his child conceived out of marriage,[10] equalizing the status of a biological child with that of a legitimate child (Gabon, Guinea, Congo),[11] or suppressing any distinctions between children with different links to the father (Côte d'Ivoire). Unless the child has been relinquished, the parents are obliged to maintain and educate the child, in principle until marriage or legal majority. The minimum legal age at marriage, 14 years for a young woman and 16 to 18 years for a young man during the colonial period, has increased now to 17 to 18 years for a woman and 20 to 21 years for a man. (These legal ages of marriage, of course, are generally higher than the mean levels actually observed.)

The Status of the Wife after the Death of her Husband

Because married women are legal minors, the fate of a woman after the death of her partner with respect to remarriage and rights to the husband's property constitutes a vital element of family law. In customary practice, marriage comprised one of the most important elements of alliance strategies between groups. Hence, the death of a partner did not end relations between the groups. The levirate and sororate offered means of both continuing the alliance and permitting the individual woman or man (respectively) who suffered a loss to regain a new partner quickly. Concerning inheritance, the wife is regarded as having had no other property than personal belongings; goods such as furniture and buildings are presumed to have belonged to the husband. Custom tends to give the inheritor of the deceased husband—usually a close male relative—the entire inheritance, composed of his property, children, and widows. In theory this relative manages his inheritance in trust for the extended family.

Not belonging to the husband's family, and being unable to join it, the wife naturally has no right to the goods of her deceased husband. Nowadays, particularly in the changing international climate, ideals of equal rights for women and realizations that the wife contributes increasingly to the accumulation of household goods sometimes lead to legal innovations. Legislators can only look beyond custom and try to innovate. The following provide some examples of legal steps that most francophone countries have taken:

1. According freedom to a widow to remarry a person of her choice, once the period of widowhood has passed:[12] four months in Guinea, 180 days in Gabon, and 300 days in most of the other countries. Efforts to prohibit the levirate (as well as the sororate), of course, can backfire. They can erode kin networks that rest on marital links; they can also make it difficult for a woman to find a partner at all.
2. Prohibiting inhumane physical treatment of the surviving partner.
3. Requiring conjugal partners to select a way to deal with their property at the time of marriage in order to avoid conflict at the time of eventual dissolution. Lacking such a declaration, the community is automatically assigned to allocate the property, but only those items acquired during the marriage.
4. Giving the widow power over the children of the marriage and including her among privileged heirs, as long as she was a partner to a clearly defined marriage and had not divorced or separated from the husband (Cameroon, Gabon, Congo, Senegal, Zaïre).
5. Granting the widow the use of the house inhabited by the spouses, the furniture in it, and eventually, the lands belonging to it that the occupant personally farmed as well as the commercial funds relating to it (Cameroon, Zaïre, Gabon).

In these issues, proposed innovations unavoidably demand a revolution in people's thinking in so far as they threaten very long-standing practices, and affect the vital interests of kinship groups as corporate bodies. Giving such rights to widows who, in the eyes of everyone, belong to entirely different families, will undoubtedly constitute a source of conflict for years to come.

Legislation on Marriage in Africa: How Realistic?

The goal of every law is to try either to model a new behaviour, or to confer a legality upon a new behaviour adopted by an important fraction of people in a country. Yet in Africa, external influences so thoroughly dislocated indigenous structures that national cultural identities are still in flux and being created. Especially now that national harmony is such an important goal, initiatives that seek to codify behaviour to create some sense of order and regularity are essential.

Despite African states' stated wishes to update marriage laws to meet changing circumstances, setbacks appear on many fronts. Efforts to educate the public about new laws and make them commonly accepted are no easy task. We would hardly expect to see quick, spectacular results simply by instituting laws on marriage. State jurisdiction, the cornerstone for national identity, often functions badly if at all. Even in countries where virtually the entire population speaks a common language and is literate and aware of current national events, this would be difficult. African populations, of course, are still largely illiterate and rural; large proportions of citizens are spread out over great expanses of remote territory. They often speak different languages and do not understand the national lingua francas.

Besides these more predictable disadvantages of communication, equally serious obstacles to creating and enforcing marital legislation arise from differences in social status and cultural practices. In most countries, options to register unions accord advantages primarily to the few citizens who have remunerated work; hence, few people feel the need to register their unions. Various legislative bodies, moreover, have found that the concessions they must make to local customs often hark back, ironically, to the legal dualism that developed during the colonial period. Yet because there is such diversity among marital customs, even within well-defined geographical areas, certain groups could not identify their own practices in the resulting compromises. Legal concessions to diverse groups also give certain élites grounds for deviating substantially from the spirit of the legal code by manipulating laws for their own advantage.

A very substantial problem in creating and implementing new marriage laws revolves around inequalities between husbands and wives. Codes on marriage in sub-Saharan Africa remain fundamentally patriarchal, as we can deduce from the primacy accorded to the husband: the head of the family, main supporter of the household, principal authority over the wife and children, etc. It is the husband who can decide to make the marriage monogamous or polygynous. And both bridewealth and the notion of the woman as an instrument of alliance between groups continue to dominate marriage practices. Hence, the obligation to refund a man's bridewealth in case of divorce originally represented an attempt to stabilize a marriage. But it may have the opposite effect of forcing a woman to endure an unfortunate marital situation or search for more suitable partners outside marriage. As a result, ideals of liberating women do not always get applied either in legal codes or in practice.

As long as individuals see little concrete advantage to be gained by complying with new national codes, most marriages will continue to remain tacit contracts between individuals or families requiring nothing more than customary ceremonies. Many legislators go so far as to argue that attempting to implement and enforce any legislation at all for civil marriage is premature, given the problems of tenuous state control over vast, heterogeneous populations. Yet restricting legal protection solely to Western-style marriages prescribed by civil law—and treating informal unions, polygyny, and customary unions as second-class arrangements—is inherently discriminatory. Hence, to treat these unions as marginal is to confer legitimacy on only a minority of the population.

How legislators have dealt with polygyny is the issue most fraught with contradictions. Because African societies base their survival on links between groups, the protective role of the state remains limited, and social security measures are in their formative stages, attempting to suppress polygyny runs the risk of undermining the basic structures of local communities. Despite economic pressures as well as legal attack, polygyny seems to have resisted the restrictions imposed by colonial legislation very well. Recent studies have pointed out the persistence, or even the resurgence of this practice (see e.g. Ngondo a Pitshandenge, 1980, 1988; and Sala-Diakanda, Mpembele, 1980). Indeed, contemporary polygyny is not exclusive to rural areas. In urban areas, one finds new adherents among the new class of bourgeois. Because polygyny involves so many people or their close relatives in one way or another, resistance to abolishing it, though seldom explicit, is strong. To quote Tunc (1966: 11), 'all of the customs are favorable to polygyny, as the African man is attached to it and women are not united in attacking it'.

The logical inconsistencies inherent in attempts to suppress polygyny deepen when we turn to the issue of children. Requiring men to give equal recognition to all their children, those by wives married under civil law as well as by custom, implicitly suggests that their mothers should be given status as recognized wives. Such concessions, if consistently carried out, would contribute to a rapid increase in bigamy. On the other hand, to take pity on the fate of natural children and to pretend to ignore that of their mothers is playing politics with the vision of an ostrich.

Polygynous marriage will continue to both challenge and change law. Practical measures for suppressing polygyny remain timid and the sanctions mere gestures. When an individual grows up in a polygynous environment, as most current African leaders did, when one or another parent was in such an arrangement, and when individuals themselves flirt with it, it is extremely difficult to take up the crusade against polygyny with any conviction.

Efforts to establish measures that promote equity and protect rights in the realm of family and marital law inevitably encounter problems. Legislation in Africa represents a constellation of elements from the precolonial past and the colonial era as well as modern international currents and changing power relations within countries. Because contemporary legislative efforts must build on long histories of tradition as well as debate and compromise, they often effectively intensify contradictions and inequities. As a result, it is impossible to speak of one African marriage system; rather, we must speak of numerous African marriage systems that reflect the multiple, often inconsistent, kinship and legal norms that underlie them.

Acknowledgements

The body of this paper was written during my stay (August 1988) in the Centre Français pour la Population et le Développement (CEPED) in Paris. I wish to express my gratitude to Mr Francis Gendreau and all his colleagues at CEPED for the very kind welcome that they showed me. I would also like to share my feelings of gratitude and friendship towards Mr Gilles Pison for the long discussions we had together in Paris; and to Citizen Lusamba Senda, Secretary of Project ZAI/86/PO1, for his re-reading of my manuscript.

Notes

1. One example among many of this collaboration: the extension of the legal protection accorded to legal marriage in the Belgian Congo by virtue of the decree of 5 July 1948.
2. Law 71–002 of 12 June 1971 published in *Le Journal Officiel*, provisional edn., 15 Dec. 1971, p. 8.
3. In 'The Rights of the Surviving Partner in Senegalese Law', *Acts of the Conference of the Federation of African Lawmakers*, Kinshasa, 20–5 July 1987, p. 2.
4. Article 20 of the code of the Côte d'Ivoire on the family stipulates: 'No minister of a religious group may direct the religious celebration of marriage without its having been justified by the presentation of a certificate showing its legal validity.'
5. Cf. in particular Article 4 of Law 62–17, 3 Feb. 1962.
6. We might conclude this from Articles 344, 345, 346, and 347 of the Family Code of Zaïre, p. 89.
7. In the presentation of the themes of Law 87–010, the main body of family code of Zaïre, 'conventional separation' is rooted in customs which demand that when the home is threatened by a serious conflict, the husband send his wife to his parents' house, or to his parents-in-laws' house while waiting for time to play a stabilizing role for the marriage (cf. *Le Journal Officiel*, special edn., Aug. 1987, p. 14).
8. Article 7 of Law 62–17 of the Mali family code stipulates, however, that 'the man having opted for monogamy may revise his contract *with the explicit consent of the wife*'.
9. Angola is one of the few countries to have dealt with the matter through legislation. A woman in a free union can receive the social and legal benefits of the union while it lasts, and after the death of the man with whom she lived, as long as the cohabitation lasted at least three years. Legislators from Gabon recognize only those free unions linking two persons who had never had any marital link (cf. *Compte Rendu Synthetique*, 004/FJA/87, Conference of the Council of the Federation of African Jurists, Kinshasa, 20–5 July 1987, pp. 9–10).
10. In Zaïre, the principle of 'affiliation' in Article 619 of the family code requires the father to recognize not only every child regardless of the conditions under which

it was born, but also the mother's family. This code has even specified that the maternal family of a child born outside marriage may demand indemnity payments and presents from the father by virtue of custom.

11. There are several restrictions to this principle of equality of rights. The consent of the wife is necessary for a man to recognize a child born of an adulterous union (Côte d'Ivoire) or to bring it into the conjugal household (Zaïre). Also, a child born out of marriage is eligible to inherit if it was affiliated with the father while he was alive.
12. Article 16 of the Gabon Law of 7 July 1966 states, e.g. that 'in case of the dissolution of a marriage by the death of the husband, the heirs . . . have no right to either the person of the widow or her property'.

References

Bernard, G. (1968), *Ville africaine, famille urbaine: Les Enseignants de Kinshasa* (Paris: Mouton).

Hulstaert, G. (1961), *Les Mongo: Aperçu général* (Tervuren: Académie Royale des Sciences d'Outre-Mer), ser. 4.

Lacombe, B. (1987), 'Les Unions informelles en Afrique au Sud du Sahara: Le Cas du deuxième bureau congolais', *Genus,* 43/1–2: 151–63.

Mertens, J. (1942), *The Crowned Chiefs among the Western Bakongo* (Brussels) cited by G. Bernard, in *Ville africaine, famille urbaine.*

Ngondo a Pitshandenge, I. (1980), *De la nuptialité et fécondité des polygames: Le Cas des Yaka de Popokabaka (Zaïre)*, Cahiers de Sciences Humaines, 109 (Tervuren: Musée Royal).

—— (1988), 'De la nuptialité dans les grands centres urbains du Zaïre', text of a seminar presented at the Department of Demography, University of Montreal, August 1988.

Pauwels, J. M. (1972), 'La Réforme du droit civil au Zaïre: Comment concilier la tradition et le développement?,' Colloque de l'Association Internationale de Droit Africain, Bordeaux.

Phillips, A. (1953), *Survey of African Marriage and Family Life* (London: Oxford University Press).

Sala-Diakanda, M. D. (1980), *Approche ethnique des phénomènes démographiques: Le Cas du Zaïre*, CABAY edn., Recherches démographiques, Université Catholique de Louvain, Cahier 4, Louvain la Neuve.

Tunc, A. (1966), *The Legal Aspects of Economic Development* (Paris: Dalloz).

Verdier, R. (1972), *The Evolution of the Legal Status of the Child in Ancient and Modern Law in Black French-Speaking Africa* (Strasburg: no pub.).

6 Types of Marriage and Marital Stability: The Case of the Moba-Gurma of North Togo

MARC PILON

Whereas anthropological research has documented considerable diversity and complexity of marriage systems in African societies, demographic studies of marriage have fallen short. Demographers are particularly handicapped because their concepts and methods were developed for European populations and the kinds of marriage they recognized were quite limited in scope. Because European marriage is monogamous, for example, such studies did not need to distinguish among the statuses of wives or children in polygynous households.

As a result of more diverse research on the family, the study of marriage has drawn renewed interest from demographers. Our study, conducted among the Moba-Gurma of Togo, builds on this increasing interest in marriage trends. It also builds on a series of projects concerning the populations of Togo. In his study of the populations of the Dayes plateau, for example, Vimard recorded marital ruptures by type of marriage: traditional, consensual union, religious, and civil marriage (Vimard, 1983). And Locoh's (1984) work on south-eastern Togo developed a typology for unions, particularly taking into account the fact of separate spousal residence. The findings presented here stem from a rural demographic survey done in 1985 among the Moba-Gurma of north Togo. The data come from retrospective questionnaires which attempted to retrace the marital histories of men and women of all ever-married men and women present in the sample households (896 men and 1,501 women).

This article aims to show the importance of taking specific characteristics into account for studying marriage and marriage change: in particular, type of marriage and dissolution, whether the union is the first for the individual, and whether the woman has the status of first wife. After presenting the different types of marriage observed and examining their positions in the marriage system taken as a whole, the paper will focus on their distribution and how it has changed. Describing the different ways in which unions may end, the article then attempts to shed light on how marital stability can be affected by the type of marriage and whether a woman is the 'first wife' in a polygynous marriage.

Ethnographic Background

The Moba and the Gurma are concentrated in the extreme rural north of Togo. At the 1981 census, 88 per cent of those in Togo resided in the Prefecture of Tône, where they comprise most of the population. The Moba, generally recognized as the prior group, live predominantly in the western part of the district; the conquering Gurma, who began arriving in the eighteenth century, live largely in the east. Though conquerors, Gurma seem to have adopted many of the social and religious practices of the Moba, and together these groups became sufficiently homogenous that we can safely speak of them as one.

Moba-Gurma society is patrilineal and patrilocal. The only visible grouping in space is the residential compound (*naag*), a focus of production and consumption. Living in domestic groups with an average size of nine people, people farm collectively under the authority of an older male head.

Fertility is a central value. Numbers bespeak the power of one clan *vis-à-vis* another. The system of production, relying on human work, requires an ample body of kin to assure the subsistence of the group and later of parents. Social status for the woman therefore rests on her fertility. For men, prestige rests on the number of wives and children. A man, if his descendants are numerous, will be guaranteed the status of a revered ancestor whose name will be pronounced often (Agounké *et al.*, 1991). Male status and well-being ultimately depend on gaining dependants. Men, therefore, need wives who can bear children. Marriage is based on the principle of exogamous exchange among clans and polygyny. Polygyny is extremely important (Pontié and Pilon, 1990). Early female age at marriage, almost inevitable remarriage among still-fecund widows and divorcees, virtual non-existence of unmarried women, and the practice of polygyny all lead to a high fertility rate among both men and women.

Since the beginning of French colonization, the Moba-Gurma area has seen temporary labour migration directed essentially towards Ghana and Côte d'Ivoire, and involving mostly young men and single people (Pilon and Pontié, 1991). For those who remain, extensive subsistence agriculture continues to occupy more than 90 per cent of the population. Principal crops include millet, sorghum, and beans. Although cash crops such as groundnuts and cotton constitute a small part of the crop, they also constitute the principal source of monetary revenue, particularly for women and young men, who cultivate them on their personal fields. The combined effect of a rapid demographic increase and the absence of an actual agricultural intensification leads to an increasing imbalance among population and resources (Pilon, 1991*b*).

Marriage Types among the Moba-Gurma

The logic of exchange of women, the rule of clan exogamy, the practice of polygyny, and the levirate determine the basic structure of marital life among the Moba-Gurma. Several types of marriage may be identified:

marriage by exchange, *puokpedu*
brideservice marriage, or marriage by work, *puokuul*
child betrothal or marriage by promise, *puopaab*
marriage by 'abduction' or consensual union, *puotugnu*
leviratic marriage, *pekual fallu*
marriage by reimbursement/payment, *taalon* (or *puoton*)
bridewealth marriage

Marriage by Exchange (or marriage by work). For various reasons (friendship, alliance strategies, etc.), two heads of domestic groups or lineages (*yaamul*) not belonging to the same clan may decide to exchange young women. This is reflected in the formulaic statement: 'I give you my sister, you give me yours.' The principal justification for marriage by exchange, generally voiced by the elders, is that it enables all men, most notably physically handicapped men, to obtain a wife, as long as they have a woman to exchange (Pontié, 1978). This system of exchange might seem to pose a problem for groups with more eligible males than females. However, the system is in fact quite flexible. Although the woman being exchanged should be in the clan of the man who exchanges her, their kin relationship can be classificatory. Eligibility may extend therefore to many members of the same lineage; there may also be a generational difference between the two members; and the exchange may be simultaneous or deferred.

Brideservice Marriage. For a man who does not have an actual or classificatory sister to exchange, another means of getting a wife entails going to work each year in the fields of a family from whom he eventually expects to obtain a wife. It is said that this is 'a wife earned by the hoe' (*puokuul*). The completion of such a marriage can take years, depending on the age of the girl when the prospective husband begins to work and the length of time her parents can draw out the labour requirement. Though in-laws obviously value this route to marriage, it is the most costly for the suitor in terms of his labour, gifts of millet to the woman's mother, and even financial assistance.

Child Betrothal or Marriage by Promise. In gratitude for services rendered (agricultural work, financial assistance, loans of land, etc.) or in witness to a friendship, a head of a family or lineage may decide to make a gift of a woman to the man. To do so, he may promise a young daughter (either actual or classificatory) in marriage or perhaps even one not yet born. The middle class translate this practice in French as 'marriage by promise', though we use the phrase 'child betrothal' here, the more common referent in English. *Tigpuo* designates one particular kind of such promise: if a healer manages to cure an infertile woman or one whose children all died young, the first girl to be born to the woman may be promised to him or to someone in his family, as long as he is not of the same clan.

Marriage by 'Abduction', or Consensual Union. Translating this kind of union as 'abduction' sounds quite archaic. Hence, some explanation is in order. First, this type of marriage must not be confused with a real practice of abduction of women (*puosuolu*), which has now disappeared, in part because of colonial influences. Previously, certain clans had the reputation of being 'woman stealers'. The abduction of a woman took place during armed conflicts, which were themselves often brought on by 'women's affairs'.

Second, the term nowadays is largely metaphorical. *Puotugnu*, 'to take away a woman', now refers to consensual unions, the only situation in which a woman has a decisive say in choosing her husband rather than having her marriage arranged by her family. To retain the flavour of the literal expression, however, we translate it as 'marriage by abduction'. More importantly, this translation, rather than 'consensual union', more aptly captures the sense, as we will see, of a woman being taken, usually from an arranged marriage, by another man.

A woman who is 'abducted' may be a single woman, already promised in marriage or not, or she may be married. The two parties agree to meet one day in an arranged place; and the woman lets herself be 'abducted'. Particularly in the past, when this form of marriage was considerably more frowned upon, the couple might leave the village and go to live in neighbouring Ghana, to avoid the anger of their parents and to wait for feelings to be soothed before returning to 'negotiate the affair'.

As we can see, the term 'abduction'—a literal translation of the Moba word used by the educated class—seems inadequate since the woman is consenting, and willingly leaves the conjugal or parental home to join her new partner. But in Moba-Gurma society, which is both patrilineal and patrilocal, and where power resides very much with men, people have difficulty attributing this initiative to the woman's own will: it is not the woman who leaves; it is the man who takes her away. It is from him or from his family that recompense will be demanded for the act. Compensation may be made in two ways: either a girl is given to replace the one who was taken away, or the pledged fiancé or former husband is reimbursed for what he paid for the marriage.

For a married man, abducting a woman can be an expeditious means of acquiring a new wife. For women, getting abducted is the only way to have a clear voice in choosing their own husbands. Yet marriage by abduction is also important for young men. For a single young man who should ideally wait to marry until all his older brothers have been wed, access to women is, in effect, controlled by the fortunes of his older siblings. Hence, abduction brings flexibility to a rigid marriage rule which can necessitate a long period of bachelorhood.

In economic terms, marriage plays a key role. For the head of a domestic group, the marriage of his dependants can enlarge the amount of farm labour available to the family, but it also poses a risk of segmenting the unit of production. Yet for a young man, such segmentation is advantageous. In effect, marriage allows him to escape a situation in which his labour is used by his father or older sibling, and to create his own household, in his turn, as the head of an independent economic unit.

Leviratic Marriage. The levirate, a common practice in Africa, requires that when a husband dies, his wife (or wives) be inherited or taken in by his younger brother or another family member. Women who have passed menopause are commonly allowed to stay with one of their sons; widows who are still in their reproductive years sometimes refuse to marry the designated agnate and marry someone else instead. But in such cases, the husband's family usually pressures them to continue to fulfil their conjugal duties by expanding the family of the deceased, depending on the age of the widow and the importance of her own family line.

Marriage by Reimbursement. Although I have described the exchange of women as a discrete type of Moba-Gurma marriage, women are exchanged in a larger systemic sense as well. Thus, even when a man obtains a wife by brideservice or by child betrothal, he (or his family or clan) are expected to give back a woman to the clan of his wife at a later date to replace the one taken. Similarly, in exchange for a widow who has left to remarry into another clan, the family of the new husband is expected to give back a widow.

No matter what form the original marriage took, families enter exchange relationships, all of which are expressed in terms of a 'woman-debt to repay'. Such a debt may take years to pay back, and may be passed on to subsequent generations, but in principle it will be honoured. According to our survey, only 8 per cent of men married by brideservice or by gift declared that they would not give back a woman from their family or lineage after taking one from another group.

This exchange process follows a logic of reciprocity or equilibrium, which forms the basis of social, economic, and political life of Moba-Gurma society. A proverb says that 'the clan that is great in number is also great in strength'. Through her fertility, a married woman contributes to the growth of her husband's clan; thus it is necessary that a woman from his clan come to replace her to re-establish the equilibrium. When an extant marriage is broken by an abduction, the husband's family will seek to re-establish equilibrium by creating a new marriage through a 'replacement woman'. When an exchange relationship already in place is upset when one of the two women who was exchanged returns to her own family, the injured husband's family has the right to take back the woman it originally gave in exchange or 'reimbursement' (the term *puopiidi* is used). In this case, the equilibrium is re-established by the annulment of the exchange relationship.

This logic of 'reimbursement' or 'replacement' places considerable pressure on women. A woman who marries but for whom no replacement is given is perceived badly because her departure has created a demographic and political imbalance between clans, to the detriment of her own. Ritual measures can give her family ways of pressuring her husband's family to provide a replacement. When her child becomes ill, the diviner or soothsayer (*jaba*) may be consulted to determine the origin of the illness and the appropriate curative measures to carry out. These may include acquiring *tigban* (personal protective medicine) for the mother, which can be given only by her father or an older sibling. So when a woman without replacement has an ill child, her family may refuse to give her *tigban*, trying to force the husband's family to provide the replacement if its sick child is to survive.

The logic of marriage exchange is thus observed in the four following types of marriage: marriage by exchange, by brideservice, by child betrothal, and by reimbursement. Each of these consists of an arrangement between families, where the future bride must accept the husband chosen for her, without her input. Although marriage by abduction can also result in establishing an exchange relationship, this type of marriage is different from the other types in that the bride gives her consent, and thus escapes active control of the older siblings and of the families. Though I do not want to overemphasize this distinction, the term 'consensual union' may be used in place of 'abduction', and the four other types of marriage may be regrouped under the rubric of 'arranged marriage'.

Bridewealth Marriage. In Africa, the term bridewealth (translated by francophone middle class people as 'dowry') applies to marriages for which the families arranged, beforehand, the amount, the composition, and the payment schedule for what some call the 'price of the fiancée'. If a man can produce an adequate bridewealth, in theory his obligations towards his in-laws end.

This type of marriage did not traditionally exist among the Moba-Gurma. Certainly, except for levirate and marriage by abduction, the suitor will have to make a certain number of outlays (gifts of millet to the mother, animals, kola nuts for the ceremonies, eventually financial aid), even aside from the labour he may have been required to perform before the woman was delivered. One may even hear it said that a man has 'dowried' his daughter, although situations vary tremendously. Moreover, marrying by bridewealth does not replace the exchange relationship. The future husband or his family is still expected eventually to give a woman to his wife's family, thus 'replacing' her.

TABLE 6.1. *Types of marriage by marriage order and sex, for marriages entered during 1980–1985 (%)*

Types of marriage	Marriage order						
	Men				Women		
	1	2	3	Total	1	2	Total
Exchange	35.8	26.9	24.1	30.9	34.2	14.9	30.0
Brideservice	4.2	2.3	10.8	4.9	5.8	—	4.6
Child betrothal	17.2	25.4	27.7	21.7	26.7	7.4	22.6
Reimbursement	8.8	13.1	4.8	9.3	11.3	—	8.9
Abduction	31.2	20.8	13.3	24.6	22.0	31.9	24.1
Bridewealth	—	0.7	—	0.2	—	1.1	0.2
Levirate	2.8	10.8	19.3	8.4		44.7	9.6
TOTAL	100	100	100	100	100	100	100
(No.)	(215)	(130)	(83)	(428)	(345)	(94)	(439)

The rare Moba-Gurma men who practise marriage by bridewealth are often those who marry women of other ethnic groups (most notably the Koussasi of Ghana, the Mossi and Boussance of Burkina-Faso) by offering four or five bulls to the bride's family. Sometimes in Moba-Gurma land one hears of problems of bridewealth repayment, an expression which may confuse the issue (Lare-Kombate, 1982). This concept usually applies only in the case of divorce or the abduction of a woman who is already betrothed or married. The reimbursement mentioned in such cases refers to all that the husband (or fiancé) has spent to acquire and maintain the woman who left him, and is therefore cast as a compensation for the 'suffering' he has endured.

The Current Distribution of the Types of Marriage and its Change over Time

Table 6.1 displays the actual distribution of types of marriage and their change in prevalence through time, as measured through the sequences of marriages for individual men and women. It reveals enormous diversity, particularly in terms of order for the two sexes. For men, the proportion of all marriages that are arranged remains approximately the same, whatever the temporal order in which they were contracted. Yet within this broad category, the internal distribution is quite different, according to order. The frequency of marriage by exchange decreases with order, while those of marriage by brideservice and by child betrothal increase. Marriage by abduction is common especially for the first marriage (31.2 per cent of the total), its frequency declining for higher orders. Leviratic marriage is the opposite: very low for first marriages (2.8 per cent), but comprising nearly one in five (19.3 per cent) marriages of order three or higher, as men's mortality risks increase.

For women, the situation is very different. For their first marriages, nearly 80 per cent of women entered arranged marriages, the rest marrying by abduction. Subsequent marriages (in case of widowhood or dissolution) in contrast, are dominated by the practice of levirate (44.7 per cent of the total). Next comes abduction, making up nearly a third (31.9 per cent) of remarriages; and arranged marriages constitute the minority (22.3 per cent).

If the way of starting married life differs little by sex (except in the case of the levirate, since women cannot inherit a sister's husband), ways of remarrying vary considerably. (The practice of polygyny creates the potential for even more diverse marriage histories for conjugal partners in terms of experiences and strategy.) For men, marriage by abduction is done mostly by single men; for women, by contrast, it is more common among those who are remarrying. It is certainly more difficult for a young woman to refuse an arranged marriage, because it is, to a certain extent, forced. Instead, a common strategy for a young woman and her parents is to collude: in order to avoid putting her family in an awkward situation, a young woman will accept the chosen (that is, imposed) husband; then after just a few months, get herself 'abducted'. Her family is thus relieved from any obligations, and the burden of settling on compensation is shifted to her husband and the man who abducted her. It is clearly very difficult to have a statistical measure of this type of situation.

The current distribution of types of marriage seen in Table 6.1 also suggests changes that have occurred over time. I do not have access to any statistical knowledge of this distribution or evolution during the precolonial period, though considerable changes undoubtedly accompanied colonization. As mentioned earlier, for example, marriage by real abduction disappeared in part as a result of the colonial influence in the country.

In any event, taking all the marriages recorded during the most recent period, between 1980 and 1985, we can make the following observations:

- The first four kinds of marriage taken together, which constitute 'arranged' marriages, make up two-thirds of the total, with a net predominance of marriage by exchange (about a third) and by child betrothal (a bit more than one-fifth); brideservice marriage is the least likely to be recorded (a bit less than 5 per cent).
- Marriage by abduction (that is, consensual union) represents nearly a quarter of the total, which shows its importance in contemporary marriage.
- The practice of leviratic marriage makes up a bit less than 10 per cent of marriages.
- Bridewealth marriage is very marginal indeed; only one case was recorded. For this reason, I eliminate this category from subsequent tables.

The data from the 1985 survey, which retraced the lifetime marriage experience of men and women, permit us to look more systematically at changes over time in the types of marriage, at least for the most recent decades. Table 6.2 illustrates this by looking at first and second marriages of men. Since the type of union entered may vary according to its order, taking this into account is necessary to avoid the risk of bias in the data, which overrepresent first marriages.

While the table dates changes only in relative terms (from men's first to their second marriages), it suggests the following trends: a decline in the levirate and of marriages by brideservice and reimbursement, a stabilizing of marriages by child betrothal, a recent increase in marriage by abduction, an increase in marriage by exchange, and a decline in the overall proportion of arranged marriages.

Men are now more hesitant to initiate a long period of brideservice in hopes of marrying, because young women increasingly refuse to have husbands imposed on them. This type of marriage is the most costly for the man, and many years may pass before the marriage actually takes place. Men do not want to risk 'suffering for nothing', in the local parlance. For these same reasons the inverse may be true for marriage by exchange which is now quite desired: it is more quickly concluded if the young women are willing, and it requires fewer expenses.

The declines in all types of arranged marriages and the levirate and the parallel increases in marriage by mutual consent or abduction support the informal observations by local people that the rationales of the old marriage system are being set aside. At least at an individual level, changes are assuredly taking place in the direction of greater personal freedom, especially for women. Although one cannot measure the real impact, Togo's legal code of 1980 concerning individuals and the family is probably intensifying this change by requiring the mutual and free consent of the future spouses and by discouraging monetary payments for women at marriage.

A parallel analysis by marriage order for women (not shown here) supports similar conclusions. The increase in marriage by abduction—that is, consensual unions—is more marked among young single women: from a level of 10 per cent before 1960 (and only 4 per cent before 1950), its relative proportion rose to 22 per cent in 1980–5.

Impact of Type of Marriage on Marital Stability

Standard demographic analyses of marital instability generally limit themselves to two ways marriages may end—the death of one of the spouses, or divorce—with no distinctions among the causes and forms that might reveal some important patterns specific to men and women and about marital dissolution in general. Our survey recorded only the type of dissolution and the identity of the person who claimed to have made the decision or to have taken the more active role. From these data, however, we were able to create a typology reflecting four forms of dissolution:

The husband sends the wife home (*puonyanu*). For various reasons, the man may decide to repudiate his wife, whatever the type of marriage, declaring that he owes nothing to his in-laws. Interestingly, a wife's sterility is not a legitimate reason to send a wife away since polygyny is seen as a means of solving this problem. (By contrast, the Togolese Code on individuals and the family recognizes sterility, either the woman's or the man's, as legitimate cause for divorce.) If the man himself is sterile, his wife is seen as justified in becoming pregnant by his brother, but she usually does so without the husband's knowledge. As in most African societies, 'social fatherhood' is more important than 'biological fatherhood'.

The wife's departure to her family (*u tun u baa nag*). In this case it is the wife who leaves her husband and returns to her parents. The causes may be many: mistreatment by the husband, lack of support for her and her children, rivalries between co-wives, or sterility. The latter is seen as especially justified because of the high value of fertility for both women and men. A woman without children is poorly regarded; when she dies she will not have the right to a normal burial. Leaving a sterile husband, therefore, is considered legitimate; but even when the soothsayer (*jaba*) attributes sterility to her, the wife, in her desperation for children, may prefer to leave and take her chances elsewhere. In either case, her family must give compensation in the form of money or another wife.

The wife's departure for another man (*u nyi jab*). This usually falls under the category of abduction.

The wife's family takes her back (*puopiidi*). This form of dissolution is linked to the logic of marriage exchange. If a woman involved in a marriage exchange leaves her husband, her in-laws are considered justified in 'taking back home' the woman they had previously given to her family in the initial exchange. Since this dissolution is the direct consequence of another dissolution or abduction, carrying it out is beyond the control of the spouses, who may thus even find themselves separated from one another against their wishes.

The data in Table 6.2 on men's and women's reports of how their unions dissolved, taken from individuals' marital histories, reveal relatively consistent reports between the sexes. One interesting difference, however, emerges when we compare men's versus the women's reports of the dissolutions that result from the man sending his wife home and those resulting from the wife's decision to leave. Women are more likely to declare that they left than men are to admit that their wives left (47 versus 32 per cent); by contrast, men are more likely to declare that they sent their wives home than women are to admit being sent (16 versus 7 per cent). Although such ambiguities pose obvious problems for interpretation, this disagreement is not surprising.

TABLE 6.2. *Forms of marital dissolution, by sex* (total marriages, %)

Form of dissolution	Men	Women
Sending wife home	16	7
Wife leaves	32	47
Abduction	15	14
Family takes wife	37	32
TOTAL	100	100
(No.)	(99)	(131)

TABLE 6.3. *Dissolved marriages, by type of marriage (excluding marriages ended by the death of one spouse)*

Type of marriage	%	(No.)
Exchange	8	(862)
Brideservice	12	(134)
Child betrothal	8	(531)
Reimbursement	8	(328)
Abduction	11	(584)
Levirate	6	(241)

We would expect individuals to prefer to say that they rejected the spouse, rather than admit being rejected.

Besides determining the frequency of various types of dissolution, we also need to ask how likely it is that marriages of various kinds will dissolve. Marriages by levirate, as Table 6.3 shows, are the most stable. Since these are not first marriages for women, they generally involve older women, many of whom may be beyond their fertile years. Although her dead husband's family may pressure a widow to stay among them (particularly if she is still fertile), marrying her dead husband's brother allows her to remain in an environment that is already familiar to her, probably near some of her children. Many of these women, indeed, are regarded as burdensome; many leviratic 'marriages' consist of little more than charitable caretaking for old women. Still, such unions are extremely stable because these women have few other marital options at this point.

The fact that the greatest instability is found among marriages initiated originally through abduction seems to run counter to common sense since these unions reflect mutual consent of the partners, when compared with the rigidity imposed by the collective family will in arranged marriages. Yet their instability is not surprising when we realize that the same liberty and flexibility that brought them into being also facilitates their dissolution. Dissolving

TABLE 6.4. *Type of marriage by forms of dissolution (%)*

Type of marriage	Action by husband	Action by others			% (No.)
	Wife evicted	Wife abducted	Wife left	Family took wife	
Men's reports					
Arranged	9	23	28	40	100 (77)
Abduction	25	7	29	39	100 (31)
Women's reports					
Arranged	5	16	46	33	100 (71)
Abduction	17	9	44	30	100 (33)

contractual marriages, that is, involves breaking political alliances as well as the loss of heavy financial investments by two families, while conjugal partners who marry by abduction find it easier to break a union that only they themselves created. Moreover, a man who has spent less on behalf of an abducted wife worries less about parting with her. As for the woman, unless her abduction is followed by reimbursement to her family, she is not subject to the family pressures that would loom large in other cases.

For men, of course, marriage by abduction does not preclude a subsequent contractual marriage or even polygyny. A married man can always become polygynous by abducting another woman. And if his first marriage was created by abduction, there is nothing to prevent him from obtaining subsequent wives through contractual marriage. Thus, for men aged 50 and over, the retrospective rate of polygyny[1] (that is, the maximum number of spouses a man has had simultaneously at any point in his life) is higher among men who have been married at least once by abduction (80 per cent) than among those who have entered only contractual marriages (61 per cent). The mean numbers of marriages entered are 2.85 and 1.99 respectively.

One of the most important findings of the study is that how unions begin is related to how they end. Because of the small sample size, I focus only on the distinction between marriage by abduction and the various marriages that I have classified as arranged (marriages contracted through exchange, brideservice, child betrothal, and reimbursement). This distinction, shown in Table 6.4, thus reflects two quite different kinds of marriage: marriages by the partners' mutual consent and those controlled by the families. The reports of men and women, though shown separately, are quite similar. For facility, we will describe only the figures that reflect the men's reports.

To be sure, type of marriage and type of dissolution are not wholly independent of each other. A couple whose marriage began with abduction generally will not go through a formal 'divorce' procedure—for the simple reason that there was no state of official marriage from which to divorce, unless the arrangement was later legitimated or reimbursed by the families. Yet this fact makes the following results all the more significant.

Table 6.4 helps us to discern whether having an arranged marriage or a marriage by abduction affects how the marriage ends. The first thing to notice is that whether a man began with arrangement or abduction seems to have little effect on whether the wife left or her family took her back (the two variables on the right). Whether the man sent the wife home or another man abducted her do seem to have important effects, though. Reading down the first column, we see that a husband is much more likely to send his wife home if he acquired her by abduction than if he acquired her by formal arrangement. A likely reason is that a man who married his wife formally but later sends her back does not in principle have the right to reimbursement by her family. Hence, the small number of cases of eviction among arranged marriages may reflect the fact that a man who married a woman formally ('arranged' marriage) risks the loss of a substantial investment by evicting her. Clearly, a man who initially paid nothing for a woman has less to lose by evicting her.

The second column shows, by contrast, that a marriage that begins by arrangement is much more likely to end by another man abducting the woman than a marriage that itself began by abduction. Three factors probably explain this result. First, as we saw above, a husband is reluctant to evict a wife he acquired through a formally arranged marriage. Second, a man who wishes to marry a woman who is already married through a formal arrangement would be expected to provide a heavy reimbursement to the husband. Seeking to escape this obligation, he may elope with her. Finally, women are increasingly resistant to having husbands imposed on them in arranged marriages. Since 1972, when a 'colloquium on the social problems of the district of Dapaong' brought local leaders together, arranged marriage has comprised a topic of much public debate, and the term 'forced' marriage has often been used in such contexts (Collectif, 1972), a phrase probably introduced by the Catholic missions during the colonial period. A woman, therefore, may be anxious to escape such a marriage by agreeing to elope with—that is, be abducted by—a lover.

Somewhat ironically, then, while marriages that begin by arrangement are likely to end in abduction, marriages that begin by abduction are likely to end in eviction.

Marital Stability and the Status of the First Wife

Type of marriage not only influences marital stability and dissolution type among all couples, it also appears to have a great deal of impact on marital stability among polygynous wives. As in other societies that practise polygyny, a Moba-Gurma 'first wife' (*naakpel*)—or, perhaps better, 'head wife'—

TABLE 6.5. *Women in polygynous unions who became 'first wives', by type of marriage (%)*

Type of marriage	
Brideservice	43
Reimbursement	20
Child betrothal	14
Exchange	9
Levirate	5
Abduction	2

enjoys a privileged status in the household. When the husband is absent, it is she who represents him. She plays a central role when ceremonies take place. It is to her that the husband gives the daily ration for the whole household to divide among all the wives. And she organizes and delegates the collective work for which women are responsible in the household.

However, the woman who becomes *known* as the 'first wife' is not necessarily the first to be married. Indeed, among women who were the first to contract a marriage with a particular man and were followed by other wives, about one in five (21.5 per cent) do not become the first wives. Conversely, among those who marry a man who is married already, 14.2 per cent obtain the status of first wife. Appointment to the privileged status of 'first wife' depends on the type of marriage, for these types imply a hierarchy. A woman obtained by brideservice has priority over all the others: she will be the first wife even if she is her husband's second or third wife, and even if she is younger than her co-wives. Next in the hierarchy of eligibility for 'first wife' status come women married by reimbursement, then those by the other kinds of contractual marriages (exchange and gift), and finally those by abduction and levirate.

Our survey data on women's marriages do not allow a complete assessment of how rigorously these customary understandings are applied. None the less, there is a hierarchy among the types of marriage, as we can see by examining Table 6.5, showing the proportions of women married by different kinds of arrangements who became 'first wives'. (The distribution is independent of the sheer number of marriages of each kind.) The list is restricted to women who wed men who were already married.

The most accurate measure of the relationship between the status of the first wife and marital instability focuses on the proportion of marriages that end because the woman appears to take the initiative: she leaves for her parents' home or gets herself abducted.

As Table 6.6 shows, whether a woman becomes the first wife clearly influences marital stability. Marital dissolution is nearly three times as common among women who did not become first wives as among those who did (8 per

TABLE 6.6. *Dissolved marriages by access to first-wife status for women who were polygynously married**

		%	(No.)
Woman was monogamous then polygynous in the same union:			
Total women who actually married first		4	(309)
Became 'first wife'?	yes	2	(242)
	no	9	(67)
Woman married an already-married man:			
Became 'first wife'?	yes	6	(70)
	no	8	(418)
Total women who became 'first wife'		3	(312)
Total women who did not become 'first wife'		8	(485)
Total women who married an already-married man		7	(488)

* Excluded are dissolutions resulting from a death, the husband's act of sending a wife home, and the woman's family taking her back.

cent versus 3 per cent; $p < .01$). Moreover, taking into account the order of the wife, independent of whether she became the first wife, we see more marital instability among women who married an already-married man (7 per cent versus 4 per cent; $p < .05$); these would most likely not be first wives. Marital instability is highest (9 per cent) among women who were displaced as first wives although they were married first. This, obviously, is the most difficult situation for a woman to accept. In contrast, the most stable marriages are those in which the women who married first remained first wives.

Type of marriage and the status of first wife in polygynous marriages, then, have a strong influence on marital stability. While these results demonstrate that it is important to take into account the characteristic features of a marriage system in order to understand marital stability, it is also necessary to place this understanding in a temporal perspective.

Elements of Stability and Change in the Moba-Gurma Marriage System

Elsewhere, increasing levels of education have been found to be associated with rises in the age at first marriage, and with greater freedom of partner choice and greater likelihood of marrying monogamously.[2] In 1981, the year of the last population census, the Moba-Gurma area had one of the lowest rates of school attendance in Togo, 28 per cent versus 51 per cent, the mean rural national level. School enrolment is particularly low for women (15 per cent for women versus 39 per cent for men). Yet levels of school attainment have remained low among Moba-Gurma women precisely because of marriage.

Marriage is the principal reason given by parents for not sending their daughters to school; and for those few girls sent to school, marriage is the main reason given for withdrawing them from school before they have finished: 'It is to avoid her getting a hard head which will lead to her refusing a husband later on', say many fathers.[3] Over time, of course, such attitudes may diminish as increasing education influences behaviour.

We must point out, however, that although imposed marriages would seem to be anathema to women, women actually have quite different opinions about imposed or forced marriages depending on whether it is they or others who are to marry. A young single woman is considerably more opposed to an imposed marriage when it involves her, rather than someone who must marry in her place. She is also more opposed to the practice than a mother who needs a son-in-law and tries to impose a husband on her daughter to obtain one.

A conversation with an older Moba-Gurma woman, aged about 60, yielded a complex combination of these positions. F. relates that she refused to marry the husband selected for her and arranged to have herself kidnapped instead by the man she really wanted to marry. In response, the man originally chosen for her asked for and received another woman in her place. For the second woman this meant a 'forced' marriage, but F. was unconcerned since the second woman was from another clan. F. is a mother today, and has daughters who are still not married. She would like very much to arrange their marriages to men who would perform brideservice, since the future sons-in-law would be obliged to come regularly to help her and give her millet. The economic advantage of obtaining sons-in-law is uppermost in the mother's thinking in this instance, rather than the marital constraints such an arrangement would impose on her daughters. Such desires are so strong among older women that men sometimes accuse their wives of trying to arrange marriages secretly, promising their daughters to young men who will come and work in their fields.

Refusing to marry men imposed by their families, then, does not mean that women necessarily reject the marriage system as a whole. It may simply mean that one option within the entire cultural array of options is better for a woman under some conditions than other options. Still, there are clear signs of change in Moba-Gurma marriage, as observed, for example, in women's increasing refusal to accept 'imposed husbands', the rise in marriages by abduction, and the diminution of expensive marriage ceremonies.

The 1980 Togolese Code on individuals and the family may be contributing to this process of evolution, since it is biased against 'imposed husbands' and bridewealth. The district chiefs in theory occupy a pivotal place in its execution. But while the judgements of some rely heavily on the Family Code, those of others remain closely tied to customary law. (For an analysis of matrimonial conflicts and the ways in which they are set right among the Moba-Gurma, see Pontié and Pilon, 1990.) Furthermore, the Code is not applied systematically—many verdicts are determined by who pays the most. And most marital conflicts are resolved not through district-chief's courts but through reconciliation by heads of family, clan, or district. The heads of many of these bodies play only a limited role in enforcing national laws. Only if a marital dispute cannot be resolved at these levels will it be taken to the tribunal by one of the parties, where it is tried under civil law and where the national family code may be invoked. A young woman who refuses a husband imposed on her might in theory have an advantage in this court. Still, few women even now dare to bring their cases to this level because the costs are quite steep, and they often include bribes. Women also fear family reprisals. These are particularly likely since a dispute judged under the tribunal is not necessarily resolved that way once everyone returns home to the village.[4]

Conclusion

Like those of other researchers, these results have confirmed the need to examine the characteristics and logic of a particular marriage system to understand marital dissolution. Among the Moba-Gurma, the types of the logic of marriage exchange remain important particularly for women, on whom pressures to stay out of school and marry remain very strong. On the other hand, the increasing practice of marriage by 'abduction' indicates an increasing tendency to exercise freedom in choice of partners as well as mutual consent. It is important to stress, however, that individuals' refusal to conform to certain rules of the marriage system does not reflect a more global rejection of the marriage system. Changes are coming about in individual strategies, but a real break with the logic and force of the system is not in evidence.

One point of methodological interest is the study's effort to combine quantitative methods with a sensitivity to qualitative issues of meaning. Certainly it is important to measure the real importance of a certain phenomenon by looking at its prevalence in the population. Doing so gave us, for example, a more precise idea of change over time in the relative importance of the different types of marriage and marital dissolution. Another important quantitative finding was whether women married under different types of union become 'first wives', and the resultant effects of this factor on marital stability.

Yet as this study has shown, applying standard categories of marriage and divorce in a superficial way can mask many complexities. The definition of marriage, the identification of different types of marriage and marital dissolution—all pose problems that require in-depth qualitative methods which the demographer is not particularly inclined to employ. On the other hand, of course, qualitative methods are not perfect; they suffer problems of representativeness. For such reasons, closer collaboration must be developed between demographers and anthropologists to develop new methodologies that are better suited to the issues that scholars in the two disciplines now find of mutual interest.

Notes

1. For a more detailed presentation of this index, see Pilon (1991*b*).
2. See on this subject Chs. 1 and 7, and Antoine and Nanitelamo (1991).
3. In the past, people claim, there was an extreme treatment for rebellious girls. A husband rejected by a woman who was meant to be his wife could put her in *condi*: to hide her away in a cell for several months. She became the responsibility of an older person in the family who administered drugs to her to 'break' her or 'make her docile'. The husband could take advantage of this period to try to make her pregnant. Although undoubtedly rare nowadays, the fact that this practice existed underscores the importance attached to marriage.
4. For an analysis of matrimonial conflicts and the ways in which they are judged among the Moba-Gurma, see Pontié and Pilon (1990).

References

Agounké, A., Levi, P., and Pilon, M. (1991), 'Evolution contemporaine des schémas de reproduction en Afrique sub-saharienne: Le Cas des Moba-Gurma du Togo', Paper presented at the International Seminar ENSEA-ORSTOM, 'Croissance démographique et stratégie de développement en Afrique', 26–28 November 1991, Abidjan.

Antoine, P., and Nanitelamo, J. (1991), 'More Single Women in African Cities: Pitine, Abidjan, and Brazzaville', *Population: An English Selection*, 3: 149–69.

Collectif (1972), 'Colloque sur les problèmes sociaux de la circonscription de Dapaong', manuscript.

Lare-Kombate, D. (1982), 'La Dot en pays moba', mémoire de Maîtrise Es-Sciences Juridique, Université du Bénin, Lomé.

Locoh, T. (1984), 'Fécondité et famille en Afrique de l'ouest: Le Togo méridional contemporain', *Travaux et documents*, Cahier 107, INED, PUF.

Pilon, M. (1991*a*), 'Contribution à l'analyse de la polygamie', *African Population Studies*, 5: 1–17.

—— (1991*b*), 'Génèse du déséquilibre entre population et ressources en pays moba-gurma (Nord-Togo)', in *Les Spectres de Malthus* (Déséquilibres alimentaires, déséquilibres démographiques, International Colloquium, Paris, 1990 (Paris: EDI-ORSTOM-CEPED), 117–36.

—— and Pontié, P. (1991), 'Développement inégal et mobilité: Le Cas des Moba-Gurma du Nord-Togo', in *Migration, changements sociaux et développement*, Troisièmes Journées Démographiques de l'ORSTOM, Paris, 20–22 Sept. 1988 (Paris: ORSTOM), 103–25.

Pontié, D. (1978), 'Les Moba de Lomé', doctoral thesis, Paris.

Pontié, G., and Pilon, M. (1990), 'Un exemple de justice au quotidien: Les Conflits matrimoniaux au Nord-Togo', *Afrique contemporaine*, 156, 4/40: 93–105.

Vimard, P. (1983), 'La Nuptialité sur le Plateau de Dayes', *Études togolaises de population, mariage et nuptialité au Togo*, 2/4: 158–97.

7 The Effects of Education and Social Stratification on Marriage and the Transition to Parenthood in Freetown, Sierra Leone

ANASTASIA J. GAGE AND CAROLINE BLEDSOE

Introduction

Nearly twenty-five years ago, Evans-Pritchard (1965) described marriage in 'simple' African societies as a given: there was no such thing as an unmarried adult woman or one who was childless by choice; companionship in marriage was weak and the concept of romantic love non-existent; and women could not choose a career instead of marriage. Although considerable deviation from these patterns arose in different times and places, the last quarter-century has seen truly dramatic changes (see Parkin and Nyamwaya *et al.*, 1987, for an anthropological assessment). Among educated Africans, some scholars argue that corporate lineages are being replaced by elementary families, kin-group selection of spouses by personal choice, and distant formality by love and companionship as the basis of marriage (Bird, 1963; Little and Price, 1973; Oppong, 1974; Harrell-Bond, 1975). Other apparent trends include greater instability of marriage, increasing fluidity and flexibility of sexual relationships, and women's increasing use of relationships with men to obtain resources and opportunities for furthering their own social and economic ambitions (Schuster, 1979: 75–82; Bledsoe, 1980*b*: 118–72; Obbo, 1980: 87–142; Oppong, 1980: 197–209; Mann, 1985).

The evolving status of multiple conjugal partners is one of the most striking changes. Missionaries frowned on polygyny, and policy-makers as well as scholars often view it as an anomaly in urban areas, where plural wives may drain rather than augment household resources (Lombard, 1954; Goode, 1970). In some African countries, polygyny has become illegal, at least on paper; in many others, the outward form of urban marriage has shifted to monogamy especially among educated people (Omari, 1960; Little and Price, 1973; Little, 1973; Harrell-Bond, 1975). But new forms of polygyny are emerging that do not require the same degree of familial commitment. These are labelled variously as 'informal arrangements' (Pellow, 1977), 'serial monogamy' (Comaroff and Roberts, 1977), and 'outside wives' (Baker and Bird, 1959;

Harrell-Bond, 1975; Karanja, 1987) or, in francophone countries, as a situation of 'la femme de cœur' or 'le deuxième bureau' (Lacombe, 1987).

Whether such women are accorded the same social recognition as the 'legal' wives is an important issue, one that is linked to increasing stratification in the wider society. We examine this question by looking at marriage change in Sierra Leone with special emphasis on its relationship to education and the transition to parenthood. The quantitative data analysed here come from Gage's retrospective survey conducted from July to September 1984 in Greater Freetown, Sierra Leone. Information was obtained on 2,654 women: their socio-economic characteristics, educational attainment, age at first marriage, current marital status, and age at first birth if any.[1] Qualitative data come from Gage's own knowledge of Sierra Leone and from Bledsoe's anthropological studies conducted in 1981–2 and in 1985.

We show that, unlike older cohorts, many younger women nowadays become mothers before marrying. Education is a crucial factor: it has a powerful positive effect particularly on the reported age of first marriage, though less so on first birth. Among uneducated women marriage seems to precede parenthood, while among highly educated young women, motherhood precedes marriage. We provide two interpretations for these results: they can be taken at face value as reflecting enormous changes in marital ages, or they may reflect educated women's increasing tendencies to focus on church or civil marriage ceremonies (which often come late) as defining marriage, despite the fact that these women may have undergone customary rites already. We focus on the changing nature of polygyny as it has evolved alongside increasing education and stratification. Such pressures induce women to strategically construe their marital statuses according to new emergent norms of monogamy and church marriage, whereas according to local norms, their nuptial status may be quite different.

Historical and Legal Aspects of Marriage Change in Sierra Leone

Any discussion of marriage in the African context brings up the age-old definitional controversy: what is marriage? Within the extended African family, a church or civil marriage that is not preceded by customary rites is rarely recognized because the elders did not officially give their consent before the marriage was contracted. For highly educated urbanites, on the other hand, a customary marriage is considered incomplete until it has been formalized by a religious or civil ceremony.

Just as marriage has different meaning for different social groups, it has changed over time. Differences in marriage among Freetown groups stem from economic and political changes over the past two centuries. The Western Area of Sierra Leone was founded in 1787 by the British as a colony for freed slaves from the United States, Canada, the West Indies, and England (for an analogous setting, see Mann, 1985). These settlers were later joined by people called Liberated Africans, who were rescued from slave ships captured on the high seas after the abolition of legal slavery. In the Colony or Western Area, the culture that evolved for the Creoles[2], as the freed slaves came to be known, drew on Western education and Christianity. The rest of the country was governed less directly as a protectorate from 1896 to 1947. In 1947 the British drew up a new constitution uniting the Western Area or Colony with the rest of the country.

As Barbara Harrell-Bond (1975) notes, the impact of colonialism led to the coexistence of three legal systems of marriage in Sierra Leone. First was customary marriage under which indigenous ethnic groups, including those living in the Western Area, could marry. The second type, statutory marriage, applied to non-natives. The third was Islamic marriage. Since independence in 1961, national policy-makers have modified these laws to take into account the changing political and social situation. Although non-Creoles were not legally allowed to marry under the Civil Marriage Act governing statutory marriage, this Act was amended in 1965. It set a minimum age at marriage of 18 for 'Natives'—persons governed under customary or indigenous law—and a minimum age at marriage of 21 for 'Non-Natives' (that is, Creoles and other groups). Another legal modification was the Christian Marriage Act, which recognized marriages contracted under customary law by forbidding marriage if one or both of the would-be spouses were already married under customary law. The other attempt to adapt marriage practices to changing social conditions was the banning of mandatory widow inheritance in 1963, since it was increasingly realized that many women did not want to marry one of their deceased husbands' relatives, and that with modernization and the spread of education, women should consent to marriages (Harrell-Bond, 1975: 93).

Despite these carefully designed changes, laws governing marriage remain remote from, if not irrelevant to, the cultural understanding of the people to whom they were meant to apply. Today, few people seem to know about legal age stipulations, widow inheritance is still widely practised, most civil or religious marriages begin as customary marriages before being 'formalized' by a religious or civil ceremony, and married Christian men sometimes marry additional wives by customary law or by taking additional women informally as 'outside' wives.

Amid these changes, new forms of sexual unions are emerging. In the urban environment, a distinction is often made between a girlfriend (in Krio, the *fejeh*: 'the woman who has the man') and the legal wife (*marrade wef*: 'married wife'—that is, 'the woman who has the ring'). Consensual unions are also socially differentiated from marriages. Locally referred to as *tap-to-me* (live-in arrangements), they are usually frowned upon because they indicate minimal interest on the part of a man. Among the most important trends are the unions of older schoolgirls (Bledsoe, 1990). Declining economic conditions have put pressure on parents to let their daughters accept school fees from *sponsors* or 'supporters': men of wealth and influence. These school fees imply an investment of either a marital or a sexual nature. 'Sponsors' may include schoolteachers, a phenomenon also observed in Ghana by Bleek (1976), or older wealthy bureaucrats in the urban area, called *sugar daddies* (see also Dinan, 1983). Although such support often excludes marriage plans, girls use such relationships to further their education or make job contacts, and family elders keep their daughters in school against the risk that they will get pregnant or choose men of whom the elders do not approve. In any case, to bemoan the demise of African marriage as an institution is to misunderstand its fundamental dynamism. As Burnham (1987: 50) argues,

> it seems analytically valuable to consider all forms of conjugal relations, from consensual unions to 'ordinance' or 'church' marriages, as representing a bundle of interactional possibilities with associated political, economic, legal and other implications. How these interactional possibilities are utilized by male and female actors is a central, if very complex, question for future research.

Age at First Marriage

Demographers are concerned to establish a definite date of marriage, since this usually marks the inception of childbearing. Yet as generations of anthropologists have stressed, marriage in Africa is more a series of events—a process over months or even years—than a single, definitive event (Evans-Pritchard, 1951; Kuper, 1970; Comaroff and Roberts, 1977). Aryee and Gaisie (1979: 287) point out that it is 'difficult to determine whether a relationship is merely a prelude to, or an actual formalization of, marriage or in a stage of transition somewhere between the two'. A series of events, then, together define a marriage, although one more than others may do so for a particular society. Such events may include preliminary visits, the exchange of gifts or symbolic tokens such as kola nuts, initial sexual relations, cohabitation, a series of bridewealth payments, the birth of the first child, and so on. Thus, one way to ask an uneducated Mende woman whether she is married is, 'Did he give kola for you?' But since the word 'kola' can refer to any gift given ceremonially as an appeasement or a token of thanks, giving kola for a woman may simply refer to one of numerous symbolic gift exchanges along the process of an evolving conjugal relationship. Even for educated women whose local rites may occur months or even years before the civil or church ceremony, which they regard as more valid, these various arrangements create enormous ambiguity. (One of the most insightful treatments of such ambiguities is that of Comaroff and Roberts, 1977, on the Tswana.)

If the customary marriage ceremonies preceded the Western ceremony for all women, then we could still use the age at which a particular ceremony occurs as the valid age at first marriage for all women. But the same ceremonies do not occur for all women. Nor do they necessarily imply the same kind of

TABLE 7.1. *Cumulative proportions marrying by age X, by birth cohort: Freetown, 1984*

	Birth cohort				
Age	1935–9	1940–4	1945–9	1950–4	1955–9
15	.01	.06	.08	.09	.06
16	.10	.11	.15	.16	.15
17	.18	.15	.28	.25	.24
18	.33	.33	.40	.34	.32
19	.36	.40	.48	.40	.39
20	.45	.52	.59	.47	.46
21	.55	.61	.68	.53	.49
22	.61	.67	.69	.54	.54
23	.71	.74	.74	.58	.57
24	.77	.76	.74	.58	.59
Median	20.5	19.8	19.1	20.6	21.1

Source: Greater Freetown Survey, 1984.

familial responsibilities and marital commitment, particularly for educated versus uneducated women. Among some educated people, the customary marriage ceremony is called the 'engagement' and the church or civil wedding, the 'marriage'.[3] Such divergences between customary norms and the national legal code make finding a common definition of marriage virtually impossible. In fact, an uneducated woman who has been married through local rites is generally regarded by her peers as being in a more valid relationship than the educated woman who has completed a customary marriage process yet has not undergone a religious or civil ceremony marked by signing a marriage certificate. In the Greater Freetown Survey, marriage was defined quite broadly: a union marked by an indigenous, religious, or civil ceremony, or a consensual union of a duration of six months or more. Ultimately, the survey used women's own conceptualizations of their marital status according to the norms and mores of their own social groups.

Factors such as the advent of education appear to decrease the proportions of Freetown women who are married by any given age: that is, education appears to raise the age at first marriage. Table 7.1 shows a life table analysis of the age at marriage by birth cohort in Greater Freetown, Sierra Leone. The data for the three oldest cohorts suggest that the age of marriage seemed to fall slightly. Those born in 1935–9 had a median age at marriage of 20.5, compared to 19.1 for the 1945–9 cohort. After that, the median age appears to rise, reaching 21.1 for the 1955–9 cohort. This upward trend in the age of marriage beginning with the 1950–4 cohort appears to parallel increasing levels of schooling.

TABLE 7.2. *Median ages at first marriage and first birth for women aged 30 years and over, by completed years of schooling: Freetown, 1984*

Education	First marriage	First birth	No. of women
No schooling	16.3	17.9	416
1–9 years	17.4	18.1	126
10 or more years	23.5	20.3	211

Source: Greater Freetown Survey, 1984.

Since first marriage, schooling, and first birth are more likely to have been completed by older women, we will focus on those aged 30 and over. For these older women, the incompatibility of being both a student and a wife is widely acknowledged; it is no surprise, therefore, that Table 7.2 shows education to be associated with an increase in the age at first marriage. Uneducated women married much earlier than educated women, and this difference widens with an increase in the number of completed years of schooling. There is an enormous difference of about 7 years in the median age at marriage between uneducated women and women with 10 or more years of schooling.

The Effect of Education on Marriage and Transition to Parenthood

Most African societies regard marriage as the ideal state for the commencement of childbearing. In some East African communities (Hahn *et al.*, 1966: 33; and Edel, 1957: 65), premarital childbearing was reportedly punishable by death. Elsewhere, alleged sanctions varied from abortion (Massam, 1927: 159) to infanticide for male children (Orde-Browne, 1925: 70). In Sierra Leone the Mende conducted purification rites for unmarried pregnant girls (Little, 1951).

Unmarried motherhood, however, is increasingly accepted. If a woman becomes pregnant and there is no talk of marriage, the elders of the man's family may come to claim responsibility for the woman's condition, in effect legitimizing the child but not the union. At this ceremony, the unborn child is given the family name of its father (see also Little and Price, 1967). This ceremony, known in Krio as *answer belleh* ('answering to the pregnancy'), is quite distinct from the marriage rites wherein the man's family claim the woman as their *yarwoe* (bride). African societies continue to place a strong emphasis on childbearing. Even Creole society greatly values fertility. Men are warned against marrying *cocks*: infertile women. Some men, in fact, have required proof of fertility before marriage (Harrell-Bond 1975: 263; see also Obbo, 1987, for Uganda). As indigenous forces lose their hold on women's conjugal behaviour, the prevailing urban attitude is that it is better for an unmarried woman to have a child than to be childless. Indeed, although they prefer their daughter to marry, desperate kin may pressure an educated woman who has passed the socially defined age at first birth to begin having children, even though no permanent husband is in sight.

Apart from the question of legitimacy, what are the implications for fertility of the apparent delay in marriage among educated women? Since marriage usually marks the transition to socially sanctioned childbearing, we might expect the rise in the age at marriage with increasing education to produce a corresponding rise in the age at first birth. However, education seems to have surprisingly little effect on timing of first birth. Table 7.2 shows that for women who are now 30 or over, women with either 0 or 1–9 years of schooling have their first births about the same time. This probably reflects the fact that most women with 1–9 years of schooling have had less, rather than more, schooling. By contrast, women with 10 or more years of schooling generally had their first births later than women in the other education categories. But the differences are considerably less than we might expect—certainly, compared to the differences in the age at first marriage for the three educational groups. The impact of education on reported age at first birth may be even smaller than our figures show. Some schoolgirls who get pregnant leave their babies with their families in order to return to school, and have little to do with them thereafter. Some infants may even become fictionalized as the offspring of the girls' own mothers.

The timings of first marriage and first birth are shown in Figure 1 for the individual educational groups of women over 30. Uneducated women and women with 1–9 years of schooling, respectively, have higher proportions married than proportions having their first births at every age. By contrast, for women with 10 or more years of schooling, the curve representing the probability of having a first birth lies above the marriage curve. That means that highly educated women are more likely to have a first birth than to marry. The proportion of first births that are apparently premarital increases with education from 16 per cent among uneducated women to 21 per cent for women with 1–9 years of schooling, and to 36 per cent for women with 10 or more years of schooling. Education, therefore, appears to have considerably more impact on the timing of first marriage than on the timing of first birth.

The fact that motherhood does not have to follow first marriage raises an important question. Does education affect fertility or does fertility affect education? Conventional wisdom holds that educated women have lower fertility *because of* their education. But fertility may be an equally plausible independent variable in areas where schoolgirls commonly engage in sex (see also Cherlin and Riley, 1986). That is, those women who manage to *avoid* pregnancy (whether because of birth control or higher infertility, or whatever) get more education.[4] A pregnancy, on the other hand, may terminate a school career and hasten a marriage or lead to one that would not have

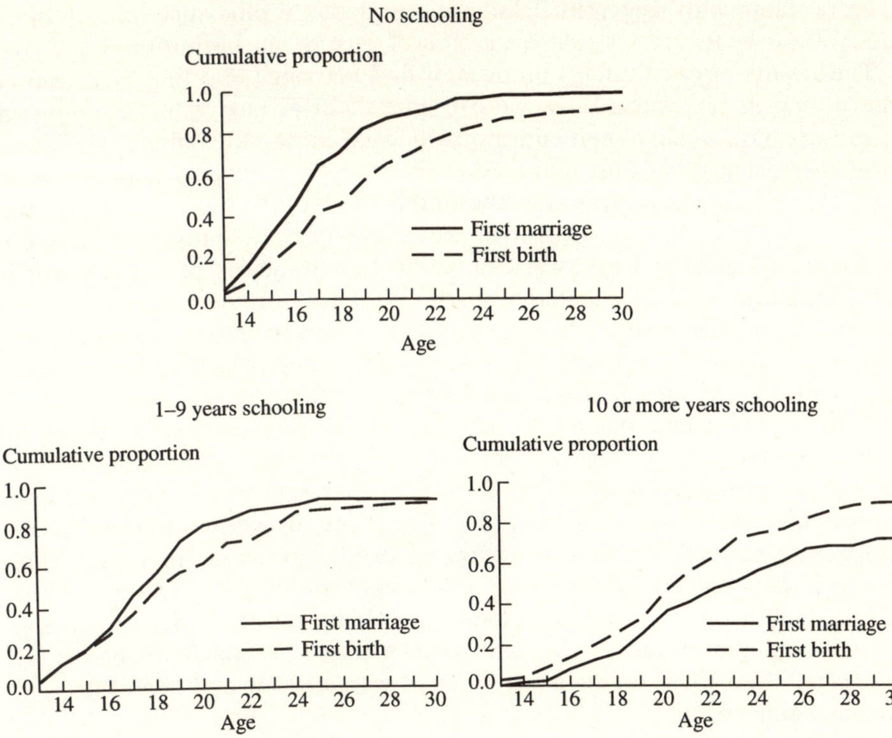

Source: Greater Freetown Survey, 1984.

FIG. 7.1. Cumulative proportion of women 30 years and over marrying and having a first birth, by education: Freetown

occurred otherwise. In fact, women may use pregnancy to try to get husbands. Therefore, whereas getting married can precipitate childbearing and determine children's legitimacy, childbearing itself can precipitate marriage and make children legitimate.

On the other hand, pregnancy does not necessarily represent the failure of a girl's educational career. It may represent her successful strategy to use her education to marry a high-status man. If a girl decides that the man who made her pregnant will make a good husband, she may drop out of school and marry him. But if he loses interest in her, she may attempt abortion or bear the child and marry someone else later. Like marriage, therefore, pregnancy has become a process of testing relationships with potential partners. There is no necessary causal relationship between pregnancy and marriage, nor are they necessarily motivated by the same factors. For many of the indigenous groups in Sierra Leone, in fact, initiation into the women's secret society is far more important than marriage in legitimating childbearing.

(The problem with using ritual initiation in demographic surveys as a likely precipitator of fertility is that even girls of 7 or 8 can be initiated.)

The results of our findings on ages of first marriage and first birth can be interpreted in two ways. First, we can take them at face value as empirical fact: there have indeed been enormous changes among the cohorts as women have increasingly become more educated and have taken on professional jobs. While many uneducated women are still given to husbands soon after puberty and have little control over the choice of their spouses, marriage is no longer a logical and necessary stage for the attainment of social status in the life course of an educated woman. With support and encouragement from kin or outside men, a growing number of young women—even some who get pregnant while going to school—may continue their schooling and get a better job than if they had married (Bledsoe, 1990).

Yet although some women are undoubtedly delaying marriage, it is hard to believe that such a large proportion have delayed so long. Rather than taking the recent apparent rise in the age at marriage too literally, we must ask whether these figures partly reflect the efforts of women to strategically construe their marital statuses, according to the norms of the social levels to which they would like to aspire. To address this difficult question, we now return to confront squarely the labelling problem raised earlier: the lack of a common definition of marriage can affect efforts to establish the age at first marriage. How, then, do different kinds of women view and label their own conjugal statuses?

The two significant trends observed in Table 7.1 give us some clues. First, in the three oldest cohorts. all those born between 1935 and 1949, the oldest women seem to have married later than the younger ones. This could stem from sampling or recall problems. But because marriage is a process marked by a series of events, a woman may reconstruct the last significant event she experienced as the one that defined her marriage. Because older women have undergone more of these events than younger ones, they may be more likely to define their entry into marriage at a later age than slightly younger cohorts. If this interpretation is correct, it means that while identifying more culturally sensitive stages of marriage helps us to understand more of the process, it may not be as useful as we think for predicting the onset of childbearing. Particularly, uneducated women may select the last major event in the marital process as the definitive one. This conclusion becomes even more plausible, in the light of rapid changes in the definition of marriage even from one generation to the next.

The second trend was the apparent turnaround—a gradual increase in the ages of marriage, beginning with the 1950–4 cohorts—which probably reflects a growing number of educated women. While educated women may put off marriage, as we acknowledged above, it is impossible to clearly separate this from a more phenomenological problem: their definitions of marriage may differ from their uneducated forbears and peers. This means that the ranks of the younger cohorts are swelled by women who tend to downgrade the significance of customary rites—which uneducated women might offer as proof of their marriage—and to report being unmarried unless they have undergone church or civil ceremonies. Moreover, such women are less likely to call themselves married if their partners are already married. A union's validity, that is, depends on who is assessing it. Many women are undoubtedly in unions of various kinds, but fewer call themselves married.

Changes in the normative rules for labelling unions give better perspective as well to the apparent increase in premarital childbearing. While the age at first birth may be largely unrelated to age at first marriage, as we stressed above, it is significant to recall that by contrast to reported ages at first marriage, ages at first birth have changed little among educated women. This suggests that to women, legitimating children follows from how their own unions are defined. Paradoxically, defining their births as premarital is preferable to admitting being married through customary rites.

The subtleties extend even to the contexts in which such questions are asked. In informal settings, young women (even schoolgirls) may refer to their partners as 'husbands', whereas they may deny being married when queried by educated strangers such as surveyors to signify that as 'civilized' individuals, they recognize only church or civil marriages. For schoolgirls, it is important not to admit being married because of the possibility of being turned out of school. Anecdotal support for this comes from one of Bledsoe's surveys in a large rural town. A teenaged girl, answering a written questionnaire administered to students at a local secondary school, reported in a typical manner that her father had paid most of her school fees that year. However, before the survey form could be collected, it had been 'corrected'. The girl's teacher, who knew her well, crossed out her answer and wrote 'husband'.

In sum, it is possible to attribute some of these dubiously large numbers of late marriages and premarital births to 'misreporting'. We prefer to recognize them as artefacts of our own insistence on labelling women definitively as married or unmarried, in contrast to the more indigenous sense of accepting marriage as a process and perhaps of defining a relationship as a marriage at an earlier stage in that process.

Increasing Stratification and the Position of Women in Marriage

Although we have posed two apparently mutually exclusive interpretations of the results—interpretations that might be labelled simplistically as demographic and anthropological—neither can be understood without the other. Both make sense only when placed into a larger context of economic competition and intensifying social hierarchies. Despite the apparent decrease in legal polygyny, the most important changes in marriage revolve around the changing nature of polygyny in the midst of increasing stratification. Education plays a pivotal role in these changes.[5]

One likely source of the changes in polygyny is declining mortality rates. The observations of Caldwell *et al.* (1982) in south India offer important insights for this phenomenon. Because Indian men are usually older than their wives, marriage comes under pressure when decreasing mortality rates enlarge younger cohorts. More women are available to marry a smaller number of older men, whose ranks were thinned by higher mortality rates when they were children. By increasing the numbers of surviving children and decreasing the likelihood of widowhood, mortality decline may therefore raise the ratio of 'marriageable women' to 'marriageable men'. Rises in fertility may have a similar effect.

Similar trends are undoubtedly affecting Africa as well. Whereas the notion of the 'marriage squeeze' implies demographic imbalances in the number of men and women in the marriageable age range, in Africa, social and economic factors are probably more important than demographic influences in creating an artificially imbalanced marriage market. Both mortality declines and migration may lead to numerical imbalances between the sexes, thereby creating a marriage squeeze that is intensified by a large gap in the age at first marriage between men and women, which may be as high as 8–10 years. Although women can marry younger men, they could also join polygynous unions, which are quite rare in South Asia. Indeed, polygyny depends on excesses of women, created largely by marital age inequalities.

Yet polygyny is increasingly unsatisfactory as a solution to the marriage squeeze because female education has made legal polygyny a highly brittle institution. Polygyny is often seen as an equalizer in Africa, spreading resources among a man's wives and children (e.g. J. Goody, 1971, 1976). We would qualify this, arguing that despite an ideology of equality, polygynous wives were differentiated even in the past by their family statuses and their order of precedence into the household (see also Murphy and Bledsoe, 1986). But nowadays education appears if not to create then to intensify differences among women. Women who progress beyond primary school, often because of their families' wealth and connections, are increasingly unwilling to accommodate themselves to a polygynous household. They particularly detest becoming junior wives to women with less education. Furthermore, an educated man who is already married to an educated woman is unlikely to marry a young woman with whom he contracts a liaison. She is often willing to forsake legal marriage for affiliation with a man of status, as a former social worker pointed out to Harrell-Bond (1975: 141): 'Well, if they don't want to get married to the kind of man who would like to get married to them—you see it may be a man of the same or lower status to themselves—they think it is better to be the mistress of Honourable so-and-so, than to be the wife of a carpenter.'

If a man does marry his mistress, she will probably displace, rather than join, the first wife. Therefore, although some highly educated professional women may not want to marry, others find it increasingly difficult because they face an 'educational marriage squeeze': there are too few unmarried educated men to meet their new aspirations. Underscoring these points, a respondent explained that being a polygynous wife 'is a role an educated woman cannot easily play, especially a secondary school girl, because at that level of education she can probably easily find a job, and is probably moving with a man who is probably highly placed. So she really doesn't have to marry, except for social standing'. Asserted another:

> I don't know of any woman who was in secondary school who is married polygamously. Any woman who has finished Class 8 will probably be a sort of mistress, rather than a polygamous wife. . . . Many of the women working in the offices now who are educated and unmarried are somebody's mistress. But especially now, people don't look down on this so much. As for me, I don't see anything too different about having a mistress or a second wife: it's just that one would be here, and the other there.

In sum, female education contributes to a growing marriage squeeze, counteracting the tendencies of African polygyny to accommodate excess young women in marriage.

The dynamic forces through which education and polygyny interact bring us back to the point about women's propensity to construe and label unions strategically. Despite the apparent decline in legal polygyny among educated individuals, competitive pressures stemming from a *polygynous model* of relations among women have not diminished. A man who marries an uneducated wife through customary procedures is likely, as his urban career spirals, to marry an educated woman under the modern civil or religious code. He keeps his first wife in the background: perhaps hidden from élite view in the rural area. To his urban friends, he may neglect to mention her; alternatively, he may deny his marriage to her, referring to her as his 'outside wife', although he may conduct normal rural marital relations with her when he goes on 'trek' upcountry (see also Clignet, 1987: 204). But if he returns to shore up rural ties, he treats her as a legitimate wife, married by customary procedures. On the other hand, he may redefine low-status women as non-wives by divorcing them, labelling them 'outside' or 'country' wives, or eventually choosing one legal wife from several partners, as Baker and Bird (1959: 115) discerned as long as 30 years ago for urban Africa: if a woman cannot fulfill the stringent qualifications normally required for a statutory marriage, namely a high degree of education and social status, she may herself be an 'outside' wife, that is, a partner to an unregularized conjugal union. . . . she stands, as it were, in a 'second class' position' (see also Harrell-Bond, 1975: 141).

Indeed, the new ideals of monogamy, coupled with increasing female education, may be exacerbating the extant hierarchies among the women attached to one man (Bledsoe, 1993). Men almost invariably allocate more resources to the most educated women with the best family connections (see Comaroff and Roberts, 1977, and La Fontaine, 1962, for analogous arguments). If anything, female education is exacerbating stratification within polygynous units.

Implications of New Conjugal Forms for the Statuses of Children

Just as education creates clear differences among the women affiliated with one man, a woman's success in becoming educated is increasingly crucial to the life chances of her children. Educational differences among polygynously affiliated women—whether unofficial or legally married—intensifies inequalities among children of the same man. As La Fontaine (1962: 114) perceived in Kenya, the changes associated with marriage and education 'have created a class structure which appears to be more rigid in recruitment and sharper in wealth differences than that of the traditional system'.

Ideally, fathers should invest more in older children, who can then contribute toward their siblings' education, or in 'clever' children. But for a man, marriage and fatherhood in the sense of willingness to claim a child are increasingly disjointed. When actual resource allocations must be made, a man cannot offend a high-status woman's family by sending her child to a rural 'granny', an elderly female caretaker, while paying the educational expenses of the child of a low-status 'outside' wife at an élite urban school. A man who privately favours his 'outside' wife over a dour 'official' wife may quietly educate the former's son. But it is rare for such children to have better opportunities than those of the 'official' wife's children. Children of high-status women almost invariably receive the benefit of the doubt when an educational choice must be made, making education the basis of inequality in successive generations. Harrell-Bond (1975: 134), in fact, suggests that women who were born as 'outside children' are themselves likely to become 'outside wives'.

An example illustrates the efforts of upwardly mobile men to slough off the children of uneducated, low-status wives (see Bledsoe, 1993).[6] A young rural man with a few years of primary education married an illiterate wife. Within a few years, his small trading business grew, and a local family removed one of their daughters from school to become his second wife. Both women bore children. In time he moved to Freetown, bought several large cargo trucks and a supermarket, and acquired a Freetown wife much more educated than he. From then, little more was heard about the first two wives or their children. Indeed, the new wife insisted that she was the only wife. Her children were educated in Freetown, while the others managed to attend school upcountry by drawing on the meagre resources of their own maternal families.

Of utmost significance is that the same man's children can inherit vastly different life chances by virtue of their maternal status. Although most Sierra Leoneans would claim they do not discriminate against children born in non-marital unions, these children face severe discrimination at inheritance time (see also Mann, 1985: 121–2, for Lagos). Attempts to ban the notion of 'illegitimacy' have failed, primarily because of opposition from legally married élite women who saw the bill as a threat to their social positions and those of their children (Harrell-Bond, 1975). For these reasons, the well-intended efforts to restrict legal polygyny while recognizing a man's illegitimate children often, ironically, work *against* children of informal unions. Fathers, feeling pressure from their legally married wives and their children, may deny paternity (Bledsoe, 1980*a*). Although a man may support children by low status women if he wants to maintain ties with their families, these children generally receive fewer advantages, and they and their maternal families become increasingly marginalized, geographically as well as socially.

Notes

1. The quality of the data is affected by a failure to report first births, especially if the child died young or was illegitimate, or misreporting the date of first birth or the age at first marriage. However, as we will show, 'misreporting' of these variables is itself linked to socio-cultural factors.
2. We use the term 'Creoles' to refer to the descendants of the former slaves, and the term 'Krio' to refer to the language they speak.
3. This interplay between customary expectations and marriage strategy has been observed also in southern Ghana. Crabtree (1950) and Oppong (1974) reported that educated men in Southern Ghana and élite Akan civil servants usually marry according to African custom although they may also have a church or civil ceremony. Tetteh's (1967) analysis of the 1960 post-enumeration survey of the Ghana Census revealed that 86 per cent of individuals sampled who had ever married had done so under customary law only. Those who chose to marry by Ordinance or church, primarily urban élites, did so after carrying out the customary rites.
4. This issue has clear parallels elsewhere. In the USA, some analysts have argued that early fertility leads to poor school performance, and others that poor school performance leads to early fertility. See Rindfuss *et al.* (1980) and Furstenberg *et al.* (1987) for summaries of this debate.
5. Men as well must certainly be taken into account. Like an educated woman, an educated man may wait longer to marry in the church, because once he does marry, he cannot take another legal wife. He also fully understands the implications of church marriages for disenfranchising his children of other women, past, present, or future.
6. Although mobility strategies induce husbands and their 'official' wives towards more 'nuclear' family structure and class division, Mann (1985: 70) demonstrates that such changes are not linear. Monogamous men take on 'outside' wives as easily as polygynous men become monogamous. This implies that at the societal level, educated people do not simply abandon polygyny by deciding at the outset of their conjugal careers to marry monogamously and have one set of children:

rather, once an educated man marries a highly educated wife, he can define their marriage as legitimate monogamy, while reconstruing other women and their children as 'country' or 'outside' wives and children. Therefore, polygyny itself can become more 'monogamous' through an active process of sloughing off wives and children, rather than a man deciding at the outset of his marital career to marry only one woman.

Acknowledgements

We thank Phil Morgan, Bill Murphy, and Étienne van de Walle for their helpful comments on the paper. For generous institutional support of our projects, we are grateful to the National Science Foundation, the Population Council, the Rockefeller Foundation, the Regional Institute for Population Studies in Accra, the Population Studies Center at the University of Pennsylvania, and the Department of Anthropology at Northwestern University.

References

Aryee, A. F., and Gaisie, S. K. (1979), 'Fertility Implications of Contemporary Patterns of Nuptiality in Ghana', in L. T. Ruzicka (ed.), *Nuptiality and Fertility: Proceedings of the IUSSP Seminar on Nuptiality and Fertility, January 1979* (Bruges: Ordina Éditions).

Baker, T., and Bird, M. (1959), 'Urbanisation and the Position of Women', *Sociological Review*, NS 7: 99–122.

Bird, M. (1963), 'Urbanisation, Family and Marriage in Western Nigeria', in *Urbanisation and African Social Change*, Proceedings of the Inaugural Seminar held at the Centre for African Studies, University of Edinburgh.

Bledsoe, C. H. (1980*a*), 'The Manipulation of Kpelle Social Fatherhood', *Ethnology*, 19: 29–45.

—— (1980*b*), *Women and Marriage in Kpelle Society* (Stanford, Calif.: Stanford University Press).

—— (1990), 'Schoolgirls and School Fees among the Mende of Sierra Leone', in P. Sanday, and R. Goodenough (eds.), *Beyond the Second Sex* (Philadelphia: University of Pennsylvania Press), 283–309.

—— (1993), 'The Politics of Polygyny in Mende Education and Child Fosterage Transactions', in B. D. Miller (ed.), *Gender Hierarchies* (Cambridge: Cambridge University Press), 170–92.

Bleek, W. (1976), 'Sexual Relationships and Birth Control in Ghana: A Case Study of a Rural Town', Ph.D. Dissertation, University of Amsterdam.

Bott, E. (1955), *Family and Social Network: Roles, Norms and External Relationships in Ordinary Urban Families* (London: Tavistock Publications).

Burnham, P. (1987), Changing Themes in the Analysis of African Marriage, in D. Parkin and D. Nyamwaya (eds.), *Transformations of African Marriage* (Manchester: Manchester University Press), 37–54.

Caldwell, J. C., Reddy, P. H., and Caldwell. P. (1982), 'The Causes of Demographic Change in Rural South India: A Micro Approach', *Population and Development Review*, 8: 689–727.

Cherlin, A., and Riley, N. (1986), 'Adolescent Fertility: An Emerging Issue in Sub-Saharan Africa', PHN Technical Note 86–23 (Washington, DC: World Bank).

Clignet, R. (1987), 'On dit que la polygamie est morte: vive la polygamie!', in D. Parkin, and D. Nyamwaya (eds.), *Transformations of African Marriage* (Manchester: Manchester University Press), 199–209.

Comaroff, J. L., and Roberts, S. (1977), 'Marriage and Extra-Marital Sexuality: The Dialectics of Legal Change among the Kgatla', *Journal of African Law*, 21: 92–123.

Crabtree, A. L. (1950), 'Marriage and Family Life amongst the Educated Africans of the Gold Coast', Master's thesis, University of London.

Dinan, C. (1983), 'Sugar Daddies and Gold-Diggers: The White-Collar Single Women in Accra', in C. Oppong (ed.), *Female and Male in West Africa* (London: George Allen & Unwin), 344–66.

Edel, M. M. (1957), *The Chiga of Western Uganda* (New York: Oxford University Press).

Evans-Pritchard, E. E. (1951), *Kinship and Marriage among the Nuer* (Oxford: Clarendon Press).

—— (1965), *The Position of Women in Primitive Societies and Other Essays in Social Anthropology* (New York: The Free Press).

Furstenberg, F. F., Brooks-Gunn, J., and Morgan, S. P. (1987), *Adolescent Mothers in Later Life* (New York: Cambridge University Press).

Goode, W. J. (1970), *World Revolution and Family Patterns* (New York: The Free Press).

Goody, J. (1971), 'Class and Marriage in Africa and Eurasia', *American Journal of Sociology*, 76: 585–603.

—— (1976), *Production and Reproduction: A Comparative Study of the Domestic Domain* (Cambridge: Cambridge University Press).

Hahn, C. H. L., Vedder, H., and Fourie, L. (1966), *The Native Tribes of South-West Africa* (London: Frank Cass & Co.).

Harrell-Bond, B. (1975), *Modern Marriages in Sierra Leone* (The Hague: Mouton Publishers).

Karanja, W. W. (1987), ' "Outside Wives" and "Inside Wives" in Nigeria: A Study of Changing Perceptions in Marriage', in D. Parkin, and D. Nyamwaya (eds.), *Transformations of African Marriage* (Manchester: Manchester University Press). 247–61.

Kuper, A. (1970), 'The Kgalagari and the Jural Consequences of Marriage', *Man* NS 5: 466–82.

Lacombe, B. (1987), 'Les Unions informelles en Afrique au Sud du Sahara: Le Cas du deuxième bureau congolais', *Genus*, 43/1–2: 151–63.

La Fontaine, J. (1962), 'Gisu Marriage and Affinal Relations', in M. Fortes (ed.), *Marriage in Tribal Societies* (Cambridge: Cambridge University Press), 88–120.

Little, K. (1951), *The Mende of Sierra Leone* (London: Routledge & Kegan Paul).

—— (1973), *African Women in Towns: An Aspect of Africa's Social Revolution* (Cambridge: Cambridge University Press).

—— and Price, A. (1967), 'Some Trends in Modern Marriages among West Africans', *Africa* 37/4: 407–24.

Lombard, J. (1954), 'Cotonou, ville Africaine: Tendances évolutive et reaction des coutumes traditionelles', *Bulletin de l'Institut Français d'Afrique Noire*, 16: 3–4.

Mann, K. (1985), *Marrying Well: Marriage, Status and Social Change among the Educated Élite in Colonial Lagos* (Cambridge: Cambridge University Press).
Massam, J. A. (1927), *The Cliff Dwellers of Kenya* (London: Seeley, Services & Co.).
Murphy, W. P., and Bledsoe, C. H. (1986), 'Kinship and Territory in the History of a Kpelle Chiefdom (Liberia)', in I. Kopytoff (ed.), *The African Frontier: The Reproduction of Traditional African Societies* (Bloomington, Ind.: Indiana University Press), 121–47.
Obbo, C. (1980), *African Women: Their Struggle for Economic Independence* (London: Zed Press).
—— (1987), 'The Old and the New in East African Élite Marriages, in D. Parkin and D. Nyamwaya (eds.), *Transformations of African Marriage* (Manchester: Manchester University Press), 263–80.
Omari, P. T. (1960), 'Changing Attitudes of Students in West African Society towards Marriage and Family Relationships', *British Journal of Sociology*, 11: 197–210.
Oppong, C. (1974), *Marriage among a Matrilineal Élite: A Family Study of Ghanaian Senior Civil Servants* (Cambridge: Cambridge University Press).
—— (1980), 'From Love to Institution: Indicators of Change in Akan Marriage', *Journal of Family History*, 5/2: 197–209.
Orde-Browne, G. St. J. (1925), *The Vanishing Tribes of Kenya* (London: J. B. Lipincott Co.).
Parkin, D., and Nyamwaya, D. (eds.) (1987), *Transformations of African Marriage* (Manchester: Manchester University Press).
Pellow, D. (1977), *Women in Accra: Options for Autonomy* (Algonac, Mich.: Reference Publications).
Rindfuss, R., Bumpass, L., and St John, C. (1980), 'Education and Fertility: Implications for the Roles Women Occupy', *American Sociological Review*, 45: 431–47.
Schuster, I. M. (1979), *The New Women of Lusaka* (Palo Alto: Mayfield Publishing Company).
Tetteh, P. A. (1967), 'Marriage, Family and Household', in W. Birmingham, I. Neustadt, and E. N. Omaboe (eds.), *A Study of Contemporary Ghana*, ii. *Some Aspects of Social Structure* (Evanston: Northwestern University Press), 206–16.

Part IV

The Changing Face of Polygamy

8 The Historical Roots and Cultural Logic of Outside Marriage in Colonial Lagos

KRISTIN MANN

Introduction

Outside marriage, the practice whereby men marry one woman by statutory law and also form extra-legal domestic and sexual unions with other women, is widespread in African towns and cities (Baker and Bird, 1959; Little, 1973; Aronson, 1978; Karanja, 1987; Obbo, 1987; Clignet, 1987). A modern form of polygyny with implications for women and children, nuptiality, and fertility, the institution has been an important agent of change in African domestic life (Parkin and Nyamwaya, 1987). The phenomenon is not new, contrary to what much anthropological and sociological research suggests, but dates to the advent of missionary and colonial contact. This essay probes the origins of outside marriage within the educated élite, in a specific historical context, early colonial Lagos, to uncover the religious, ideological, material, and legal forces that produced it.

Multiple and changing forms of marriage have long existed in African societies. In a recent discussion of shifting themes in the analysis of African marriage, Burnham has argued that all forms of conjugal unions represent a range of possible ways that men, women, and their respective kin interact (Burnham, 1987). Domestic and sexual behaviour is the product of choices, themselves shaped by moral as well as material imperatives. Following this perspective, the question regarding outside marriage is why men took outside wives and why women became them. What was the logic of the system, given the cultural and historical setting in which it emerged? The answers to these questions shed new light on the relative experiences of statutory wives, outside wives, and their children with respect to moral authority, legal status, resource allocation, and domestic interaction. This information in turn illuminates the marriage and fertility strategies of statutory and outside wives. When coupled with Karanja's research (this volume), the data presented here show how outside marriage has changed over the past century.

The article is of more than historical interest. Africanist social scientists are increasingly aware of the past in considering the present (see, for example, Peel, 1983; Moore, 1986; Joseph, 1987). Not only have historical processes shaped contemporary institutions and relationships; the experiences of their forebears also affect the understanding and behaviour of contemporary actors. Few places is this second phenomenon truer than in domestic life, where individuals carry with them vivid memories of lessons learned from their progenitors. If scholars want to understand African marriage in the present, then they must study it in the past. Africanist demographers are keenly interested in changes in nuptiality, both for their own sake and because of their implications for the persistence of high fertility on the continent (Lesthaeghe, Kaufmann, and Meekers, 1989; Caldwell and Caldwell, 1987). This essay shows that changes in African marriage that they usually identify with the contemporary period are in some places much older. Before demographers can draw accurate conclusions about the character and causes of changes in African nuptiality, and about their impact on fertility, they need better data about the history of marriage. By demonstrating the possibility of historical scholarship on African marriage, this essay points the way to vital future research.

The Educated Élite and Christian Marriage

In the second half of the nineteenth century, the community of Lagos, on the West African coast, developed into a leading centre of international trade in manufactured imports and agricultural exports, primarily palm produce. Christian missionaries of European and African descent arrived to establish churches and schools, and soon they began attracting local converts. Great Britain annexed the settlement in 1861 and commenced building a colonial bureaucracy and legal system, the machinery of direct colonial rule (Newbury, 1961; Hopkins, 1964; Ajayi, 1965; Ayandele, 1966; Burns, 1969; Tamuno, 1972; Adewoye, 1977).

Over the next forty years the town changed dramatically. The population grew and diversified as liberated slaves returned from Brazil, Sierra Leone, and Cuba, immigrants flocked to the coast from the interior, and merchants and officials arrived in increasing numbers from Europe (Verger, 1968; da Cunha, 1985; Kopytoff, 1965; Mann 1977). Originally a small Yoruba settlement that had in the precolonial period been affected by contact with the Aja-Ewe peoples to the west and the Benin kingdom to the east, the community continued to absorb heterogeneous influences, while remaining predominantly Yoruba (Aderibigbe, 1975; Law, 1983; Agiri and Barnes, 1987; Smith, 1988). The growth of international trade brought fresh opportunities in commerce and new individuals were able to take advantage of them, generating upward social mobility. The authority of the local *Ọba* (king) and chiefs waned, although it by no means disappeared, as Britain imposed hegemony and new individuals of wealth and power emerged in the local community. Indigenous communal land tenure survived, but the alienation and commoditization of land occurred, and individual land ownership developed into an important new basis of wealth. New mechanisms for mobilizing credit took hold, and new means of organizing labour were forged as British colonialism and commercial capitalism undermined slavery (Hopkins, 1980; Mann, 1987, 1989, 1991*a*). For reasons that I have examined elsewhere, women were less able than men to take advantage of changes in land tenure, credit mobilization, and labour organization and to profit from new possibilities in trade. Inequality in men's and women's access to resources and opportunities widened in the early colonial period (Mann, 1985, 1991*b*).

During the second half of the nineteenth century, a small but influential educated élite developed in Lagos, drawn primarily from repatriated slaves from Sierra Leone and the New World and secondarily from the indigenous population. The members of this group were Christian and highly educated relative to the population as a whole. Its men made their living in commerce, the professions, and the colonial service, while its women aspired to be housewives and avoid work outside the home. The élite enjoyed wealth derived from trade and modern occupations, which took the form of privately owned land, capital, and consumer goods. Members of the group exercised considerable influence with Europeans and other Africans, owing to their education and occupations. They developed a distinctive culture, shaped by Christianity and Western education, that identified the élite, distinguished it from the rest of the population, and contributed to group maintenance.

Marriage and domestic life lay at the heart of élite culture (Mann, 1977, 1985).[1] The mission churches, Anglican, Wesleyan, Baptist, and Roman Catholic, regarded marriage as central to Christianity. Investing domestic behaviour with great significance as an outward sign of inward religious conversion, they made marriage a pivotal testing ground of individual religious faith by contrasting what they saw as the Christian ideal with local practices. Whereas the local culture was polygynous, for instance, the missionaries identified Christianity with monogamy. They required polygynists to renounce all wives but one as a condition of baptism and treated backsliding among converts as grounds for excommunication. Besides demanding monogamy, missionaries spread Victorian domestic culture. They insisted that Christian marriage united two individuals, not two kin groups, as among local peoples, and they stressed that it should be based on the Western ideals of love and companionship.

Missionaries also transmitted to educated Christians middle-class Victorian values about the sexual division of labour in society. Local women pursued economic activities outside the home, independently of their husbands, and contributed to the support of themselves and their children as well as to the needs of their affines and kin. The missionaries taught, however, that Christian husbands ought to support their wives and offspring economically, while Christian wives should remain at home caring for the hearth and children.

In local domestic culture spouses fulfilled obligations to one another, their children, and their affines, but retained separate resources and interests. Wives did not become members of their husbands' families, although they commonly lived in their husbands' households, but remained members of their families of origin. Moreover, spouses did not inherit from one another. Men's and women's property passed to their siblings and children (Ajisafe, 1924; Folarin, 1939; Fadipe, 1970; Hopkins, 1969; see also Coker, 1958; Lloyd, 1965; Eades, 1980). Contrary to these local practices, Christian missionaries taught that marriage should form a conjugal estate and create an identity of interests between partners.

The authority of the colonial state reinforced Christian teaching about marriage. Little more than a year after the first governor arrived in the colony, the local legislative council passed ordinances for licensing and registering Christian marriages.[2] In 1884, the government enacted a new marriage ordinance that repealed the older laws and made detailed provisions for licensing, celebrating, and registering unions. The new legislation provided for the appointment of marriage registrars, and it licensed churches for the celebration of marriage (The Marriage Ordinance, 1884). Persons wanting to wed under the Ordinance were required to give written notice to a marriage registrar, which he was then obliged to publish by entering it into a Marriage Notice Book and posting it on the outer door of his office. After twenty-one days, the registrar could issue a marriage certificate, provided he was satisfied by affidavit that (1) each party was twenty-one years old or had the consent of a parent or guardian, (2) there was no impediment of kindred or affinity to the marriage, and (3) neither of the parties was married to another person by 'Native law or custom' (ibid. 447). Anyone who knew a reason why the marriage should not take place could block the issue of the certificate by writing 'forbidden' opposite the relevant entry in the Marriage Notice Book and recording the grounds for objecting to the union. If this happened, the registrar was to refer the matter for hearing to the English judge of a colonial court.

Under the 1884 Ordinance, marriages could be celebrated only in licensed churches by recognized ministers or in marriage registries by registrars. Following their celebration, such marriages were registered in the principal registrar's office. The law declared ordinance marriages invalid on the grounds of kindred or affinity or of prior marriage to someone else by native law and custom. It also declared persons married under the Ordinance incapable, during the continuance of the union, of contracting a valid marriage to someone else under native law and custom. Thus the law made ordinance marriage and polygyny mutually exclusive.

The Marriage Ordinance of 1884 contained a further provision that altered the inheritance rights of spouses, their children, and their children's children. It specified that when locals married under the Ordinance or had contracted a valid Christian marriage prior to the passage of the Ordinance and died intestate (without having written a will) leaving a spouse or children, then their personal and real property which could be disposed by will should be distributed according to English law regulating the distribution of the personal estates of intestates.[3] This section of the Ordinance did not alter the ability of persons to distribute their estates by writing wills. Moreover, it affected only freely disposable personal and real property, not collectively owned family land, which could not be disposed by will. Finally, the section of the Marriage Ordinance on inheritance governed intestate succession not only to the property of couples themselves but also to that of their children. At the time the Ordinance was written, English law of intestacy held that a wife's personal property passed to her heirs, subject to her husband's right to use it for life. One-third of a husband's personal property devolved to his wife for life, while the remaining two-thirds went to his heirs. For the purpose of intestate succession, heirs included legitimate children or next of kin (Geldart, 1963; Bromley, 1962; Crane, 1957).

In the eyes of the churches and state, and of many educated Africans themselves, Christian marriage gave men, women, and their children a new status. Justice Brandford Griffith expressed this view in his decision in the celebrated case *Cole* v. *Cole*: 'Christian marriage imposes on husbands duties and obligations not recognized by native law. The wife throws in her lot with the husband, she enters his family, her property becomes his. . . . In fact a Christian marriage clothes the parties to such a marriage and their offspring with a status unknown to native law' (1898, 1 NLR, 22).

By his decision, Justice Griffith overturned the ruling of a lower court that customary laws of inheritance should govern the distribution of the estate of an African who had contracted a Christian marriage in Sierra Leone and died intestate in Lagos. Because the marriage in question had taken place outside the colony, it was not regulated by the 1884 Marriage Ordinance. As a consequence of the lower court ruling, the brother of the deceased had inherited the estate, rather than the lone child, a lunatic son. The widow appealed, and Griffith held in the higher court that to decide questions of inheritance among Africans who made Christian marriages outside the colony on the basis of native law and custom would be contrary to 'principles of justice, equity, and good conscience' (ibid). He ruled that the law in force in England ought to apply, and declared the son heir, lunatic or not. The widow retained her right to dower and to one-third of her husband's personal and real property. The decision in *Cole* v. *Cole* underscored the legal, cultural, and moral distinction between Christian and African marriage. It attracted considerable public attention, suggesting that the local community appreciated the case's significance.[4]

Christian teaching about marriage deeply affected the educated élite. Yet it was at odds with local domestic culture, which continued to influence the members of the group, nearly all of whom were of Yoruba descent. To compound the tension and ambiguity surrounding marriage, élite men and women responded differently to the problems of how to marry and what marriage should mean.

Most élite women embraced the Christian regime for both religious and instrumental reasons. These women had internalized the Christian message about the spiritual, moral, and cultural superiority of monogamy and of Victorian conjugal relations. They were also drawn to Christian marriage as a mark of élite status and a means to alliances with élite families. In addition, élite women had special economic concerns that shaped their thinking about marriage. Having embraced an ideology that emphasized retreat from the workplace into the home, they sacrificed the economic autonomy of women in local culture. Élite women depended on their husbands for their social status and economic well-being. If women born into élite families or educated to élite standards failed to make Christian marriages to men who would support them in the manner they desired, then the consequences were devastating (Mann, 1983). Because changes spawned by the growth of foreign trade and imposition of British colonialism had adversely affected women's access to resources and opportunities, élite women could not have competed with élite men economically, had they wanted to. Well-educated women faced no alternative but to secure their status through Christian marriages to élite men.

Christian marriage gave women a standing in the eyes of the churches and state that they believed would protect them and their children. Christian wives had a moral and legal right to monogamy. They and their children were supposed to enjoy exclusive access to their husbands' time, affection, and resources. Monogamy freed women and their offspring from the competition and conflict often associated with polygyny. Moreover, Christian marriage gave women new rights to their husbands' estates in cases of intestacy, as it did children new rights to their fathers' estates. In time, Christian wives discovered that they could not necessarily enforce their rights and those of their children, despite the backing of religious and secular authorities. None the less, élite women saw in Christian marriage a host of moral and instrumental advantages.

Élite men responded to the Christian marriage regime more ambivalently than did élite women. Like their female counterparts, many were drawn to it out of a deeply felt desire to be good Christians. All recognized it as a mark of élite status and a path to it. But élite men, much more than élite women, also felt the pull of local domestic culture (Mann, 1985). Of the 200 educated, élite men who resided in Lagos between 1880 and 1915, roughly two-fifths were monogamous and three-fifths were polygynous.

Monogamy confronted élite men with a major obstacle to the Christian marriage regime. They argued that it prevented wives from observing the long period of sexual abstinence following the birth of children, which local culture regarded as essential to the health of mothers and babies. It was one thing to expect women to forgo sex during this interval, but it was another to require men to exercise the same restraint. Without other wives to whom a man could turn for a sexual outlet following the birth of a child, he would be driven to have sexual relations with the new mother, which locals believed would spoil her breast milk. Or he could fornicate, a more grievous sin than polygyny in the eyes of many Christian men. Violation of the post-partum sexual taboo, contemporaries argued, also interfered with birth spacing. They believed that too frequent childbirth was weakening the wives of monogamists and their offspring.[5] From the point of view of many élite men, monogamy offered two unacceptable reproductive alternatives: infertility, when Christian wives could not have children, and high fertility accompanied by high maternal and infant mortality when they could.[6]

Élite men expressed two other reasons for objecting to monogamy. First, they thought that it deprived some women of marriage, woman's natural calling. Second, they maintained that it created a problem of what to do with widows. In polygynous cultures, commentators observed, all women could marry and widows of childbearing age would almost certainly remarry. Widows too old to bear children would find a home with offspring, kin, or affines. But in cultures that insisted on monogamy, they noted, some young women were deprived of the care and protection of a husband, while a man's death could leave his widow destitute for life.[7] Élite men blamed monogamy for such Western social problems as female poverty and prostitution.[8]

The discourse about polygyny centred on the evils of monogamy rather than the attractions of polygyny. Apart from stating the general proposition that polygyny was preferable because it reflected local culture, writers said little about what drew men to plural marriage. Had élite men expounded on the advantages of polygyny, they probably would have emphasized the greater access it provided to female labour and marriage alliances and the larger number of offspring it enabled men to have. Almost one-third of the élite men between 1880 and 1915 were merchants, who had to move in local circles in pursuit of trade. Much more than élite women, they needed the alliances with local families that polygyny made possible. Moreover, the merchants sometimes used wives in organizing their commercial activities. The most honest élite men might also have mentioned the freedom that polygyny gave them to pursue openly emotional and sexual relationships with multiple women.

Other aspects of Victorian domesticity suited local culture little better than did monogamy. Some élite men resisted the obligation to assume full financial support for their wives and children.[9] Many found the companionate conjugal ideal impossible to achieve and better left untried (interview: Archdeacon J. O. Lucas). As primary breadwinners, they felt that their economic activities outside the home entitled them to maintain control over their earnings. Few, therefore, liked the claims that Christian wives could make on their resources during life or the rights that these women enjoyed to their property after death.[10]

Despite the opposition that men felt to the Christian marriage regime, and despite the myriad ways it conflicted with local culture, roughly nine out of every ten men among the 1880 to 1915 élite made Christian marriages.

Many did so without fully appreciating the difficulties it would cause them. Others entered these unions knowingly, because they were committed to Christian ideals or desired the benefits associated with Christian marriage: educated wives, élite affines, and status recognition. But roughly six out of every ten élite men who made Christian marriages also entered unions with other women, in spite of religious and secular prohibitions against such behaviour. As Guyer argues in her essay (this volume), neither the church nor the state could effectively regulate the moral culture of family relationships. The pattern among élite men of making a Christian marriage to an élite woman who aspired to a Western conjugal relationship and at the same time of forming multiple unions with non-élite women who accepted indigenous domestic relationships gave birth in the 1850s and 1860s, the moment of missionary and colonial contact, to the institution of outside marriage.[11]

The Educated Élite and Outside Marriage

Although it involved élite men, outside marriage differed from Christian marriage in fundamental respects. Like local customary marriages, it allowed men multiple unions. Moreover, outside wives came from different backgrounds from Christian wives, and they enjoyed different opportunities. Outside wives were less well-educated than Christian wives, many having had no schooling at all, and they came from families with less education. Their fathers were more likely to trade or farm and less likely to work in the colonial service or a profession. Ineligible for Christian marriage, outside wives invariably engaged in some economic activity outside the home, usually trade. Absent from their lives were fathers, uncles, brothers, and husbands willing to support them if they withdrew from the workplace. Some outside wives were Christian, but others were Muslim or practised an African religion. A few came from powerful local families, but others were of humble origin. A number of such women had been slaves and accepted outside unions as a way out of bondage (interviews: Mrs Lelia Abiodun, Mrs H. F. Pereira, Mrs R. A. Wright, Mrs Ayo Adeshigbin).

Outside marriage also had a very different standing from Christian marriage in the eyes of the various churches and of the state. Élite men formed outside unions both before and after Christian ones. If before, then the men had no legal right to enter Christian unions with other women. Mechanisms existed, as we have seen, for blocking the Christian marriage, but outside wives rarely used them. An examination of Supreme Court records to 1900 uncovered only one judicial hearing into the issue of a marriage licence. Adel Coker, the customary wife of Dr Oguntola Ṣapara, stunned the élite community in 1911 by opposing the issue of a licence for her husband to wed Ore Green under the Ordinance; as did Herbert Macaulay by objecting to the union when the banns were read in a local Anglican church (interview:

Archdeacon J. O. Lucas; 'In the Matter . . . of an Intended Marriage . . .' (*In re Ṣapara*), 1911).

If men formed outside unions after Christian marriages, then the unions had no standing in the eyes of the churches or state. Religious and secular authorities regarded outside wives as adulteresses, with no rights to inherit unless provided for in wills and certainly with none to monogamy. The mission churches and colonial state viewed outside children as illegitimate, the state denying them inheritance in cases of intestacy and the churches baptism except in special circumstances (*Report of . . . Proceedings*, 1907: 129–37).

Throughout the 1880s and 1890s the churches and state clarified and tightened their position on marriage, in partial response to the growing incidence of outside unions. The move began with the passage of the 1884 Marriage Ordinance. Religious debate and action focused on polygyny in general rather than outside marriage in particular. In December 1884, James Johnson, pastor of St Paul's Anglican church, Lagos's largest congregation, disciplined a communicant who had given his daughter in marriage to a polygynist by suspending him from communion for six months and dismissing the daughter until her husband 'reformed his life' (Lucas, 1946: 32–4). When members of the congregation protested, Johnson publicly rebuked 700 of his parishioners by pronouncing them unworthy and refusing them the Eucharist (Ayandele, 1970: 107). In 1887, Bishop Ingham instructed the Diocesan Conference in Lagos, 'The Church of Christ does not tolerate polygyny. The system is bad . . .' (quoted in Okoro, 1887: 3). As if to ratify the position of these local religious authorities, the pan-Anglican Lambeth Conference in 1887 confirmed the obligation of Christians to sexual purity and monogamous marriage, with a series of widely publicized resolutions endorsed by the first Diocesan Synod of Western Equatorial Africa in 1906 and 1907 (Davidson, 1889: 277–8, 292–5, 298–300; *Report of . . . Proceedings*, 1907: 19–20). While the thoughts and deeds of Wesleyan and Baptist missionaries are less well documented, they adopted the same position as the Anglicans with respect to polygyny in any form (Hewitt, 1971: 43). At the very end of the century, the judicial decision in *Cole* v. *Cole* affirmed and publicized the state's commitment to protecting the inheritance rights of Christian spouses and children against other claimants.

Outside unions differed from Christian ones in an additional fundamental regard. Although often preceded by protracted courtship rituals (Mann, 1985), Christian marriage was celebrated definitively at a precise moment in a church and registered so that there could be no question about its existence. The status of outside unions, on the other hand, was often ambiguous, sometimes deliberately so. As the editors of this volume emphasize and as much anthropological research has shown, marriage in much of Africa is not a discrete event but a process enacted over time as men and women live out relationships with one another, their children, and their affines (Comaroff and Roberts, 1977, 1980; Bledsoe, 1980; Guyer 1981). In nineteenth-century Yorubaland, the process at its most ritualized and formalized began with betrothal which was negotiated by kin, continued through the transfer of bridewealth and brideservice and the ceremonial movement of the new wife into her husband's home, and culminated in the birth of the first child. Perpetuating the union required fulfilling conjugal, parental, and affinal obligations throughout its lifetime (Mann, 1985). However, there were different kinds of marriage in Yorubaland, as in other African cultures, many of them less highly ritualized and formalized. Multiple domestic categories existed, instead of only two—married and unmarried. Each kind of union entailed different conjugal and affinal relationships and imparted a different social value. Moreover, unions were fluid and negotiable; their character changed as parties redefined relationships.[12] As a result, unions were sometimes perceived differently by different persons or at different times.[13] Finally, marriages were often attenuated. Women in particular found changing partners one of the few means within their control of responding to their shifting needs.

Outside marriage emerged within this conjugal culture. True of Yoruba marriage in general, these traits were doubly so of outside unions that involved élite men and non-élite women. Such relationships were often begun without a clear outcome in mind by men who were drawn to the women and to polygynous, customary marriage, but felt bound by Christian beliefs and status considerations, as well as by prior or expected commitments to Christian wives and their children. The trajectory of outside unions often became clear only as the relationships unfolded. Because these unions were forbidden by the churches, outlawed by the state, and resented by Christian wives, men had good reason to want to keep them hidden and ambiguous. Their very ambiguity made it that much easier for men and women to manipulate them, claiming one sort of relationship on one occasion and another on another.

In some instances élite men admitted marriage to outside wives in local circles, but denied it in élite ones. However, the factors influencing the representation of these unions were often more complex than a simple élite/non-élite divide. They included the changing character of the relationship, the circumstances of the man, the Christian and the outside wife, and the beliefs of the man and the Christian wife. In a context in which divorce between Christian couples was socially unacceptable and difficult to achieve, the ambiguity of outside unions enabled élite men to withdraw from Christian marriages as they became disadvantageous and to emphasize the legitimacy of other domestic relationships.

Two examples, both documented in court records, illustrate some of these points.[14] In 1904, Dr Oguntola Ṣapara personally negotiated a customary marriage with Adel Coker's kin. His own relatives opposed the union because the woman had borne a child after a fleeting sexual liaison with her cousin. Adel moved to Ṣapara's house in ceremonial fashion, and he paid a modified form of bridewealth to her relatives. The couple lived together for six years, during which Adel bore two children for him, one of whom died. Adel desired the status of 'wife' in the sense in which the élite understood the term. She sat at the head of the doctor's table, kept his purse, went with him to chapel, and called herself 'Madame Ṣapara'. Ṣapara, however, had other ideas. Although he twice referred to himself in official contexts as married by 'native law and custom', to wed Ore Green, a member of the élite community, in church, the doctor in 1911 denied that what he had paid for Adel was bridewealth and claimed that she was only his mistress (*In re Ṣapara*).

A second example shows the ambiguity and fluidity of outside unions even more clearly. In 1892, Q. F. Gomez, a member of Lagos's Brazilian community, went to the Niger River to trade. There he met a slave woman, Lanlatu, who had been hired as a carrier from the slave's mistress in Ilorin by a female trader named Awo. When it came time for Lanlatu to leave the Niger, she pleaded with Gomez to allow her to stay with him, but he refused.[15] In 1896, Gomez returned to the River, where he again met Lanlatu, still in Awo's service. By Gomez's later account, the slave woman 'stayed in my house during the day, but went away at night'.[16] When Awo was ready to leave, Lanlatu again pleaded to remain with Gomez. This time he redeemed her, to keep as his 'maid servant', he told the court.[17] Gomez admitted giving the woman cloth and beads, but he denied paying bridewealth. Awo, on the other hand, claimed that he gave Lanlatu a looking glass and tin bowls, items that commonly formed part of bridewealth in northern Nigeria. In 1897, Gomez returned to Lagos, and soon after he helped Lanlatu follow him there.

Gomez had married in the Roman Catholic church in 1889. In Lagos, he resided with his brother, his Christian wife, and several male and female 'servants'. Lanlatu did not enter his household immediately on her arrival; she resided instead with Gomez's sister, who agreed to teach her to wash and sew to make her 'useful in the house'.[18] During this time Gomez gave Lanlatu 'money for her expenses'.[19] Soon she protested that she did not like washing and sewing and wanted to trade. To raise capital or precipitate a change, Lanlatu stole onions from the sister and sold them. Gomez then gave her thirty shillings to begin trading.

In December 1898, Lanlatu moved into Gomez's household. When cross-examined about the relationship, Gomez testified, 'she was not my wife, as I had not married her. She may have been my sweetheart or friend. I used to have connection with her sometimes'.[20] His lawyer subsequently admitted that Gomez had fathered a child by Lanlatu, which had died.

In May 1899, Lanlatu left Gomez to live with another man, Oke. We have no way of knowing why she went, but the record suggests a possible explanation. She had had a fight with another 'maid servant', possibly one whose relationship with Gomez was as complex as her own. Lanlatu believed that Gomez was partly to blame for the quarrel. This unpleasantness may have convinced Lanlatu that she no longer wanted to live with Gomez. He tried without success to persuade her to stay. After she had gone, he sent people to bring her back. One of them, finding her in the street, claimed to have been greeted with, 'I have got married. My husband is Oke. Come and see him tomorrow'. This witness reported that Lanlatu was then wearing native cloth, silver bracelets, seggi beads, and silver chains, in the style of 'newly married women'.[21] Soon after, Gomez initiated a case against Lanlatu and Oke in the Lagos Supreme Court.

Gomez's claim in the case appears in the court record as follows: 'return of money and goods [taken by Lanlatu] value £24.3.8 . . . and . . . against defendant Oke £40 damages for enticing and harbouring plaintiff's wife'.[22] The second part of the claim is inconsistent with Gomez's representation of his relationship with Lanlatu to the court. Throughout the trial he and his witnesses framed their testimony to make it appear that Lanlatu was his 'maid servant'.

It is difficult to know how to explain this inconsistency. Gomez was represented at the trial by an attorney, who must have known that because his client had married another woman under the Ordinance in 1889, the court would not recognize his relationship with Lanlatu as a valid marriage. Quite likely, Gomez began the legal proceedings on his own by talking to the court registrar and only later hired a lawyer. Gomez may have articulated his grievances to the registrar much as they appear in the claim, or the registrar may have imposed this formulation on complaints that Gomez expressed differently. Once he hired a lawyer, Gomez must certainly have been made aware of his dilemma. He and his attorney then adjusted the presentation of the case as best they could, but from a position of weakness. It is also possible that Gomez knew when he initiated the case that the court would not recognize Lanlatu as his wife, but that he sued Oke for damages none the less because he was angry with the man, as well as with Lanlatu, and wanted retribution from them both. Finally, the contradiction may reflect the ambivalence that Gomez himself felt about the relationship and the likelihood that he was unused to referring to Lanlatu in a consistent way.

Whatever the explanation of the contradiction, Gomez had no leg upon which to stand so far as his second charge was concerned. He failed to prove his first: that Lanlatu had taken anything from him. The judge threw the case out of court. The British official must have felt some sympathy for Gomez, however, because he partially vindicated him by charging Oke with court costs.

This case demonstrates many of the characteristics of outside unions. They were ambiguous, negotiable, and changing; presented differently on different occasions and were understood differently by different people. Here we see Gomez in the process of constructing such a relationship. In certain ways and on some occasions, he behaved toward Lanlatu as though she were his wife. On the Niger he paid something resembling bridewealth for her. In Lagos he gave her money for expenses and capital to begin trading, as befitted a new husband. The couple cohabited and bore a child. When Lanlatu left him, Gomez tried to persuade her to stay. When she refused, he sent people to bring her home. On discovering that she was living with another man, he sought damages from the culprit in court. Yet the evidence suggests that even before the trial Gomez presented Lanlatu as his 'maid servant' to his sister and probably also to his other Catholic relatives and his Christian wife. Gomez managed Lanlatu's entry into his matrimonial home by first placing her with his sister to be taught domestic skills and then bringing her into his household to assume responsibility for specific tasks. Gomez's sister was astonished to learn at the trial that her brother had fathered a child by Lanlatu, saying he had never told her so. In court, except for his initial charge, Gomez and his witnesses steadfastly maintained that Lanlatu was his maidservant. Construing the relationship in this way enabled Gomez to avoid violating the letter of the Marriage Ordinance and to minimize censure by his relatives and the church.

It is impossible to know how Lanlatu understood the relationship. She did not speak in court, because after hearing from Gomez and his witnesses the judge decided that the plaintiff had no case against the defendants. Hence, he did not call them to testify. It is very unlikely that before Lanlatu arrived in Lagos she understood the ambiguities of outside marriage as it existed in the colony. She may not have known that Gomez had a Christian wife; or if she knew, she may not have appreciated just what that meant. It is clear, however, that Lanlatu played an active role in initiating and defining a relationship with Gomez. She used it to obtain redemption and make her way to Lagos. Once there, she pushed Gomez to give her a start in trade. When the relationship no longer suited her, she escaped it by 'marrying' another man. Whatever her understanding of the relationship, at this point she evidently realized that she could leave it with relative impunity because of the very ambiguity of her union with Gomez.

The treatment of outside wives and children within the African community was quite different from their position in the eyes of the church and state. Beyond the educated, Christian population, outside wives and children were from the beginning accepted and supported by both men and women, consistent with the values of a polygynous culture. This support was essential to their long-term welfare. Within the educated, Christian community, the responses to outside wives and children varied during the second half of the nineteenth century from ostracism to toleration. The experiences of outside wives and children depended heavily on the beliefs, interests, and commitments of the man, his Christian wife, and their kin. Also important was the ability of the outside wife to win support for herself and her children.

Most élite men who had inside and outside families during the decades from 1850 to 1900 privileged the inside wife and her children, either because they thought that was the right thing to do or because they were forced into it by their Christian wives, affines, and kin. Men expended on inside wives and children greater resources, which translated into better food, housing, clothing, and education. Moreover, they had closer contact and more regular interaction with their Christian wives and children. These relationships were open and public in a way that those with outside wives and children usually were not. A strong feeling existed among the early élite that outside wives and children ought not be brought into polite society (interviews: Lady Bank-Anthony, S. O. Foresythe, Mrs Ayo Adeshigbin, Mrs R. A. Wright). In addition, élite men favoured their Christian wives and offspring when it came to inheritance. Of the fifty-four such men who lived in Lagos between 1880 and 1915 and made Christian and outside marriages, only eleven wrote wills that provided for their outside wives and only twenty left wills that provided for their outside children. The rest either made no such provisions in their wills or left no wills.[23] What those who died intestate intended is difficult to know. Prior to *Cole* v. *Cole*, a few may have failed to appreciate the legal implications of intestacy. Others may have provided for outside wives and children before dying or trusted their Christian wives and children to share their estates. But some undoubtedly chose intestacy as a way of effectively disinheriting outside wives and children.

In general, outside children received better treatment than their mothers. Few men cut outside children off entirely, even when they ended their emotional and financial relationships with the children's mothers. Shortly before his death in 1894, John Langley went to see his minister, the Revd James Johnson, and confessed that he had a child whose paternity he had denied and whom he had kept from his Christian wife. Langley told Johnson that before he died he wanted to persuade his Christian wife to raise the boy. He begged the clergyman to intercede with the woman on his behalf.[24]

A minority of élite men treated their outside wives and children more equally, either because they were determined to do so despite the objections of others or because their Christian wives and kin accepted their outside families. In the 1870s, Hezekiah Africanus Caulcrick fathered two daughters, each by a different woman (interviews: Mrs Lelia Abiodun, Mrs H. F. Pereira). In 1878, he wed another woman, Matilda Beatrice Scale, in the Anglican church (Marriage Register, St Paul's Church). The following year, Caulcrick fathered his first son by a fourth woman, which offended Matilda. Moreover, he began resisting his Christian wife's claims to full support. Soon the couple fell to quarrelling. Eventually Matilda moved out of her husband's house. Caulcrick set up his outside wives in residences of their own, and gave each a weekly allowance. Throughout his life, H. A. Caulcrick accepted responsibility for all his children, but he favoured his first son, by an outside wife, over his second son, by the Christian wife. His will bequeathed landed property to all the children and made the following provision for his wives:

> [A]s . . . trustees I direct [my male children] to support the women that might be under my guardianship as wives at the time of my death . . . during such time that each might remain without a husband or during lifetime. I direct that Mrs. Matilda Beatrice Caulcrick though now away from my premises under mutual separation should share in such maintenance and support during her widowhood and lifetime.[25]

In a handful of cases Christian wives accepted outside wives. All the wives and offspring of merchant B. C. Vaughan met for lunch on Sundays after church (interview: Michael Ayo Vaughan). It was more common, however, for Christian wives to embrace outside children but to maintain a distance from their mothers. Colonial servant Edward Sogunro's outside son lived with his mother, but visited regularly in the house where his father lived with the Christian wife (interview: F. O. Sogunro). Isaac Nathan's Christian wife served as godmother to his outside son, born after their Christian marriage. She allowed the boy, but not his mother, to live with her, her husband, and their two children.[26]

No significant change occurred in the legal position of outside children until the middle of the twentieth century. A judicial decision in 1946 (*Philip* v. *Philip*, p. 102) upheld the right of an illegitimate son whose paternity had been acknowledged by his father to share communally owned family property equally with legitimate sons. But not until 1954 and 1955 did landmark decisions rule that offspring of polygynous and/or outside marriages were legitimate, provided their paternity had been acknowledged, and entitled to share their fathers' estates equally with descendants of ordinance marriages.[27] These decisions affected outside wives only in so far as mothers were often permitted to live on land or in houses that children inherited from their fathers (Coker, 1958: 264–83; Elias, 1962: 258–65). As widows did not inherit from their husbands in local culture, outside wives had less to gain than their children from changes in the law.

Although the legal position of outside wives and children remained the same between 1890 and 1920, the attitude of élite men towards them changed. This change in attitude improved the treatment of outside children and, to a lesser extent, their mothers. A number of phenomena explain the transformation. The first was that around the turn of the century, élite men began rethinking Christian domestic ideology. Troubled by their attraction to polygyny yet their desire to live as good Christians, several turned to the Bible for guidance and concluded, after careful study, that the scriptures nowhere forbade polygyny (Agbebi, 1903; Coker, 1904; Coker, 1915).[28] In time these men rejected the identification of monogamy with Christianity.[29]

This religious reassessment of polygyny coincided with a second watershed event: the creation of African churches, founded for political and constitutional reasons rather than over the issue of polygyny, but open to polygynous members (Webster, 1964). The African churches introduced a type of 'Native Christian' marriage, sometimes also called 'parlour marriage' because the ceremony took place at home, in which the union received Christian blessings but was not registered under the Marriage Ordinance and did not carry the expectations of monogamy and a Victorian conjugal relationship. The African churches gave Christian men an alternative. Henceforth, they could take multiple wives and remain members in good standing of a Christian congregation.

Third, the religious debate over polygyny occurred at roughly the same time as a separate but related cultural nationalist movement. At the end of the nineteenth century and beginning of the twentieth, educated Christians began to celebrate African culture and to rethink the meaning of civilization (Ayandele, 1966; July, 1967; Omu, 1978; Mann, 1985). Intellectual leaders eventually worked out a position of cultural relativism. They asserted that African customs such as polygyny were culturally determined; one was not better or worse than another, only more or less well-suited to a given cultural environment. Monogamy might be right for Europeans, but it was neither inherently superior to polygyny nor right for West Africans.[30] These Yoruba thinkers challenged European assumptions about the relationship between morality and culture. As a result of these developments, many educated Christian men and some such women became more tolerant not only of polygyny but also of other aspects of African marriage. The official position of the European denominations with respect to polygyny remained much the same as it had been earlier. As late as the 1960s, the interdenominational Christian Council of Nigeria was still agonizing over what to do about 'marital irregularities' (Hewitt, 1971: 43). Informally, however, local clergymen grew more understanding of the cultural pressures that pushed men to polygyny. Moreover, the European churches began to distinguish between Christian religious beliefs and local cultural practices such as the sexual division of labour (interview: Archdeacon J. O. Lucas).

Despite all the rethinking about marriage, most élite men and virtually all élite women continued to make Christian marriages. The women were afraid to turn their backs on the legal protection offered by marriage under the Ordinance, and both they and the men were drawn by the status advantages associated with Christian marriage (Mann, 1985; see also Oppong, 1974; Harrell-Bond, 1975). The continued appeal of Christian marriage in an age of growing acceptance of polygyny and other aspects of African marriage produced a rapid increase in the incidence of outside unions. With greater acceptance and greater incidence came new willingness on the part of men to recognize and assume responsibility for outside children and, to a lesser extent, their mothers.[31] Most men still did not share their time and resources equally with their inside and outside families, but to a greater degree than in the past they contributed to the maintenance of their outside wives and the upkeep, education, and inheritance of their outside children. Moreover, men became more willing to have a public relationship with their outside families and to encourage open and friendly relationships among their children by different mothers (interview: Archdeacon J. O. Lucas). Most inside wives continued to feel deeply ambivalent about outside wives. Yet some made peace with a situation they could not change by arguing that the welfare of children was paramount and that inside wives should not punish innocent offspring by refusing to accept them. Élite women also accommodated by tolerating outside wives so long as they could negotiate the terms of their husbands' relationships with these women (interviews: Mrs Ayo Manuwa, Mrs Ayo Adeshigbin, Mrs R. A. Wright).

The Impact of Outside Unions on Women's Strategies for Marriage and Fertility

The fact of outside unions reinforced élite women's insistence on Christian marriage, to protect their interests and those of their children and to distinguish themselves from non-élite women (Mann, 1985). Élite parents educated daughters at least through secondary school and often sent them to finish their education in England, so that they would make attractive Christian wives for élite men (interviews: Mrs T. E. Williams, Mrs Adeniyi Williams). This suggests that advanced education for a girl was in part a strategy for trying to keep her from becoming an outside wife. Educated women of the sort Karanja describes, who preferred outside unions because they were disillusioned with Christian marriage, appeared only in tiny numbers prior to the 1950s (Karanja, 1987: ch. 9; see also Dinan, 1977; Obbo, 1987). Their advent had to await significant improvements in women's educational and occupational opportunities.

Owing to prolonged education, most élite women married later than non-élite women and had shorter reproductive lives. The mean age of marriage for the Christian wives of Lagos's 1880 to 1915 élite was 22.6 years, considerably older than the marriage age of illiterate women, who wed soon after puberty.[32] Some élite women, however, offset delayed marriage by abandoning the post-partum sexual taboo, in the interest of preventing monogamous husbands from going elsewhere for sexual outlets, effectively reducing the interval between the births of their children. Christian wives sometimes also felt pressure to continue having children after they wanted to stop, to accommodate the desires of their husbands, would-be monogamists, for many offspring (interviews: Mrs E. M. E. Agbebi, Mrs Ayo Manuwa, Archdeacon J. O. Lucas).

Domestic and sexual relationships with élite men opened to non-élite women, ineligible for Christian marriage with members of this group, the possibility of access to resources and opportunities that the women would not otherwise have enjoyed. Although Yoruba women commanded good access to resources and great economic autonomy relative to women in other cultures, they remained none the less at a disadvantage relative to men in their own culture (Mba, 1982; Afonja, 1983). Economic, legal, and educational changes in colonial Lagos heightened women's disability, just as it widened the gap between the haves and the have-nots (Mann, 1985, 1991*b*). Élite men controlled property, income, influence, and information within the reach of most non-élite women only through their relationships with these men.[33] Outside marriages, as we have seen, were often informal and open-ended, subject to change and termination. They gave women no guarantees. However, they sometimes developed into stable, long-term unions. Even if only short-lived, as in Lanlatu's case, outside unions could bring non-élite women real, if limited, benefits. When longer-lasting, as in the case of H. A. Caulcrick's outside wives, they imparted more substantial rewards, extending even to care in widowhood. The hope of obtaining access to assistance, resources, and opportunities through outside unions provided an incentive for non-élite women to enter them and for their kin to encourage the women to do so.

Procreation could concretize and solidify an outside union. It made the relationship more visible and public than it would otherwise have been. Moreover, the birth of a child opened élite men's resources to non-élite women if anything would. An article in the local press referred to women who bore children for men outside Christian or customary marriage and clung 'to the bald hope of a promised sustenance from the man concerned'.[34] Procreation also gave the relationship a future that it would not otherwise have had. Few outside unions that did not produce issue lasted for long. Childbirth sometimes played a role in formalizing unions. A local chief commented, 'A man may take a woman, and they live together and get children, and [he] may make her *iyawo* [wife] by giving the *anọ* [bridewealth] afterwards.'[35] All of these factors motivated outside wives to bear children for élite men. As Karanja shows (this volume), a woman may not have wanted to bear too many children for a man if she thought the relationship was limited or unlikely to last; but if she believed it had good potential, then continuing to bear children provided a publicly visible means of perpetuating it.

As we have seen, most élite men did not expend on outside children the same resources as on inside children. Some provided for outside children by placing them with guardians or bringing them to live with their Christian wives, rather than by channelling resources for the children's care through their mothers (interviews: Mrs Adewakun, Mrs Ayo Manuwa, Mrs Ayo Adeshigbin). Even so, many outside wives benefited in the short-term from maintenance provided by fathers to their children. Women gained in the long-term from inheritance passed down to their children from fathers and paternal kin or, more commonly, from education paid for by them. There was of course a danger that men would permanently remove children from outside wives, weakening the relationship between the women and their offspring. Worried about the prolonged absence of her son at his paternal grandmother's home in Abeokuta, Age went to court to obtain a writ of *habeas corpus* requiring the boy's father, Isaac Nathan, to produce him.[36] The case was settled when Age and Nathan agreed that the child should remain in Nathan's custody, on the condition that he should be allowed to visit his mother and that he would not be sent to Abeokuta again without her consent. Few outside children, however, including those who achieved marked upward social mobility through their fathers' assistance, failed to maintain close ties with their mothers. Outside wives ultimately benefited from their children's accomplishments, and this provided a powerful incentive to have them. Although not always realized, outside unions had the potential, in Jane Guyer's terms (this volume), for short-term lateral advantages to the women themselves and for short- and long-term lineal benefits through their offspring. As outside wives and children gained greater acceptance and began to receive better treatment in the early twentieth century, the attractions of outside marriage increased, despite the continued legal disadvantages that outside families encountered.

Some outside wives limited their domestic and sexual relationships to one man. Others had multiple partners, serially as in the case of Lanlatu, or simultaneously. A newspaper article asserted that 'mothers of three or more children who owe their paternity to an equal number of fathers, can be counted by the scores if not by the hundreds'.[37] In the first two decades of the twentieth century, male commentators complained bitterly about the loss of patriarchal control over women's sexuality. They decried the decline among women of premarital chastity and marital fidelity and the emergence of 'unqualified polyandry'.[38] Men lamented that women could now leave partners at will.[39]

In the early twentieth century, some Lagos women were in fact moving in and out of multiple domestic and sexual relationships.[40] Cultural factors gave women domestic and sexual mobility, and socio-economic realities pushed them to use it. Polygyny created a large demand for women, so that they had no difficulty finding partners as long as they were of childbearing age. The informality and negotiability of indigenous unions allowed women to enter, redefine, and terminate relationships with relative ease. The fact that in local culture children's claims to resources through their fathers depended upon recognition of paternity rather than on the marital status of their mothers meant, ironically, that offspring of customary and outside marriages usually sacrificed little when their mothers terminated unions with their fathers (Fadipe, 1970; Ajisafe, 1924). If men had assisted children when unions were intact, they usually continued to do so after they had ended. Finally, Lagosians perceived, accurately, that many Europeans believed African women were frequently forced into marriage against their wills owing to the practice of childhood betrothal, and that religious and secular authorities regarded this state of affairs as repugnant. Locals thought that women could find in the colonial courts and mission churches allies in their struggles to change partners, simply by playing upon European prejudices about African marriage.[41] This belief may have given women new freedom by emboldening them to leave partners, in the face of opposition from affines and kin.

If these phenomena gave women domestic and sexual mobility, others pushed them to take advantage of it. In Lagos, resources such as land, housing, and credit could be bought on the market by those who had the means, but most persons still obtained access to them, as well as to vital information and assistance, through social relationships ranging from kinship, marriage, and affinity to clientage and slavery (Mann, 1987, 1991*a*, 1991*b*). Forms of marriage and affinity, constructed through domestic, sexual, and maternal interaction, offered Lagos women better prospects for help and support than any other relationship save kinship. Moreover, such relationships gave women one of the few means within their control of adapting their circumstances to their changing needs. Consequently, Lagos women felt great pressure to treat domestic and sexual behaviour instrumentally and to turn it to good advantage. In a changing, uncertain, and competitive world, moreover, building a wide and diverse network of relationships on which one could draw for resources and assistance provided an effective strategy for maximizing opportunities and minimizing risks. Forming multiple domestic and sexual unions, serially or simultaneously, offered women a way of constructing such a network. It allowed them to diversify their options. Bearing children of these unions enabled women to give them a value and meaning that they would not otherwise have had. In a culture that regarded polygyny as the ideal but polyandry as anathema, the ambiguity of outside unions made it easier for women to have multiple partners simultaneously, because the precise character of these relationships, and thus of the women's obligations within them, was subject to debate. Some non-élite women found outside unions attractive precisely because they gave them new flexibility, in addition to entrée into a rich network of help and support.

Conclusion

The demands of living in a rapidly developing colonial and commercial capital made outside marriage an attractive alternative for a growing number of élite men and non-élite women, despite the religious and secular prohibitions against it. Characteristics of indigenous marriage facilitated the emergence of this new kind of union, which originated among this community of élite men caught between conflicting moral and material orders and of non-élite women subordinated within both. The relationship offered élite men a way around the religious and legal prohibition against polygyny and Christian marriage and around the moral and cultural identification of Christian marriage with Victorian domesticity. For women the story was quite different. Statutory wives, outside wives, and, to a lesser extent, their respective children occupied different legal, moral, and material positions, all of which affected the marriage and fertility strategies of both kinds of women. Whereas élite women had little to gain from their husbands' outside marriages, they may have had much to lose, in terms of pride or even the possibility of conjugal dissolution, by raising public objections. For non-élite women, however, outside marriages gave them the possibility of access to resources and opportunities which otherwise they would not have enjoyed: the hope of upward mobility for themselves or their children in a highly unequal world.

Notes

1. The discussion here of Christian marriage among the educated élite draws on my book, *Marrying Well: Marriage, Status, and Social Change among the Educated Élite in Colonial Lagos*. In that work, I defined élite men as including all professionals, high-ranking civil servants, and Western-educated, import-export merchants who lived in Lagos between 1880 and 1915. For a fuller consideration of the definition and composition of the élite and a more detailed development of the argument regarding marriage, see that work.
2. Colonial Office Dispatches (CO) 147/3, 3 July 1863, Minutes of the Legislative Council.
3. Communally-owned family property could not be disposed by will.
4. *Lagos Weekly Record* (*LWR*), 6 Aug. 1898, p. 4, col. 3; 31 May 1913, p. 4, col. 2; *Lagos Standard* (*LS*), 10 Aug. 1898, p. 5, col. 2; 29 Mar. 1899, p. 3, col. 3.
5. 'Is Native Marriage Advisable or not Advisable?', 26 July 1904, Coker Papers (CP) 4/2/53; Coker, 1915; *LWR*, 22 June 1901, p. 3, col. 2; 11 July 1908, p. 3, col. 1; 7 June 1913, p. 4, col. 2; *LS*, 20 Aug. 1913, p. 4, col. 2.
6. Blyden to Chancellor B, n.d., Macaulay Papers (MP) 22/8; interview: Mrs E. M. E. Agbebi.
7. *LWR*, 9 May 1903, p. 6, col. 1; 19 July 1913, p. 4, col. 1; 11 Sept. 1915.
8. *LWR*, 10 Apr. 1897, p. 4, col. 1; 16 Sept. 1899, p. 3, col. 3; 19 Oct. 1907, p. 4, col. 2; 'Is Native Marriage Advisable or not Advisable?'; Coker to cousin, 16 July 1913, CP 2/1/10.
9. *LS*, 27 Feb. 1901, p. 5, col. 1; Agbebi, n.d.; Coker, 1917; interviews: T. A. Doherty, Dr da Rocha-Afodu.
10. *LWR*, 6 Aug. 1898; 22 Apr. 1905, p. 5, col. 2; *LS*, 10 Aug. 1898; 'Resolution Adopted at a Conference of Bishops', Mar. 1906, Church Missionary Society, Yoruba Mission Papers, G3A2/012; 'Report of the Proceedings . . . of the First Synod of Western Equatorial Africa', 1907.
11. *LWR*, 31 May 1913, p. 4, col. 2; Coker to cousin.
12. *LS*, 15 Mar. 1911, p. 5, col. 3; 23 Aug. 1911, p. 5, col. 3. See also Comaroff, 1980; Parkin and Nyamwaya, 1987.
13. *LS*, 20 Sept. 1911, p. 6, col. 1 to 13 Mar. 1912, p. 6, col. 1.
14. For discussions of Lagos court records and how they can be used in social historical research see the introduction and my essay in Mann and Roberts (1991).
15. *Gomez* v. *Oke and Lanlatu*, 19 July 1899, JNCC, xxii. 48–51, 55–61. In this case, we have unfortunately only Gomez's testimony and that of his witnesses, because after hearing what they had to say, the judge ruled that Gomez had no case to make against Lanlatu. She was not called to speak, because there was no legal need for her to defend herself.
16. Ibid. 48.
17. Ibid. 49.
18. Ibid. 58.
19. Ibid. 49.
20. Ibid. 51. The word 'connection' was used to refer to sexual intercourse.
21. Ibid. 56.
22. Ibid. 48.
23. Wills, the Lagos Probate Registry.

24. *Yara* v. *Ibiyemi*, 12 Nov. 1896, JNCC, xviii. 218–29.
25. Wills, 1908, iii. 97–100.
26. *Age* v. *Isaac Nathan*, 12 July 1899, JNCC, xxii. 13–14. See also *LS*, 25 Mar. 1911, p. 5, col. 3.
27. *Bamgbose* v. *Daniel and others*, *AC*, p. 107; *Alake* v. *Pratt*, 15 *WACA*, p. 20.
28. *LWR*, 15 June 1901, p. 4, col. 1; 28 Sept. 1907, p. 4, col. 2; *LS*, 24 May 1911, p. 4, col. 3.
29. Blyden to Chancellor B; 'Is Native Marriage Advisable or not Advisable?'; Coker, 1917; *LWR*, 29 Jan. 1910, p. 4, col. 1; 31 May 1913; 7 June 1913.
30. 'Is Native Marriage Advisable or not Advisable?'; *LWR*, 1 Sept. 1906, p. 6, col. 1; 18 May 1907, p. 3, col. 2; 4 July 1908, p. 5, col. 1; 31 May 1913, p. 4, col. 2; *LS*, 17 Mar. 1915, p. 4, col. 2.
31. *LWR*, 28 Mar. 1903, p. 4, col. 3; 18 Feb. 1911; interview: Dr Akinola Maja.
32. Marriage Registers in the Lagos churches and Marriage Registry include information on age of Christian marriage. For the age of marriage in indigenous culture see Fadipe, 1970; Ajisafe, 1924; Folarin, 1939; Hopkins, 1969.
33. For a similar analysis of the situation of women on the East African coast see Le Guennec-Coppens, 1987.
34. *LWR*, 28 Mar. 1903.
35. *LS*, 25 Oct. 1911, p. 7, col. 2.
36. *Age* v. *Isaac Nathan*.
37. *LWR*, 14 July 1906, p. 4, col. 3.
38. *LWR*, 27 Aug. 1904, p. 4, col. 1; 29 May 1909, p. 4, col. 2; 29 Jan. 1910; 11 Sept. 1915; *LS*, 13 July 1910, p. 6, col. 1.
39. *LWR*, 21 Jan. 1911, p. 4, col. 1; 9 Sept. 1911, p. 4, col. 1; 17 Feb. 1912, p. 4, col. 1.
40. *LWR*, 27 Feb. 1904, p. 4, col. 2; Moses to Coker, 28 May 1911, CP 4/1/14; Coker to Jibolu, 14 Dec. 1929, CP 1/2/3; Fadipe, 1970.
41. CO 147/145, 24 Nov. 1899, MacGregor to Chamberlain; Fadipe, 1970; *LWR*, 8 July 1911, p. 4, col. 1.

References

Unpublished

Oral

Mrs Lelia Abiodun, 207 Igbosere Road, Lagos; May 1974 and July 1980
Mrs Ayo Adeshigbin, 60A Apapa Road, Lagos; April 1974
Mrs Adewakun, 72 Abuleraro Road, Lagos; April 1974
Mrs E. M. E. Agbebi, 3 Carrena Street, Lagos; March 1974
Alhaji Atanda Balogun, Oluwole Street, Lagos; July 1974
Lady Bank-Anthony, Ikoyi, Lagos; March 1974
Dr da Rocha-Afodu, Medical Research Institute, Lagos; April 1974
Mr T. A. Doherty, 23 Odunlami Street, Lagos; April and May 1974
Mr S. O. Foresythe, 4 Wakeman Street, Lagos; April 1974
Archdeacon J. O. Lucas, 8 Turton Street, Lagos; April, May, and August 1974
Dr Akinola Maja, Aje Street, Lagos; May 1974
Mrs Comfort Maja, Garber Square, Lagos; August 1974 and July 1980

Mrs Ayo Manuwa, Antie Ayo School, Ikoyi, Lagos; May 1974
Prince A. L. A. Ojora, 14 Chapel Street, Lagos; October 1984 and January 1985
Mrs H. F. Pereira, 52 Olonode Street, Lagos; February 1974 and July 1980
Mr F. O. Sogunro, Bodeja Estate, Ibadan; May 1974
Mr C. B. Thomas, 14 Chapel Street, Lagos; October 1984 and January 1985
Mr Michael Ayo Vaughan, 99 Tokunboh Street, Lagos; March 1974
Mrs Adeniyi Williams, 2 Rotimi Williams Street, Ibadan; May 1974
Mrs T. E. Williams, Palm Grove, Ikorodu Road, Lagos; April 1974
Mrs R. A. Wright, 73 Oyerokun Road, Lagos; March and April 1974

Archival

Church Missionary Society; Yoruba Mission Papers, Church Missionary Society, London
Coker Papers (CP), Nigerian National Archives, Ibadan
Colonial Office Dispatches (CO), Lagos Colony, 147 series, Public Record Office, Kew
Judges' Notebooks of Civil Cases (JNCC), Lagos State High Court, Tafewa Balewa, Lagos
Macaulay Papers (MP), Africana Collection, Ibadan University Library, Ibadan
Marriage Registers, Marriage Registry, Ikoyi, Lagos
Marriage Register, St Paul's Church, Lagos
Wills, Lagos Probate Registry, Tafewa Balewa, Lagos

Newspapers

Lagos Standard (*LS*)
Lagos Weekly Record (*LWR*)

Published

Aderibigbe, A. B. (1975), 'Early History of Lagos to about 1850', in A. B. Aderibigbe (ed.), *Lagos: The Development of an African City* (Ikeja: Longman), 1–26.

Adewoye, O. (1977), *The Judicial System in Southern Nigeria, 1854–1954* (Atlantic Highlands, NJ: Humanities Press).

Afonja, S. (1983), 'Women, Power and Authority in Traditional Yoruba Society', in L. Dube, E. Leacock, and S. Ardener (eds.), *Visibility and Power: Essays on Women in Society and Development* (Delhi: Oxford University Press), 136–57.

Afonja, S. (1986), 'Land Control: A Critical Factor in Yoruba Gender Stratification', in C. Robertson and I. Berger (eds.), *Women and Class in Africa* (New York: Africana Publishing), 78–91.

Agbebi, M. (n.d.), *Africa and the Gospel: Sermons, Debates, and Lectures*, by David Brown Vincent (Lagos: T. A. King).

—— (1903), *Inaugural Sermon Delivered at the Celebration of the First Anniversary of the 'African Church', 21 December 1902* (New York: Edgar F. Howorth).

Agiri, B. A., and Barnes, S. (1987), 'Lagos before 1603', in A. Adefuye, B. Agiri, and J. Osuntokun (eds.), *History of the Peoples of Lagos State* (Lagos: Lantern Books), 18–32.

Ajayi, J. F. A. (1965), *Christian Missions in Nigeria, 1841–1891: The Making of a New Élite* (London: Longman).

Ajisafe, A. K. (1924), *Laws and Customs of the Yoruba People* (Lagos: Church Missionary Society Bookshop).

Alake v. *Pratt* (1955), 15 West African Court of Appeals Reports (WACA), 20.

Aronson, D. R. (1978), *The City is Our Farm: Seven Migrant Ijebu Yoruba Families* (Boston: G. K. Hall).

Ayandele, E. A. (1966), *The Missionary Impact on Modern Nigeria, 1842–1914: A Political and Social Analysis* (London: Longman).

—— (1970), *Holy Johnson: Pioneer of African Nationalism, 1836–1917* (London: Frank Cass).

Baker, T., and Bird, M. (1959), 'Urbanisation and the Position of Women', *Sociological Review*, NS 7: 91–122.

Bamgbose v. *Daniel and ors.* (1955), Appeals Cases (English) (AC), 107.

Bledsoe, C. H. (1980), *Women and Marriage in Kpelle Society* (Stanford, Calif.: Stanford University Press).

Bromley, P. M. (1962), *Family Law* (London: Butterworths).

Burnham, P. (1987), 'Changing Themes in the Analysis of African Marriage', in D. Parkin and D. Nyamwaya (eds.), *Transformations of African Marriage* (Manchester: Manchester University Press), 1–34.

Burns, Sir A. (1969), *History of Nigeria* (London: George Allen & Unwin).

Caldwell, J. C., and Caldwell, P. (1987), 'The Cultural Context of High Fertility in Sub-Saharan Africa', *Population and Development Review*, 13: 409–37.

Clignet, R. (1987), 'On dit que la polygamie est morte: Vive la polygamie!' in D. Parkin and D. Nyamwaya (eds.), *Transformations of African Marriage* (Manchester: Manchester University Press), 199–209.

Coker, G. B. A. (1958), *Family Property among the Yorubas* (London: Sweet & Maxwell).

Coker, J. K. (1915), *Polygamy Defended* (Lagos).

Coker, S. A. (1904), *Three Sermons on the Christian Ministry* (London: T. Fisher Unwin).

—— (1917), *The Right of Africans to Organize and Establish Indigenous Churches, Unattached to and Uncontrolled by Foreign Church Organizations* (Lagos: Tika Tore Press).

Cole v. *Cole* (1898), 1 Nigerian Law Reports, 15.

Comaroff, J. L. (1980), 'Introduction', in J. L. Comaroff (ed.), *The Meaning of Marriage Payments* (New York: Academic Press), 1–47.

—— and Roberts, S. (1977), 'Marriage and Extra-Marital Sexuality: The Dialectics of Legal Change among the Kgatla', *Journal of African Law*, 21: 97–123.

Crane, F. R. (1957), 'Family Settlements and Succession', in R. H. Graveson and F. R. Crane (eds.), *A Century of Family Law* (London: Sweet & Maxwell), 227–55.

da Cunha, M. C. (1985), *Negros, Estrangeiros: Os Escravos Libertos e sua Volta a Africa* (São Paulo: Editora Brasiliense).

Davidson, R. T. (1889), *The Lambeth Conferences of 1867, 1878, and 1888* (London: Society for Promoting Christian Knowledge).

Dinan, C. (1977), 'Pragmatists or Feminists: The Professional Single Women of Accra', *Cahiers d'Études Africaines*, 19: 155–76.

Eades, J. S. (1980), *The Yoruba Today* (Cambridge: Cambridge University Press).

Elias, T. O. (1962), *Nigerian Land Law and Custom* (London: Routledge & Kegan Paul).

Fadipe, N. A. (1970), *The Sociology of the Yoruba* (Ibadan: Ibadan University Press).

Folarin, A. (1939), *The Laws and Customs of Egbaland* (Abeokuta, Nigeria: E.N.A. Press).

Geldart, W. (1963), *Elements of English Law* (London: Oxford University Press).

Guyer, J. I. (1981), 'Household and Community in African Studies', *African Studies Review*, 24: 87–137.

Harrell-Bond, B. E. (1975), *Modern Marriage in Sierra Leone: A Study of the Professional Group* (The Hague: Mouton).

Hewitt, G. (1971), *The Problems of Success: A History of the Church Missionary Society, 1910–1942*, I (London: SCM Press).

Hopkins, A. G. (1964), 'An Economic History of Lagos, 1880–1914', Ph.D. thesis, University of London.

—— (1969), 'A Report on the Yoruba, 1910', *Journal of the Historical Society of Nigeria*, 5: 67–100.

—— (1980), 'Property Rights and Empire Building: Britain's Annexation of Lagos, 1861', *The Journal of Economic History*, 40: 777–98.

'In the Matter of the Marriage Ordinance, and in the Matter of an Intended Marriage between Dr. O. Ṣapara and Sarah E. O. Green; Adel Ṣapara, Caveatrix' (1911), in P. A. Renner (ed.), *Reports, Notes of Cases, and Proceedings of Judgments in Appeals . . . Relating to the Gold Coast Colony and the Colony of Nigeria from 1861–1914* (London: Sweet & Maxwell), I 605–14.

Joseph, R. A. (1987), *Democracy and Prebendal Politics in Nigeria: The Rise and Fall of the Second Republic* (Cambridge: Cambridge University Press).

July, R. W. (1967), *The Origins of Modern African Thought* (New York: Praeger).

Karanja, W. W. (1987), ' "Outside Wives" and "Inside Wives" in Nigeria: A Study of Changing Perceptions in Marriage', in D. Parkin and D. Nyamwaya (eds.), *Transformations of African Marriage* (Manchester: Manchester University Press), 247–61.

Kopytoff, J. H. (1965), *A Preface to Modern Nigeria: The 'Sierra Leonians' in Yoruba, 1830–1890* (Madison, Wis.: University of Wisconsin Press).

Law, R. (1983), 'Trade and Politics behind the Slave Coast: The Lagoon Traffic and the Rise of Lagos, 1500–1800', *Journal of African History*, 24: 321–48.

Le Guennec-Coppens, F. (1987), 'L'Instabilité conjugale et ses consequences dans la société swahili de Lamu (Kenya)', in D. Parkin and D. Nyamwaya (eds.), *Transformations of African Marriage* (Manchester: Manchester University Press), 233–45.

Lesthaeghe, R. J., Kaufmann, G., and Meekers, D. (1989), 'The Nuptiality Regimes in Sub-Saharan Africa', in R. J. Lesthaeghe (ed.), *Reproduction and Social Organization in Sub-Saharan Africa* (Berkeley, Calif.: University of California Press), 238–337.

Little, K. (1973), *African Women in Towns: An Aspect of Africa's Social Revolution* (Cambridge: Cambridge University Press).

Little, K., and Price, A. (1973), 'Some Trends in Modern Marriage among West Africans', in C. Turnbull (ed.), *Africa and Change* (New York: Knopf), 185–207.

Lloyd, P. C. (1965), 'Yoruba Inheritance and Succession', in J. D. Derrett (ed.), *Studies of the Laws of Succession in Nigeria* (London: Oxford University Press), 139–73.

Lucas, J. O. (1946), *Lecture on the History of St Paul's Church, Breadfruit, Lagos* (Lagos: Church Missionary Society Press).

Mann, K. (1977), 'A Social History of the New African Élite in Lagos Colony, 1880–1915', Ph.D. thesis, Stanford University.

Mann, K. (1983), 'The Dangers of Dependence: Christian Marriage among Élite Women in Lagos Colony, 1880–1915', *Journal of African History*, 24: 37–56.

—— (1985), *Marrying Well: Marriage, Status, and Social Change among the Educated Élite in Colonial Lagos* (Cambridge: Cambridge University Press).

—— (1987), 'Trade, Credit, and the Commoditization of Land in Colonial Lagos', Paper presented at a conference, New Perspectives on Colonial Africa, African Studies Center, University of Illinois.

—— (1989), 'Redefining Servitude: From Slavery to Tenancy-Clientage in Late Nineteenth-Century Lagos', Paper presented at the African Studies Association Annual Meeting, Atlanta.

—— (1991*a*), 'The Rise of Taiwo Olowo: Law, Accumulation, and Mobility in Early Colonial Lagos', in K. Mann and R. Roberts (eds.), *Law in Colonial Africa* (Portsmouth, NH: Heinemann).

—— (1991*b*), 'Women, Landed Property, and the Accumulation of Wealth in Early Colonial Lagos', *Signs*, 16: 682–706.

Mann, K., and Roberts, R. (eds.) (1991), *Law in Colonial Africa* (Portsmouth, NH: Heinemann).

The Marriage Ordinance, 1884, in E. H. Richards (ed.), *Ordinances, Orders, and Rules in Force in the Colony of Lagos, 1893* (1894) (London: Stevens & Sons), 443–57.

Marris, P. (1961), *Family and Social Change in an African City* (London: Routledge & Kegan Paul).

Mba, N. E. (1982), *Nigerian Women Mobilized: Women's Political Activity in Southern Nigeria, 1900–1965* (Berkeley, Calif.: Institute of International Studies, University of California).

Moore, S. F. (1986), *Social Facts and Fabrications: 'Customary' Law on Kilimanjaro, 1880–1980* (Cambridge: Cambridge University Press).

Newbury, C. W. (1961), *The Western Slave Coast and its Rulers* (Oxford: Clarendon Press).

Obbo, C. (1987), 'The Old and New in East African Élite Marriages', in D. Parkin and D. Nyamwaya (eds.), *Transformations of African Marriage* (Manchester: Manchester University Press), 263–80.

Okoro, K. (1887), *Views of some Native Christians of West Africa on the Subject of Polygamy* (Lagos: The General Printing Press).

Omu, F. I. A. (1978), *Press and Politics in Nigeria, 1880–1937* (Atlantic Highlands, NJ: Humanities Press).

Oppong, C. (1974), *Marriage among a Matrilineal Élite: A Family Study of Ghanaian Senior Civil Servants* (Cambridge: Cambridge University Press).

Oppong, C. (ed.) (1987), *Sex Roles, Population and Development in West Africa: Policy-Related Studies on Work and Demographic Issues* (London: Heinemann).

Parkin, D., and Nyamwaya, D. (1987), 'Introduction', in D. Parkin and D. Nyamwaya (eds.), *Transformations of African Marriage* (Manchester: Manchester University Press), 1–34.

Peel, J. D. Y. (1983), *Ijeshas and Nigerians: The Incorporation of a Yoruba Kingdom* (Cambridge: Cambridge University Press).

Philip v. *Philip* (1946), 18 Nigerian Law Reports, 102.

Report of the Proceedings of the First and Second Sessions of the First Synod of the Dioceses of Western Equatorial Africa (1907) (Exeter: Townsend & Sons).

Smith, R. S. (1988), *Kingdoms of the Yoruba* (Madison, Wis.: University of Wisconsin Press).

Tamuno, T. N. (1972), *The Evolution of the Nigerian State: The Southern Phase, 1898–1914* (London: Longman).

Verger, P. (1968), *Flux et reflux de la traite de nègres entre le Golfe de Bénin et Bahia de Todos os Santos, du XVIIe aux XIXe siècle* (Paris: Mouton).

Webster, J. B. (1964), *African Churches among the Yoruba, 1888–1922* (Oxford: Clarendon Press).

9 The Phenomenon of 'Outside Wives': Some Reflections on its Possible Influence on Fertility

WAMBUI WA KARANJA

Introduction

Despite high levels of infant and child mortality and low life expectancy in sub-Saharan Africa, most of the continent has been experiencing unprecedentedly high levels of population increase in recent years. As the former President of the World Bank points out, 'the population problem in Africa . . . is not that there are too many people there. There are not. The problem is that the population growth rate is explosively high' (McNamara, 1985: 11). This rapid population increase has had profound effects on African social structures culminating in high dependency ratios, especially of those under 15 years. All these factors have spawned a proliferation of literature attempting to establish the determinants of African fertility. Such determinants include age at marriage, family structure, education, migration, labour force participation, post-partum abstinence, religion, and so on (e.g. Brass, 1968; Caldwell, 1975; the contributors to Caldwell, 1968; UNECA, 1979; UN, 1978; and Oppong, 1987*a*, to mention only a few).

Using data obtained primarily from Lagos and Ibadan (Nigeria), I propose to examine the fertility implications of a contemporary African phenomenon I refer to as private polygyny, or the practice of taking 'outside wives', as it is more widely known in the literature (see e.g. Baker and Bird, 1959; Karanja, 1981 and 1987) and to explore its possible effects on fertility. My data are insufficient to assess whether the practice of outside wifeship has clear-cut effects on total fertility rates overall. I argue instead that outside wives, because of the fundamental instability of most outside unions, seek to keep their options open to establish new unions, if necessary, by having just enough—but not too many—children within their current unions. After presenting a brief outline of the methodology used in the study, I outline the main characteristics of outside wives. Moving to the fertility patterns of outside wives, I attempt to link the effects of this phenomenon to the larger institution of marriage and the family.

The Phenomenon of Outside Wives: An Overview

In the local parlance of West African anglophone countries, the term 'outside wife' is frequently used by the popular press and in private informal discussion to refer to girlfriends, concubines, mistresses, and 'illegal wives' of the men of means (e.g. Little, 1973; Dinan, 1983; and Bleek, 1987). In francophone countries, the term 'deuxième bureau' is sometimes used. Of the practice, Parkin and Nyamwaya (1987: 12) report, 'the Senegalese jokingly distinguish between une femme de cour (wife concerned with running the rural homestead) and une femme de cœur (a "love wife" of the town)'. How did this phenomenon arise?

Modernization theorists, especially those of convergence school (e.g. Feldman and Moore, 1962; Goldthorpe, 1984; Smelser, 1968; Kerr *et al.*, 1973) seem to view today's underdeveloped countries as inevitably moving towards a model of society that is equivalent, if not identical, to today's Western societies. As Goode (1963: 187) asserted: 'Women's increasing freedom of choice and the gradual shift from a subsistence agricultural economy to a market or industrial one has meant a *steady decline in polygyny*. Philosophical and economic factors against polygyny have been supported by the pressures from both political and religious leaders' (emphasis added). He continues, 'In Africa's urban industrial areas polygyny has become relatively rare: statutory marriage generally imposes the legal obligation of monogamy, while urban life confers on the women none of the benefit that polygyny has in tribal areas' (p. 188).

To be sure, the sort of polygyny referred to by Goode is on the decline in African societies, not only because 'in tribal areas each wife could generally support herself' (ibid.) but also because a good many of those who might have been practising public polygyny now consider themselves too 'modern' to go 'native'. As a result, reports, especially from the urban areas in many countries, indicate that people have opted for a different type of polygyny, private polygyny, which they perceive as being more in consonance with their perceptions of modernity.

A Brief Note on Methodology

My interest in the phenomenon of outside wives, as opposed to girlfriends, mistresses, or concubines, dates back to the late 1960s when I first arrived in Nigeria. During that period, one found, as indeed one would today, lively reporting and discussions of 'the other woman', outside wives, and girlfriends of 'sugar daddies', in the popular press and other dailies. Similarly, in private discussions with a wide spectrum of Nigerians, one frequently hears women in particular (who are legally married under statutory law or church auspices) talk with emotional intensity about outside wives, charging them with being recklessly and relentlessly bent on wrecking legitimate marriages (see e.g. Obbo, 1987).

By late 1973, while teaching at the University of Lagos, I felt that the phenomenon merited serious investigation. For the next four years, I collected intensive, albeit indirect, information on outside wives mostly from 'inside wives' and from a few men who either admitted having such 'wives' or had friends who did. Most 'legally' married men, understandably, are reticent to speak of their participation in this practice.

Through an informal network of individuals who, through their own networks, had friends or relatives occupying the status of outside wives, I obtained a list of outside wives, and managed to conduct a series of intensive interviews with twenty of them over a period of fourteen months (from September 1978 to October 1979). Given the sensitive nature of the subject, I felt it best to employ ethnographic approaches of intensive interviewing, informal conversation, participant observation, and tape recordings. By the end of 1979 I had recorded personal histories of the twenty women who, at the time of the interviews, I defined as outside wives. Eight of them said they had friends with similar 'marital' statuses. In addition, colleagues and friends working in Nigerian civil service, the business sector, and the various universities, provided further information of many other outside wives. A four-year teaching appointment, 1982–6, at the University of Ibadan provided an opportunity to expand my knowledge on outside wives in a different city and setting. However, the data reported here refer mostly to outside wives in Lagos.

Throughout the interviews with all the outside wives and the men who agreed to be interviewed about their outside marriages, it was understood that all the information given would be handled with extreme care and confidentiality. To this end, the discussion and reporting throughout this chapter is carried out in such a way as to protect the identity of the subjects.

The Genesis of the Concept

In élite Lagos society the extensive practice whereby a legally married man takes on a steady girlfriend dates back to the beginning of this century if not before (see Mann, this volume). Three older women I interviewed in 1979 who had married in the 1920s declared that their husbands and most of their husbands' friends, had had at least one girlfriend. The wealthier had more. These women had been members of the Lagos élite society of the late 1920s; all three had at least eight years of Western-style education and all had been born and brought up in monogamous Christian homes by parents with some Western-style education. Their husbands had graduated from British universities and were regarded as being among some of the most successful young professionals of 1930s Lagos society. According to these women, the practice of married men maintaining girlfriends had been going on long before their time.

It is important to point out, of course, that few inside wives (those married under statutory law or in the church) accept the sort of polygyny discussed here. Those who tend to acquiesce are, on the whole, substantially inferior to their husbands, by socio-economic as well as educational standards. An older woman who had been an inside wife was vehemently opposed to the practice. According to this woman the custom was hypocritical; it also caused continuous tension for wives. She cited outside wifeship as the principal reason why she had been in and out of three marriages, all of which had been legally contracted. By the time I interviewed her, she had been single for many years, but still believed strongly in the institution of marriage, though she felt that she herself had been unlucky in matters of marriage.

Asked why their husbands and other such men did not marry polygamously, since there were few legal or institutional obstacles to contracting such marriages, they argued that the Lagos élite society would not have tolerated such 'backward' or 'bush' behaviour by its esteemed sons. According to these women, what distinguished the Western-educated Christian élites from the traditional Muslim élites of Lagos (see e.g. Cole, 1974) was the fact that the former were prima facie 'monogamously' married, while the latter practised *juju* (witchcraft) openly and married polygamously, both of which were considered a manifestation of backwardness and as lacking class and taste. (Further probing indicated that the Western-oriented élites also practised *juju* extensively but only clandestinely.)

As the population of Western-educated élite men rapidly expanded a decade or so before Nigerian independence (especially in Western Nigeria where free primary education had become relatively universal), the practice of maintaining girlfriends or mistresses in Lagos élite society correspondingly gained wider currency. Indeed, it became a status symbol, since only the rich and the successful could afford to so indulge themselves. In a strong sense, outside wifeship drew on traditional practices, in which a man's prestige and influence were inseparably linked to his marital status: the more wives he had, the higher would be his position in the power structure. In short, many wives meant many powers (Clignet, 1970).

Thus, by the late 1960s and early to mid-1970s the practice had become so widespread that the media and people in Lagos referred to girlfriends or mistresses of married men as outside wives, especially if there were children born of the relationship. Later, during the oil boom, which brought wealth beyond the wildest dreams to a few Nigerians, some outside wives could now be maintained by the *nouveaux riches* in style, with shopping trips abroad and holidays to the world's major capitals.

Juxtaposing the practice of keeping girlfriends or mistresses to legal marriage practices may represent Nigerian (we might argue 'African') élite men's attempts to embody Western ideals of a modern man (i.e. monogamous, Christian, educated, and the like) while secretly admiring the African ideology of polygyny. As I have argued elsewhere (Karanja, 1987: 252) this hypothesis of desire to emulate logically inconsistent lifestyles would help explain why there is no general consensus in Nigeria as to who exactly is an outside wife. To muddy the waters even further, terms connoting adherence to statutory laws and Christian marriage are often used within the relationship between men and their outside wives, while women in the 'traditional' structures who are married under native law and custom are not considered as outside wives. Indeed, Baker and Bird (1959) writing on this subject in Nigeria could report that if the outside wife remains either with her own or her husband's kin, her position may be similar to that of a woman in the indigenous system.

Despite the prevailing confusion as to who is an outside wife, the term in Nigeria would, in all probability, connote a woman

1. who is financially maintained on a regular basis by a man with whom she has regular sexual relations, the relationship lasting for several years;
2. with children whose paternity is acknowledged by her partner;
3. who is usually not publicly acknowledged as a wife and always maintains separate residence from her 'husband';
4. with no formal legal status since her marriage involves no bride-price or marriage rites;
5. with considerably less social and political-jural recognition than an inside wife.

Outside Wifeship Compared to Traditional Polygyny

The coalescence of Western with African notions of marriage has resulted in considerable confusion as to what is meant by marriage in African urban areas. In other words, 'although marriage has a definite meaning in most African cultures, it is difficult to define in general terms' (van de Walle, 1968: 186). This confusion stems from the fact 'that on the whole there is a substantial increase in the number of unions, including many households set up without marriage ceremony' (Goode, 1963: 184).

For the men and women who participate in the practice we refer to as 'outside wifeship' or private polygyny, the tendency has been to define their situation as marriage. But for many other people, the institution is a grey area that defies a single definition. For example, mutual co-operation, residence, and even reproduction, which in the past might have been thought necessary—though not sufficient conditions—for defining marriage, now vary so widely that most people no longer think them important in talking about marriages in African cities. Perhaps this is what Burnham (1987: 46) had in mind when he wrote:

> Polygamy remains a widespread feature of African societies and one which, despite the predictions of modernization theorists, does not necessarily seem to be on the decline. We must also recognize that even in situations where monogamy is morally valued or even legally enjoined, as in so-called 'ordinance marriage' in several African

States, functional alternatives to polygamy are frequent, such as concubines or 'outside wives'.

Africa is, par excellence, the home of polygyny. The various essays in the influential volume edited by Radcliffe-Brown and Forde (1950) and numerous other studies clearly delineate the centrality of polygyny in African social structures. Yet polygyny was an ideal; as with other ideals, it was one that only a few could achieve.

Marriage in Africa was a rite of passage of great public interest, for it was within this framework that alliances and the various rights and obligations—politico-jural, socio-economic, and the like, were established and transferred. For example, the transfer of resources (especially through bride-price) from wife-takers to wife-givers has been shown to be important in forging and creating affinal ties and bonds that frequently outlive a particular union in question. The whole process also established principles of legitimacy, as evidenced by, for instance, the importance placed on premarital virginity (Fadipe, 1970) in traditional African societies. In sum, the payment of bride-price, the performance of the various rites and ceremonies, and so on, before a girl could be considered properly married, established such clear boundaries as to leave little room for speculation as to what a marriage was.

By contrast with marriage in the more indigenous setting, contemporary outside wifeship is clandestine and private; that is, the process of establishing such a relationship and the existence of the relationship thereafter are pre-eminently private. This is why, in an earlier publication, I argued for a distinction between private polygyny (outside wifeship) and public polygyny (Karanja, 1987: 257–8). The latter would then exclusively refer to two or more marriages contracted publicly by a man under native law and custom. This contrasts to an 'outside marriage' which lacks politico-jural status because it is sanctioned by neither tradition nor statutory law. To be sure, a man usually wants his friends to know he has an outside wife because of the prestige it brings; he may also tell his wife, should she become too insubordinate, to let her know that she has competition. But in important public forums, everyone is supposed to be complicit in ignoring the affair.

Again, in contrast to traditional or public polygyny, unidirectional resource flow from 'husband' to 'wife' is one of the major, indeed crucial, criteria of private polygyny. This and the advent of children are (I would argue) the most salient components of outside wifeship. The flow of resources is by definition linear because only well-to-do men can afford to have outside wives, who are almost always considerably inferior economically to their 'husbands'. Indeed, it would be a contradiction in Nigeria, and elsewhere in West Africa, to refer to a woman as an outside wife if she is *not* financially maintained by her 'husband'.

Of course, the practice of outside wifeship itself, as argued previously, is an attempt to synthesize the contradictions stemming from the introduction of a cash economy and a settler culture. Yet wives, whether monogamously or polygamously married, were not separate entities from the wider domestic or kinship mode of production who could only receive economic resources. Indeed, they were integral to it. The question of financial or economic support of wives by husbands would have been antithetical. In pre-capitalist modes of production, a woman, when she married, was allocated a piece of land with which she was to feed herself, her children, and her husband for the rest of her life. The celebrated economic autonomy or independence of Nigerian market women traders (Nadel, 1942; Bauer, 1954; Sudarkasa, 1973; Baker, 1974; Karanja-Diejomaoh, 1980), and their counterparts in West Africa (Oppong, 1974; Harrel–Bond, 1975; Robertson, 1976) is a contemporary manifestation of these African domestic modes of production.

Women's Motivations for Becoming Outside Wives

The primary consideration for becoming an outside wife is clearly economic. As many scholars (Dinan, 1983; Bleek, 1987) have argued, an African woman in town with meagre or no resources of her own can avoid poverty by attaching herself to a well-to-do man. Of the twenty outside wives reported here, four of the 'husbands' were top civil servants, eight were successful businessmen, six were senior military officers, and two were chief executives of private companies. Though the women were not randomly drawn, these numbers, together with other supporting evidence, support the proposition that it is such men who can afford the financial luxury of keeping girlfriends and outside wives. Less successful men must, perforce, be content with occasional affairs requiring considerably less capital outlay or alternatively, where bride-price is not prohibitively high, acquire 'village wives': that is, women who have less Western education and are considered to lead a less sophisticated lifestyle. These 'wives', as in traditional structures, mostly support themselves and their children after being provided with land or small capital with which they start a trade.

To be sure, it is not only the poor and the ill-educated who become outside wives. Some evidence suggests that women undergraduates, young professionals, secretaries, and the like are willing to attach themselves to men of means while they wait for the 'right' man to come along. Morgan (1975: 201), reporting on data collected in Lagos over twenty years ago, observed that '"mutual consent unions," an institution widely recognized in Lagos, in which a couple enter into a conjugal union which they and their associates define as "marriage" . . . comprised 17.6 per cent of all Lagos unions'. And of course the right man, as virtually all my informants contend, must be able to provide a wife and her children with 'reasonable' financial or economic support. In Nigeria one often hears young girls and women say 'for "love" [that is, money] for *stay*, for "suffer" [lack of money] for *go*'.

These findings suggest that most outside marriages and other similar unions are not, except at a very semantic level, cemented on love. As Bleek (1987: 141) confirms, 'most so-called lover relationships have a "commercial" component'. This is not to say that deeper affections play no role in outside relationships; deep affection may well develop eventually. It is simply to underscore the fact that the ability to regularly support and maintain a woman and her children is not only important; it implicitly defines the relationship. The prestige and the status attendant to being attached to a man of influence are important elements of the relationship as well, especially since the woman's peers and significant others will eventually be let into the 'secret'. But prestige alone is insufficient to sustain the relationship. Few women, if any, would be content to remain an outside wife to a man simply because he was influential in the community. Women informants always argued that you cannot 'chop' (eat) love.

Outside wives are aided in their quest for support by the fact that a man's ego and honour are at stake. He worries first about losing face among his peers if he cannot support his 'wife' in style. He also knows only too well that if he does support her, she will—albeit with caution—augment her income by engaging in 'extramarital' affairs. Most 'husbands' of such 'wives' attempt to avert such eventualities by making sure that they support their 'wives' to the best of their abilities. In fact, official corruption at all levels has frequently been attributed to Nigerian men's propensity to acquire outside wives, girlfriends, and mistresses who must be maintained in style.

Men marrying for the first time will, in all probability, prefer to seek fresh grounds. The same is true for older men of means. Thus, while some outside wives can command substantial support, others seem to resign themselves to the status of outside wife because, after playing the field for a while, they may find it difficult to find a suitable single man who will marry them. This is especially true if they have had children by a previous man or have been an outside wife to several men. Similarly, divorcees, especially those with several children, may find it difficult to find a man willing to take them to the altar. If they are 'fashionable', attractive, and not too old, they can easily find a 'husband' via outside wifeship. Just as Fortes (1949) observed that the Tallensi view marriage and parenthood as ultimate goals of all adult men and women, in Nigeria as well, a woman unable to contract proper marriage may readily opt for the status of an outside wife, for some kind of marriage is better than no marriage at all.

A further motivation for becoming an outside wife is disillusionment with legal marriages, either because of direct experience or from observing friends or relatives who are married: a status that outside wives often cast as turbulent. Conversely, many inside wives believe that outside wives get a better economic deal (Karanja-Diejomaoh, 1978). They also believe that outside wives have a more romantic life: they have more freedom for other liaisons, if need be, which, according to my information, are frequent. As Bleek (1987: 142) argues, there is an 'increasing disenchantment with marriage . . . [because]

the relative economic contribution of men to their marriage has decreased over the last few decades with the worsening of the general economic situation'. Inside wives in Nigeria argue, by contrast, that the decreasing economic contribution by husbands is a consequence of maintaining too many girlfriends or outside wives or both. They may have a point. In a study in Ghana on 'Responsible Fatherhood and Birth Planning', Oppong (1987*b*: 170) reports that 'more than 40 per cent of the men in [her] sample had children with more than one woman'.

As for the long term, outside wives' aspirations vary widely. They are based on a number of factors such as the current age of the woman, her fertility, and socio-economic status, the socio-economic status of the man, and the stability of the man's inside marriage. Related to this is the outside wife's perception of whether the man is likely to terminate his inside marriage and make her what is termed 'the Madam of the house' or the 'ring wife' (Aronson, 1978). If the age gap between the woman and her 'husband' is wide, and if the man is very rich, then the outside wife may attempt to have the number of children she wants by him in rapid succession 'just in case' (i.e. lest the man dies intestate). She would like, of course, to have the 'right' sex configuration in her children, for she knows that male children will help to make her claim more credible.

Also, the state of the man's inside marriage is important to an outside wife's long-term plans. If she knows that the inside marriage is turbulent, the outside wife tries to consolidate her 'marriage' by being all the things that the inside wife is not, in the hope that the man will 'see sense' and divorce his inside wife for her. Here the outside wife is not without grounds, for divorce in Nigerian cities is not rare.

On the other hand, if an outside wife who is much younger than her 'husband' sees that he is too committed to his inside wife, she may limit her fertility by having one or two children and looking around for the 'right' man for a 'proper' and long-term union. Since these 'extramarital affairs', when discovered, frequently cause acrimony and altercations, such outside wives try to hide them until they are ready to take the leap into a legal marriage.

Finally, all outside wives have a long-term goal of being 'properly' set up by the man: having a house built for them, children sent to the right schools, shopping trips abroad for them and their children, built-up savings, jewellery (mostly gold), a 'decent' car, clothes, and the like. In short, as most of my informants used to say, 'a good and easy life'. Since only a few outside wives ever realize these goals, some take on a series of men in order to meet some of them.

Men's Motivations for Acquiring Outside Wives

There are several reasons why some new élite African men in towns (see e.g. Lloyd, 1966) engage in private polygyny. Five reasons (not necessarily in order) seem to predominate: (1) pressures from peers and from the traditional sector to acquire the symbols of affluence and power, (2) as an alternative to divorce, (3) the lack of negative sanctions against the practice, (4) creating an image of virility, and (5) attempting to produce children in general and male children in particular.

Concerning social pressures, I have already pointed out that economic factors militate against maintaining more than one household in urban centres. Because only the few relatively well-to-do can afford to maintain girlfriends, mistresses, or wives outside legal marriage, the practice has come to be thought of, at least by male peers, as a status symbol. It indicates, as it were, that a man has 'arrived'. Acquiring an outside wife is akin to acquiring a Mercedes-Benz or building an elegant house, as a manifestation of success. In fact, I would posit that a successful man who does not avail himself of these status symbols would be suspect. A number of women I came to know over the years reported that their husbands had become impotent as the marriage had advanced. None the less, these same men had a string of girlfriends and one man admitted to having an outside wife—a source of continuous 'palaver' in his marriage.

For a newly successful man from a non-élite family, additional pressure to acquire an outside wife stems from pressure from his relatives to show that he has not altogether abandoned his roots by having only one wife. For such a man, acquiring an outside wife indicates to his family that their son still cherishes and upholds his primordial ties. At the same time, it indicates to the man's peers, and even to his 'monogamous' wife, that he has become modern.

Men also experience pressures from peer group and family to avoid divorce as an alternative to a statutory marriage that has proved untenable. The existence of children, admission of failure, the dim likelihood of finding a better wife, and so on, are evoked as reasons for not dissolving an unsatisfactory marriage. Taking an outside wife becomes, in effect, a convenient way of dealing with or escaping from a bad marriage.

While pressures from friends and family move men to take outside wives, society at large attaches little or no stigma to the practice. The sociopsychological costs are minimal, while the satisfaction of being considered successful or a 'big shot' in the community cannot be taken lightly. It is true, as pointed out earlier, that inside wives abhor the practice and continually fight to obliterate it. But they are also continually reminded, by peers, friends, and family alike, that given time and patience, the problem will gradually wither away. Thus, they are counselled to weather the storm; in short, to adapt to the circumstances. And since most married women in Nigeria perceive their status and prestige as inextricably tied to being what is called in Nigeria, 'respectably married', they would prefer, despite the anger and anxiety they experience, a turbulent marriage to alternative options such as divorce. One therefore finds women being officially addressed as 'Dr Mrs', 'Prof. Mrs', 'Chief Mrs', and so on.

Among the factors that prompt men to take outside wives, one of the most important, especially for an elderly man, is to prove to himself, his friends, and his family they he is still virile (perhaps as a way of dealing with increasing awareness of mortality). Related to this, having sexual intercourse with a young 'wife' (or indeed any young woman) is supposed to revitalize the blood of an old man—not an uncommon belief in Africa. In fact, in many indigenous African societies a man would not have sex with a woman whose menses had ceased either temporarily or permanently. Reporting on the Guro society in Côte d'Ivoire, for example, Deluz (1987: 126) observes that 'the ceasing of menstruation was accompanied by a ceasing of sexual activity'.

Finally, men are anxious to produce children in general and male children in particular. In terms of anxiety related to fertility issues, a woman's inability to give her husband a male child is second only to fear of barrenness, which itself is only second to the fear of death. Orubuloye (1987: 89), reporting on Nigerian data from a survey covering 2,994 respondents, found that 'about three fifths of the men and about one half of the women said they would like to have four boys and two girls in a family of six'. A woman whose first child is a girl knows only too well that it is only a matter of time before her husband goes out to look for a male issue. She prays fervently that the next child will be a boy. As for the man, he feels his masculinity is being put under public trial.

Thus, participating in either 'private' or 'public' polygyny in order to have male children is likely to be considered as legitimate by most men, and even by a number of women in Nigeria. If the inside wife does not have children, regardless of where the fault lies, then few people would look askance at a man's attempt to acquire an outside wife to bear children.

The Stalemate between Outside Wifeship and Statutory Marriages

From all the available evidence, the coalescence of private polygyny and public monogamy has been occurring for as long as Nigerian men have been acquiring higher levels of Western education. Modernization and Westernization may have won some ground to the extent that open polygyny is no longer considered trendy. But private polygyny is a manifestation of some inertia: these conflicting forces have culminated in a sort of a stalemate, which in turn has implications for marriage in African cities.

The result is a continuous tension between women's monogamous ideology and men's predilection towards polygyny. In defence of the latter, Nigerian men passionately argue that élite Nigerian women have betrayed and rejected their African roots. In a recent discussion, a distinguished Nigerian male academic is reported to have said, 'Élite Nigerian and African women do not behave like Africans, which they are, nor do they behave exactly like Western women, which they are not. Who are they?' His rhetorical question succinctly captures the dilemma of attempting to marry a new version of polygyny to Western notions of marriage.

Characteristics of Outside Wives

To address the question of how outside wifeship may be related to reproduction, we need first to examine some of the characteristics of outside wives that emerged in the study.

Age at First Becoming an Outside Wife

Although an ideal study of the relationship between this kind of conjugal union and fertility would demand fairly accurate data on age at marriage, the age data in this study are limited. The demographic world has long recognized the problems of age misreporting in trying to collect demographic data in Africa (e.g. Brass, 1968, and Caldwell, 1975). In Nigerian cities, it is not uncommon to hear people (especially the élite) talking about their official and non-official ages. The latter is usually the correct one, known only by family and sometimes close friends, while the former is the age one declares to employers and the public in general. Besides these more general problems of establishing age, dating the time when a union began is often equally problematic since many unions develop gradually over a period of time. It is sufficient to note that I was able to establish, through cross-checking with friends and intimate interaction with informants, that the approximate average age when the women in my sample first became outside wives to the men with whom they were presently associated was 25 or 26 years old. The range was between 21–32 years, with the greatest concentration being clustered around 23 years. At the time of interview, the women had been 'married' to these particular men for an average of 5.8 years, and their average age at the time of the interview was about 31. (I do not have systematic records of these women's previous conjugal links or of the women's ages when these former unions began.) Although we did not obtain parallel age information on men, it is important to point out that outside wives are usually considerably younger than their partners: sometimes by 20–30 years.

Education

Because having an outside wife is a status symbol, most men seem to look for women with considerable Western education, relative to the ill-educated women in Lagos and Ibadan. Only three women in the small non-random sample had not completed their secondary education; the average number of years spent in school was 14.2. Educational achievement ranged from less than 4 years of secondary education to doctorate degrees, held by four women. One woman was a trained lawyer and another an accountant.

Occupation and Income

At the time of the interview, seventeen of the twenty outside wives were employed as civil servants in the Nigerian Federal Civil Service. The other three women were self-employed. One of them was a commodity trader while the other two operated expensive boutiques that sold the latest in European fashion and expensive Nigerian dresses.

As is frequently the case in Africa, it was not possible to ascertain the exact incomes of the three self-employed women. People in such occupations tend to be reticent to disclose their incomes for reasons that range from questions of taxation to wanting to keep husbands and strangers ignorant of their exact income. As for the women civil servants, their incomes from their jobs were public knowledge since salary scales in civil service, whether state or federal, are standard. The average income of the outside wives who worked as typists and secretaries, i.e. intermediate cadre, was N350.00, while the senior civil servants earned about N650.00.

Income also came, of course, from the 'husbands'. The average income from 'husbands' was around N250—the lowest being N150 and the highest N600. The three self-employed women and the five typists said that their 'husbands' paid rent for their flats as well as a monthly allowance for food, commonly referred to in Nigeria as 'chop money', and occasional gifts. The other women who were senior civil servants were all housed by their respective organizations. Since they did not demand money for rent, they expected their 'husbands' to be more generous with 'chop money' and gifts.

The 'wives' emphasized that the maintenance money was not always constant, but varied with the man's fortune and his moods, both good and bad, as well as with the state of the relationship, whether tranquil or tumultuous. But unless a man wanted to terminate a relationship, it was understood that he could not go for months at a time without meeting his financial obligations to his outside marriage.

Religion

All twenty women said that they were practising Christians—e.g. Baptist, Anglican, and Methodist. All, including the ones who held doctorates, believed and practised *juju*, i.e. local ritual medicine. Consequently, they saw no contradictions between being Christians and being outside wives to men who were already married. They all attended church regularly and saw to it that their children were christened and later confirmed. Prayers and the practice of *juju* were frequently combined to achieve two objectives. The first was to channel (permanently, if possible) the 'husband's' favours and attention for themselves and their children. The second was to ensure that their economic fortunes did not suffer or fluctuate unduly.

Residence

The cities where the study was carried out, Lagos and Ibadan, are situated in the most urban parts of Nigeria. Consequently, it is not surprising to find that most of the informants (15) had been born in Lagos, Ibadan or other surrounding Yoruba cities. The other (5) women said they had been brought to Lagos from other States by parents or relatives while very young. Thus, none of these women could be considered rural or immigrants. They are, *par excellence*, city women.

Fertility Patterns among Outside Wives

Motivations for Having Children

Within the Nigerian context, reproduction remains the primary *raison d'être* of the family (Iweriebor, 1985: 243). Because few women wait till marriage to begin sexual relations, men increasingly demand evidence of fertility before marriage. Women, for their part, are keenly aware of the centrality of reproduction in cementing a relationship. Page (1975: 47) observes that 'a prospective union may not be fully accepted by the community until the birth of a child proves that the union is fertile'. Many women, being aware of this expectation, use pregnancy to force the issue. Among many southern Nigerian peoples, once a woman claims to have a child for a man, he finds it is extremely difficult to extricate himself from some sort of a future relationship with, and obligation to, her. Women know that a man who is reluctant to acknowledge paternity will receive little moral support from friends and family. His family members may, in fact, leave him little choice in the matter: they may claim the child, thereby conferring legitimacy on it. The Nigerian constitution gives women further support, stating that *no* child will suffer disabilities because of the circumstances of its birth. Given all these institutional and structural pressures, even the most recalcitrant of men eventually relent. The child, then, is a woman's link to a man's support (see also Guyer, this volume).

A wife, whether outside or other, with no children must continually worry about the ephemeral nature of her union: barrenness is a legitimate ground for divorce. Indeed, among African peoples, fertility is often integral to the definition of womanhood. In addition, having children for a wealthy man ensures some financial security for the woman during her life time. For whether the 'marriage' remains extant or not, the man must support the children while he is alive and when he dies; the constitution guarantees that the children will have legitimate claim to the man's estate. Thus, through children, a woman is always fairly certain of some financial security, for custom demands that children take care of their parents. Children are therefore used as insurance against a man's predilection towards financial default.

In sum, having children accomplishes three important objectives for an outside wife, and other wives generally. It first establishes lasting socio-economic and politico-jural ties to the father of her children. Second, because a man is not usually barren according to local beliefs, her fertility and hence her womanhood is established beyond doubt. And third, bearing children links the woman to the man's wider kinship network for the duration of her life. A woman can always seek help from her 'in-laws' to force the man to toe the line. This usually works well if the outside wife's children are males; it clinches the deal completely if the 'husband' does not have male or any children with his inside wife.

Observed Fertility Patterns of Outside Wives

Unlike the case of public or traditional polygyny, in which it is unclear whether polygyny reduces or increases fertility (see e.g. Ekanem, 1974: 117 and forthcoming), my study suggests that outside wifeship is associated with low fertility. This statement must be qualified, however. Fertility levels are usually expressed in terms of the total number of children a woman bears throughout her reproductive life. In the case of outside wives, however, we must look at these fertility goals, as the women themselves do, within the context of one conjugal relationship at a time. For each outside marriage, then, a women usually desires to keep her fertility low, although the total number of children she bears throughout her reproductive life-cycle may be substantially higher. Altogether, though, an outside wife's efforts to keep her fertility low within each relationship may mean that her total fertility is lower than a woman of comparable socio-economic status in an inside union.

The available quantitative evidence to assess fertility levels among outside wives is limited. Obviously, data from only twenty women selected in non-random ways are unlikely to be representative. Moreover, the results of tabulations of these are ambiguous. The mean number of live children that the twenty outside wives reported they bore at the time of the interviews was low: 1.9. The range was from one to four children. (At the time of this writing, most of the women in the sample are now menopausal. None has more than four children.) With an average of almost two children and an average duration of 'marriage' of almost 6 years—that is, one live birth every 3 years—a young outside wife could, in theory, have ten children in 30 years. Yet because post-natal abstinence is very short nowadays among highly educated men and women, this rate might reflect lower fertility than we would expect from an average 'inside marriage', especially one in its early years.

The qualitative evidence is much stronger in support of the proposition of low fertility among outside wives. We can look at a few individuals on a case-by-case basis. Since the study ended, I have maintained correspondence with four of the women, one of whose outside marriage has collapsed. They represent important cases because their fertility cycles are complete. Three are in their mid-forties and one is now a grandmother in her late fifties. Three of them, who are very senior civil servants approaching the peaks of their careers, had two children apiece. One woman now in her mid-forties, whose 'husband's' inside wife has remained childless, has four children. The other three had only two children each, though the woman in her mid-fifties, with uncompleted secondary education, had five children with her first legal husband.

Other qualitative evidence concerns stated fertility goals. Most of the women in the study expressed a desire to have no more than four children, regardless of the sex, unless the pressure from the man to have a boy was extreme. Even then, they would only continue trying to bear a boy if the 'marriage' were threatened. The most educated women declared that they would be quite content with one or two children; they would have no more than two or three children regardless of the circumstances. Among these women, though, were two important exceptions. One woman professional already had four children but reported that one or two more would not disturb her 'socio-economic or psychological' equilibrium. Her 'co-wife', a senior civil servant who already had three children, said she would like to have many children (could not decide the number) because financial support was not an object. I find it significant, however, that the two were 'married' to an extremely rich Lagos businessman.

The prodigious use of birth control among these women lends further support to the thesis that outside unions, individually, have low fertility. All the women (except one who said she preferred abstinence after a birth) reported that they used birth control. The most popular methods were the 'pill' and the IUD. Two women said they preferred condoms or coitus interruptus for health reasons; the IUD, they said, resulted in excessive bleeding. Nine women reported that their 'husbands' frequently insisted on using condoms, even when the women themselves were using birth control devices. They were not sure why this was so but suspected that their 'husbands' wanted to ensure that no unwanted pregnancies resulted. I suspect that guarding against infections may also have been a factor since men do not completely trust their outside wives in sexual matters and since infections carried to inside wives frequently lead to conflict in inside marriages.

Reasons for Low Fertility: Some Tentative Conclusions

The sample of twenty outside wives was too small and too selective—and the study's time-span was too short—to speculate confidently on fertility trends of outside wives in Lagos and Ibadan. A longitudinal study of outside wives throughout their reproductive life-spans would be ideal. Still, we can draw some tentative conclusions about the relationship between outside wifeship and low fertility.

It is important to stress that outside wifeship cannot be separated cleanly from other factors that also appear to reduce fertility. In the first instance, the outside wives involved in this study had, on average, substantially higher socio-economic status than most urban women in Lagos and Ibadan. With respect to education, for example, since outside wives are a status symbol, men prefer women with high levels of educational achievement, preferably women who have completed secondary education and who can confront the exigencies of a modern way of life. Since education is negatively correlated with fertility, it should be no surprise that the women in this sample had low fertility goals. Even within the sample, we can see the difference: outside wives who had at least one degree or more desired, and actually had, fewer children than their counterparts with less education. This pattern may stem from the more educated women's career aspirations (Fapohunda, 1983 and 1987; Lewis, 1982) urban residence (Boserup, 1970), lowered infant and child mortality, and less need for children for economic rewards than for emotional ones.

Higher educational levels are also associated with older ages at 'marriage'. All the graduates were around 25 years old before becoming outside wives. Those with advanced degrees tended to be three to four years older than those with only one degree before becoming outside wives. In addition to age, most of the outside wives I know report wanting fewer children than the usual 'as-many-as-God-gives' response cited so often in African fertility literature.

The practice of outside wifeship also seems to act on fertility through 'co-wife competition'. Men are aware that their various 'wives' compete either implicitly or explicitly over childbearing. My data indicate that 'husbands' attempt to contain this competition by contraception, most often with their outside wives, and even demanding at times that their outside wives abort when 'mistakes' have been made. Four of the twenty outside wives reported that their 'husbands' had forced them to abort.

An additional influence on low fertility is health problems. Women who become outside wives tend, in their search for the right man, to have multiple partners. This may expose them to a greater risk of STDs (see Ch. 12), which can lead to subfecundity and sometimes sterility. Outside wives may also have higher rates of abortion, although it is not legal in Nigeria except in special medical cases. Women who have multiple partners appear more likely to attempt to obtain abortions either through self-induction or through unlicensed practitioners. Those who do so risk gynaecological complications which, if not properly treated, can lead to infertility.

Finally, the impermanent character of outside marriage itself appears to be a primary factor underlying outside wives' desires for low fertility. Since outside wifeship is on the whole inherently unstable, men and women alike tend to minimize spousal complications with outside partners with respect to having many children. For a woman in particular, limiting her fertility seems to be the best strategy for retaining manœuvring room in case her current union fails. Finding a new man with a handful of children in tow from a previous union would be extremely difficult. This realization seemed particularly strong among the women with some independent means, or those whose economic statuses were substantially lower than those of their 'husbands'. Validation for this hypothesis came most prominently from the younger graduates, who explained that having many children with one man reduced the chances of a subsequent union in case the current one failed. By contrast, the young secretaries seemed to feel the decision was more the man's than their own; they felt that four children was a good number if that was the number the man wanted.

Summary and Conclusions

The institution we term 'outside wifeship' or 'private polygyny' has existed for a long time in Nigeria. It emerged as a way of resolving the contradictions that are intrinsic in grafting Western ideals of marriage and family on to traditional African practices. Rather than declining, outside wifeship seems to be gaining momentum as some women, in line with the contemporary ideology of rapid accumulation, find it financially expedient to attach themselves to men of means. Outside marriage coexisted with powerful antithetical forces such as Christianity, modernization, and élite women's opposition largely because men of power and means are the central actors in the drama. Because these men are the ones who set the norms, or can flout them with most impunity, the practice attracts less stigma than would be the case if lesser men were the ones primarily involved.

As to the possible influence of the practice on fertility levels, conclusions must remain tentative. While the Western norm of two or three children had not become the norm for everyone, none of the informants expressed the commonly heard African desire to have as many children as God gave them. Even those with the least amount of education felt that the question of how many children a woman has could not be left entirely to chance, and they used one form of birth control or another. The most educated took an even harder line: they vowed that they would *not* continue to give birth until they had a male child.

Finally, it is clear that the economic conditions that can potentially undermine conjugal unions *whatever* their form tend to foster instability in outside unions. Since women traditionally marry up the socio-economic hierarchy, the inclination even among those married 'monogamously' is to maximize economic gain, a pattern that exacerbates an already tense situation in which husbands feel exploited. If women's economic fortunes improve, through education and the like, the stalemate might be resolved because women would no longer benefit significantly from playing second fiddle. Until then, African marriages in towns will continue to go through crises reflecting a fundamental imbalance, with telling consequences for the subordinate partners of such unions—the women—and, by extension, their children.

References

Aronson, D. R. (1978), *The City is Our Farm: Seven Migrant Ijebu Yoruba Families* (Boston: Hall).

Baker, P. (1974), *Urbanization and Political Change: The Politics of Lagos 1917–1967* (Berkeley, Calif.: University of California Press).

—— and Bird, M. (1959), 'Urbanisation and the Position of Women', *Sociological Review*, NS 7/1: 99–122.

Bauer, P., and Yamey, B. (1954), *West African Trade* (London: Oxford University Press).

Bleek, W. (1987), 'Family and Family Planning in Southern Ghana', in C. Oppong (ed.), *Sex Roles, Population and Development in West Africa* (London: Heinemann Educational Books, Inc.), 138–53.

Boserup, E. (1970), *Woman's Role in Economic Development* (New York: St Martin's Press).

Brass, W. (ed.) (1968), *The Demography of Tropical Africa* (Princeton, NJ: Princeton University Press).

Burnham, P. (1987), 'Changing Themes in the Analysis of African Marriage', in D. Parkin and D. Nyamwaya (eds.), *Transformations of African Marriage* (Manchester: Manchester University Press), 37–54.

Caldwell, J. C. (1968), *Population Growth and Family Change in Africa: The New Urban Élite in Ghana* (Canberra: Australian National University Press).

—— (1975), 'Fertility Control', in J. C. Caldwell (ed.), *Population Growth and Socio-Economic Change in West Africa* (New York: Columbia University Press), 58–97.

Caldwell, J. C., ed. (1975), *Population Growth and Socioeconomic Change in West Africa*, New York: Columbia University Press.

Clignet, R. (1970), *Many Wives, Many Powers: Authority and Power in Polygynous Families* (Evanston, Ill.: Northwestern University Press).

Cole, P. (1974), *Modern and Traditional Élites in Politics of Lagos* (Cambridge: Cambridge University Press).

Deluz, A. (1987), 'Social and Symbolic Values of Feminine Knɛ Initiation among the Curo of Ivory Coast', in D. Parkin and D. Nyamwaya (eds.), *Transformations of African Marriage* (Manchester: Manchester University Press), 113–35.

Dinan, C. (1983), 'Sugar Daddies and Gold-Diggers: The White-Collar Single Women in Africa', in C. Oppong (ed.), *Female and Male in West Africa* (London: George Allen & Unwin), 344–66.

Ekanem, I. I. (forthcoming), *Population Dynamics and Policies in Nigeria* (forthcoming).

—— (1974), 'Correlates of Fertility in Eastern Nigeria', *Nigerian Journal of Economic and Social Studies*, 161: 115–28.

Fadipe, N. A. (1970), *The Sociology of the Yoruba* (Ibadan: Ibadan University Press).

Fapohunda, E. R. (1983), 'Female and Male Work Profiles', in C. Oppong (ed.), *Female and Male in West Africa* (London: George Allen & Unwin), 32–53.

—— (1987), 'Urban Women's Roles and Nigerian Government Development Strategies', in C. Oppong (ed.), *Sex Roles, Population and Development in West Africa* (London: Heinemann Educational Books), 203–12.

Feldman, A. S., and Moore, W. E. (1962), 'Industrialization and Industrialism: Convergence and Differentiation', *Transactions of the Fifth World Congress of Sociology*, ii. 151–70.

Fortes, M. (1949), *The Web of Kinship Among the Tallensi* (London: Oxford University Press).

Goldthorpe J. E. (1984), 'Social Stratification in Industrial Society', *Sociological Review Monograph*, 8: 97–122.

Goode, W. J. (1963), *World Revolution and Family Patterns* (London: The Free Press).

Harrell-Bond, B. (1975), *Modern Marriage in Sierra Leone: A Study of the Professional Group* (The Hague: Mouton).

Iweriebor, I. (1985), 'Women in the Family: Labour and Management—What Can Be Done?', in *Woman and the Family*, Proceeding of the Second Annual Women in Nigeria Conference (Zaria: CODESRIA), 239–53.

Karanja-Diejomaoh, W. M. (1978), 'Disposition of Incomes by Husbands and Wives: An Exploratory Study of Families in Lagos', in C. Oppong (ed.), *Marriage, Parenthood and Fertility in West Africa* (Canberra: Australian National University Press), 407–32.

—— (1980), 'Perception of Marriage, Family, and Work in Nigeria: A Study of Lagos Market Women', unpublished D. Phil. thesis, Oxford University.

Karanja, W. W. (1987), '"Outside Wives" and "Inside Wives" in Nigeria: A Study of Changing Perceptions of Marriage', in D. Parkin and D. Nyamwaya, (eds.), *Transformations of African Marriage* (Manchester: Manchester University Press), 247–61.

—— (1981), 'Women and Work: A Study of Female and Male Attitudes in the Modern Sector of an African Metropolis', H. Ware (ed.), *Women, Education, and Modernization of the Family in West Africa* (Canberra: Australian National University Press), 42–66.

Kerr, C., *et al.* (1973), *Industrialism and Industrial Man* (London: Penguin).

Lewis, B. (1982), 'Fertility and Employment: An Assessment of Role Incompatibility among African Urban Women', in E. G. Bay (ed.), *Women and Work in Africa* (Boulder, Colo.: Westview Press), 249–76.

Little, K. (1973), *African Women in Towns: An Aspect of Africa's Social Revolution* (Cambridge: Cambridge University Press).

LLoyd, P. C. (1966), *The New Élites of Tropical Africa* (London: Oxford University Press).

McNamara, R. S. (1985), 'The Challenges for Sub-Saharan Africa', Sir Crawford Memorial Lecture, Washington, DC, 1985.

Morgan, R. W., with Ohadike, P. O. (1975), 'Fertility Levels and Fertility Change', in J. C. Caldwell (ed.), *Population Growth and Socio-Economic Change in West Africa* (New York: Columbia University Press), 187–235.

Nadel, S. F. (1942), *A Black Byzantium: The Kingdom of Nupe in Nigeria* (London: Oxford University Press).

Obbo, C. (1987), 'The Old and New in East African Marriages', in D. Parkin and D.

Nyamwaya, (eds.), *Transformations of African Marriage* (Manchester: Manchester University Press), 263–80.

Oppong, C. (1974), *Marriage Among a Matrilineal Élite: A Family Study of Ghanaian Senior Civil Servants* (Cambridge: Cambridge University Press).

—— (1987*a*), 'Responsible Fatherhood and Birth Planning', in C. Oppong (ed.), *Sex Roles, Population and Development in West Africa* (London: Heinemann Educational Books), 165–81.

—— (1987*b*), 'Introduction', in C. Oppong (ed.), *Sex Roles, Population and Development in West Africa* (London, Heinemann Educational Books), 19–24.

Orubuloye, O. (1987), 'Values and Costs of Daughters and Sons to Yoruba Mothers and Fathers', in C. Oppong (ed.), *Sex Roles, Population and Development in West Africa* (London: Heinemann Educational Books), 86–90.

Page, H. (1975), 'Fertility Levels: Patterns and Trends', in J. C. Caldwell (ed.), *Population Growth and Socio-Economic Change in West Africa* (New York: Columbia University Press), 29–57.

Parkin, D., and Nyamwaya, D. (eds.) (1987), *Transformations of African Marriage* (Manchester: Manchester University Press).

Radcliffe-Brown, A. R., and Forde, D. (eds.) (1950), *African Systems of Kinship and Marriage* (London: Oxford University Press).

Robertson, C. (1976), 'Ga Women and Socio-Economic Change in Accra, Ghana', in N. J. Hafkin and E. G. Bay (eds.), *Women in Africa: Studies in Social and Economic Change* (Stanford, Calif.: Stanford University Press), 111–33.

Smelser, N. J. (1968), *Essays in Sociological Explanations* (Englewood Cliffs, NJ, Prentice-Hall).

Sudarkasa, N. (1973), *Where Women Work: A Study of Yoruba Women in the Marketplace and in the Home* (Ann Arbor, Mich.: University of Michigan Press).

UNECA (1979), *Population Dynamics: Fertility and Mortality in Africa* (Addis Ababa: UNECA).

United Nations (1978), *Fertility Behaviour in the Context of Development: Evidence from the World Fertility Survey* (New York: UN).

van de Walle, E. E. (1968), 'Marriage in African Censuses and Inquiries', in W. Brass (ed.), *The Demography of Tropical Africa* (Princeton, NJ: Princeton University Press), 12–87.

10 Social Change and Marriage Arrangements: New Types of Union in Lomé, Togo

THÉRÈSE LOCOH

The analysis of marriage change in African urban areas has long been dominated by simplistic, Western-centred views. Schooling, Christianity, new legislation, all were seen as creating new types of union that would break with the practices of polygyny and the inequality between spouses. Families were supposed to be headed, at least in the cities, towards the nuclear model centred on close bonds between partners and the privileged treatment of children. Since analysts believed that such family changes would produce more costly outlays for children, it therefore created hopes, in the international community, for the rapid penetration of family planning, another hallmark of modernization.

It is true that a small élite appears to adopt a style of family life that is closer to the Western model, either under the influence of Christianity or as a result of emulating a 'modern' family model. But closer inspection of the issue in the city of Lomé, Togo, reveals that this is the choice of only a few people. Certainly family structures change and there are lively tensions between the old and new models. But rather than the wholesale adoption of an imported model, one sees instead new forms of marriage that straddle indigenous and Christian models. Female household heads, co-wives living across town from each other and perhaps even from the husband, a man who is theoretically a monogamist but also has an informal 'outside' wife—such arrangements often pose difficulties for outsiders to comprehend. They appear to be at the crossroads between the traditional marriage based on lineage decisions and a new one, referring to the personal choice of the individuals.

To understand current conjugal dynamics better, we can glean important clues to people's actions and desires by studying forms of residence. A careful analysis of residence arrangements that initially confuse outsiders reveals that while people have not rejected the former models, neither are they merely reproducing rural patterns in the city. Instead, they are generating new patterns of marital residence to cope with economic as well as ideological pressures of urban ways of life. So doing, they regain some of the flexibility that rural polygyny offered.

TABLE 10.1. *Marital statuses of young women (15–24) and of women over 49: Lomé (L) and surrounding rural population (Région Maritime, RM), 1981 Census* (%)

Age		Single	Married	Widowed	Divorced, separated	Total
15–19	L	81	19	—	—	100
	RM	54	44	—	2	100
20–4	L	39	59	—	1.8	100
	RM	13	84	—	3	100
Total, 15–24	L	60	39	—	1	100
	RM	33	65	—	2	100
50 +	L	3	49	38	10	100
	RM	2	58	36	4	100

City and Country: Different Marital Statuses

Censuses are imperfect sources of data for analysing new forms of union that city life produces. Despite their limits, census results from 1970 and 1981 for the city of Lomé show three patterns of interest to us: a continuing high level of marriage for women (1.5 per cent of women aged 50–4 were single in 1970 and 3 per cent in 1981), a marked reduction of polygyny among men aged 40–9 (from 36 per cent down to 26 per cent), and a slight increase in the mean age at first marriage among women (from 20.8 years up to 21.5 years) though no change in the mean age at first marriage among men (around 27 years). It is unclear, of course, whether these results reflect an actual change in practices, or merely a change in the declarations that people made when questioned by the census taker, the objective situation being, in fact, the same. The truth probably lies somewhere between.

The Togolese census results also highlight regional variations. The uniqueness of urban situations is evident when we use the 1981 Census results to compare the marital status of women from Lomé to that of women in the surrounding rural population in the Maritime Region (Table 10.1). In Lomé, 81 per cent of young women aged 15–19 are single, compared to only 54 per cent in rural areas. In the age group in which women usually enter their first marriage, 15–24 years, twice as many people in Lomé (60 per cent) are single as in the rural population (33 per cent). At the other end of marital life, after age 50, there are more widows and divorced or separated women in Lomé than in the interior of the country. After age 50, married women represent 49 per cent of women in Lomé, compared with 58 per cent in the rural population of the Région Maritime. In the city, more than elsewhere, a woman may have her own home without necessarily depending on a man, whether a husband, father, or brother.

At every age, it is in Lomé that one sees not only the highest proportions of unmarried women but also, among married women, the highest proportions who are living apart from husbands: that is, their husbands live elsewhere, either as a result of migration, or because they are polygynous. One also sees, among the married women, a high proportion of female heads of household (18 per cent; 1981 census). It is not surprising, therefore, that Lomé appears to many observers to be a 'female' city. This impression probably stems from the intense commercial activity among women in the city, but also from the fact that a sizeable number of women are responsible for households, whether full- or part-time. The figures have remained quite stable: in both the 1970 and 1981 censuses, Lomé has had 25 per cent of households headed by women and 22 per cent of the city inhabitants living in households headed by women.

What the censuses are least equipped to describe is the variety of marital situations in which men and women are actually living. First of all, they record the declaration of marital status by the interviewees, as such. Secondly, they do not combine this declaration with information on where the spouse lives, which is central, as we will see, to the analysis of unions in Lomé. Individuals living in identical situations may, in fact, describe their situations in quite different ways. A small business employee may boast that he has two wives (a sign, in certain circles, of economic status) while his manager, in an identical situation, may declare himself to be monogamous, although he may in fact maintain two or three households. Alternatively, a young social service employee acknowledges that she is single, a status that ordinarily is anathema to a woman over the age of 20–5, but she adds that she 'has a child' which redeems her status. A housewife in her neighbourhood, in the same conjugal situation, might declare herself 'married' because, for her, having a child with a man is sufficient grounds to be considered his wife. Even partners to the same union might describe their marital statuses differently: a woman might say that she is married to Mr X while the man declares himself to be single.

While the census remains a rudimentary instrument, surveys can collect more detailed information. Our 1983 survey in Lomé of 3,600 women who gave birth in 1979 (the APEL Survey: 'The Arrival of the Next Child in Lomé') asked questions about the order of the current union at the time the survey was done, the status of the partner (monogamous or polygynous), and whether the spouses lived together. (A 1976 survey in a rural area near Lomé had already demonstrated the importance of defining unions based on these three criteria (Locoh, 1984).) In Lomé, the survey refers to married women who had had at least one child. In the south-east survey, it covered all women, but only 3 per cent had never had children; one may thus legitimately compare the two sets of observations. The rest of the analysis used in this paper draws on data from these surveys.

Table 10.2 distinguishes eight different situations among married women that combined these three criteria in comparing the two samples, the rural

TABLE 10.2. *Types of union among married women aged 20–4 and 35–9; rural south-east (1976) and Lomé (1983)* (%)

Type of union	20–4 years old		35–9 years old	
	South-east	Lomé	South-east	Lomé
First union monogamous				
living together	25	47	18	33
living apart	14	14	1	7
First union polygynous				
living together	38	17	47	26
living apart	7	17	3	11
Second union monogamous				
living together	5	1	6	6
living apart	3	2	1	1
Second union polygynous				
living together	8	1	21	7
living apart	—	1	3	8
TOTAL	100	100	100	100
No. of cases	250	479	240	509
% in first union	84	95	69	78
% living apart	24	34	9	28

south-east in 1976 and Lomé in 1983, for two age groups. In the rural area, the most common situation among women over age 20 is that of wife in her first union, married to a polygynous man with whom she co-resides. Women in their first unions whose partners are monogamous represent 25 per cent among young women (20–4 years old) and only 18 per cent among women aged 35–9. The proportion of women who do not live together with their husband, all unions taken together, decreases from 24 per cent at ages 20–4 to 9 per cent at ages 35–9 (Locoh, 1984).

In Lomé, the situation is different. Monogamous unions with co-residence of spouses are the most frequent situation at both ages (47 per cent among women aged 20–4, and 33 per cent at ages 35–9). Unions without co-residence are much more common than in the rural sample especially when the union is polygynous, whether at the start of married life (20–4) or later (35–9). In Lomé, common residence of co-wives is not easy to manage and may be less acceptable among certain groups. Another difference is that women in Lomé enter successive unions less often than do those in the rural area; at age 35–9, 78 per cent are still in their first union, instead of 69 per cent in the rural sample.

TABLE 10.3. *Lomé women who have had at least one live birth, by age and by marital status of the husband, APEL Survey, 1983* (%)

	20–4	25–9	30–4	35–9	40 +
Monogamous husband	55	60	58	46	36
Polygynous husband	30	37	37	51	57
No current union	15	3	5	3	7
TOTAL	100	100	100	100	100
No. of cases	563	1,275	863	524	363

TABLE 10.4. *Lomé women who have had at least one live birth, by age and order of the union, APEL Survey, 1983* (%)

	20–4	25–9	30–4	35–9	40 +
In first union	81	86	78	76	64
In second union	4	11	16	21	29
Single	12	1	—	—	—
Widowed divorced	3	2	5	3	7
TOTAL	100	100	100	100	100
No. of cases	563	1,275	863	524	363

Household, Residence, Polygyny: Towards a More Precise Typology of Unions in Lomé

Although Lomé is distinct in some ways from the rural south-east, some domains of conjugal life reflect a great deal of stability. Certainly polygyny is still alive and well in the city, as several authors have stressed for urban Africa (e.g. Clignet, 1987). Table 10.3, which divides Lomé women in each age group by the marital status of their partners, shows that among the youngest women, those 20–4 years old, 55 per cent are married to a monogamous man. After age 35, however, most women are polygynously married: 51 per cent at ages 35–9 and 57 per cent among those 40 and older. Whether this is a cohort or a life-cycle effect cannot be determined by these data.

Once they enter a marriage, many Lomé women also remain married to the same man. Table 10.4 shows that most women remain in a first marriage at all ages observed (64 per cent at ages 40–9). Single women comprise 12 per cent of the first age group, 20–4, but they are subsequently a negligible number: most women who have children either marry or call themselves married after a certain age.

More complicated conjugal permutations appear as we inquire further into

TABLE 10.5. *Lomé women who have had at least one live birth, by age and co-residence status, APEL Survey, 1983* (%)

	20–4	25–9	30–4	35–9	40 +
Spouses live together	56	74	72	70	71
Spouses live separately	29	23	23	27	22
Husband elsewhere in Lomé	(20)	(16)	(13)	(16)	(15)
No current union	15	3	5	3	7
TOTAL	100	100	100	100	100
No. of cases	563	1,275	863	524	363

TABLE 10.6. *Lomé women in union who do not reside with their Lomé-resident spouse, by age and marital status, APEL Survey, 1983* (%)

	20–4		25–9		30–4		35–9		40 +	
	%	No.	%	No.	%	No.	%	No.	%	No.
All women in union	26	427	19	1,145	15	739	19	457	17	311
Women in monogamous union	16	275	14	705	9	466	6	209	6	113
Women in polygynous union	44	152	26	440	26	273	31	248	24	198

residential arrangements. Table 10.5 shows that, at the start of married life (20–4), there are slightly more Lomé women (29 per cent) not co-residing with their husbands than among older women. At all ages, however, about a quarter of the women consistently declare themselves married but not co-residing with husbands. Among these are the wives of officials who are posted to the interior and the wives of migrants. (Lomé exports artisans and qualified labourers to neighbouring countries such as Benin, Nigeria, and Gabon.) Even among wives of male Lomé residents, a number of them live apart from their husbands (20 per cent at age 20–4, 15 per cent at age 40–9). To control for the possible effects of a male emigration, we calculated separately the unions in which the two spouses live in Lomé, but do not reside together. The proportion of married women not cohabiting with their Lomé-resident husband was still high; it varied from 26 per cent at 20–4 years of age to 17 per cent at 40–9. Whether these are arrangements that women favour or men impose, they are not unusual.

Non-coresidential unions, as Table 10.6 shows, are especially common in Lomé among wives of polygynous men (44 per cent at ages 20–4). But a surprising number also of women in monogamous unions, especially among the younger generation, do not live with their husbands. Before 25 years of

TABLE 10.7. *Lomé women married to polygynous men, by residence of the husband and the co-wives, APEL Survey, 1983* (%)

	Woman resides without co-wives	Woman resides with co-wives	All women
Husband			
co-resides in Lomé	24	40	64
lives elsewhere in Lomé	25	1	26
does not live in Lomé	8	2	10
TOTAL			100
No. of cases			1,447

age, 16 per cent of monogamous women are in this situation and 14 per cent of women between 25 and 30 years of age. The partners simply pay each other 'visits'. Among older women, such a situation may be a *de facto* separation which is not declared as such.

The fact that a high proportion of young women do not live with their partners suggests that there may be a period of progressive entry into unions that varies according to the financial means of the spouses and their families. (Our records of this phenomenon are minimal since our sample takes into account only women who had given birth, and who, having already had a pregnancy, have a higher likelihood of cohabitation than women with no children.) In such situations cohabitation might come into play only after a certain time or after a birth or two. Alternatively, we may be observing a new conjugal pattern among younger generations with new views of married life. Both explanations probably have some truth.

Polygyny appears to play an important role in residence arrangements. Evidence of this emerges in Table 10.7. For women in polygynous unions we can look not only at their cohabitation with husbands but also with co-wives. The table shows a diversity of living arrangements. The 'traditional' scenario—a household in which the husband and some or all of the wives live together—still prevails in 40 per cent of the cases. In contrast, 24 per cent of the wives of polygynous men live alone with their husbands, the husband having the rest of his wives elsewhere, and 33 per cent live apart from both husband and co-wives. These findings suggest that wives may be seeking independence from each other, instead of from the husband. Alternatively, the husband may be seeking to place distance between the wives.

Combining these factors, Table 10.8 shows a more comprehensive typology of women's marriage arrangements. (We took up only the configurations that remained significant when we grouped together all second unions, for example.) The results show that although Lomé is a city highly influenced by Christianity and Western styles of life, its forms of married life continue to

TABLE 10.8. *Lomé women by age, order of union, marital status of the husband, and living arrangement, APEL Survey, 1983* (%)

	20–4	25–9	30–4	35–9	40 +	All women
1st monogamous union	(52)	(53)	(50)	(39)	(25)	(47)
spouses live together	40	44	43	32	21	
spouses do not live together	12	9	7	7	4	
1st polygynous union	(29)	(33)	(28)	(37)	(39)	(32)
spouses live together	14	24	19	26	32	
spouses do not live together	15	9	9	11	7	
2nd monogamous union	(3)	(6)	(8)	(7)	(10)	(7)
spouses live together	1	4	6	6	8	
spouses do not live together	2	2	2	1	2	
2nd polygynous union	(1)	(5)	(9)	(14)	(19)	(8)
spouses live together	1	2	4	7	10	
spouses do not live together	2	3	5	7	9	
Widowed and divorced	(3)	(2)	(5)	(3)	(7)	(4)
Single	(12)	(1)	—	—	—	(2)
TOTAL	100	100	100	100	100	100
No. of cases	563	1,275	863	524	363	3,588

be very diverse. Of particular note is the fact that although monogamy may be more predominant than in rural areas, urban polygyny still thrives in the various manifestations that we have described here.

Marriage Arrangements, Residence, Life Cycles and Challenges to Customary Values

To discern associations between the various forms of union in Lomé, we examined the following variables through factorial correspondence analysis[1]: the status of the husband (polygynist *v.* monogamist); the order of the union; the residence of the spouses, and eventually of the co-wives. All variables were defined in discrete and non-continuous terms. By combining variables that describe women's current residence and marital situations with data representing women's stated preferences about family size and structure, we sought to discern more about the marriage system as experienced by women at various stages of the life-cycle, and of different socio-economic characteristics. The analysis revealed some unlikely associations among diverse variables. Four distinct groups emerged.

A first group, many of whose members do not co-reside with husbands, includes women who are very likely to be highly educated, have high standards of living, have married late, practise modern contraception, and be Christian (Catholic or Protestant). Women in this small group, whose ages range from young to old, may be termed 'modern'. By contrast to this group is a group of women with characteristics of a more 'traditional' sort: women who married young, married husbands chosen by their families, want many children (or at least as many as God will send them), and state that they prefer polygynous unions and desire early marriages for their daughters. Many have recently migrated from a rural setting, often from the north of the country, and profess the Muslim faith or indigenous beliefs.

Between these two groups, two new groups have emerged. One contains older women with ostensibly 'traditional' characteristics: they live with husbands, are members of polygynous unions with co-wives present, have numerous children, do not practise modern methods of family planning, and have little or no schooling. Yet these women appear to be questioning old values. Such a woman may state that she wants no more children or only one more child, would not marry if she could start over again, and wants her daughter to marry late. She appears to have built up experiences (since much of her married life is already behind her) which, rather than schooling, have encouraged her, in retrospect, to reassess critically the values of her society. This would be a quite important group to track in the future. Of particular interest would be the values these women transmit to their children, because they comprise a much larger group than do highly educated women who enjoy a high standard of living.

Finally, there is a group of single mothers. This group occupies a position like that of spouses who do not co-reside; but the situation of the single mother is more likely to be lived through, rather than wished for, as a deliberate choice of a lifestyle. It often stems from failed or non-existent contraception. Alternatively, it may represent the choice of a more traditional woman to bear a child as she enters her reproductive years in order to show that she is fertile.

In sum, by looking for associations among multiple aspects of both residence and other variables characteristic of women's aspirations in conjugal life, we gain new information that simple cross-tabulations sometimes obscure. We learn that, contrary to common-sense expectations, single mothers appear to accept traditional values of marriage more than they challenge them, and that non-coresidence may be either a modern choice by women that defies tradition or a situation that is simply imposed upon them by their partners. By contrast, polygyny without shared residence of the wives belongs on the 'modern' end of the spectrum of marriage, in the sense that such polygynous women—usually older ones—want to finish childbearing soon, and devalue marriage. These findings suggest that it is not only highly educated, well-to-do women who attempt to challenge traditions related to marriage and fertility.

Such challenges also arise among women with less schooling, fewer privileges, and numerous children.

New Forms of Union: Mediation between Cultural Models

This background sets the stage for a closer look at how ethnic groups in southern Togo have begun to use these various combinations of marriage and residence, drawn from both customary and modern practices, to create new life options and new flexibility.

Marital Practices in the Rural Societies of the Ewe-Mina

Among the ethnic groups in the south of Togo, marriage is, as in the rest of Africa, above all an alliance between two families. It is arranged by mediators between the maternal aunts of the man and the family of the woman. Custom dictates three successive visits. On the third, the aunts bring presents in kind (drinks, clothes) and a modest sum of money: it is this bride-price that concludes the alliance. There is no declaration to a village, municipal, or religious authority (Durand, 1979). In some cases nowadays, though, the sequence culminates in the eventual recording of the declaration.

Residence is generally virilocal, and kinship patrilineal, although residual matrilineal influences on family organization can be traced to the 'Akan' branch of Ghana, from which some people in southern Togo originated. Once the bride-price has been paid, the wife joins the husband in his compound. The spouses maintain a rigorous separation of property, although the man gives part of his harvest to his wife, often according to the number of children she has by him, and sometimes gifts of cloth or jewellery when he sells the harvest. The wife prepares food for her husband (each in her turn, if there is polygyny) and eventually participates in work on his land. Aside from such assistance, she often has a field of her own, assigned to her by her husband, on which she cultivates food crops, either for family consumption or for her own cash profit. If she produces something through her work (e.g. manioc flour) or by selling what her husband produces, she will be paid one way or another by the husband.

In the coastal area of Togo, women have few constraints on their movement. If they are businesswomen, they may sell in the market or even undertake long-distance trading trips. It is not unusual for a woman to leave the conjugal dwelling for some time, especially after her first birth, to go home to her mother or an aunt to acquire advice on child care. This temporary departure facilitates maintaining post-partum abstinence, while polygyny in turn facilitates these periods of absence.

After a union ends, remarriages may take place after a short time: 1.92 years for divorcees; 2.4 years after widowhood if there is remarriage, which, in the south of Togo, is very often (Locoh, 1984). Women rarely stay single and polygyny allows them to find a new domestic arrangement. For widowhood, the levirate still occurs, but this is not a strict rule.

Since the start of the twentieth century, a small urban élite, often children of Afro-Brazilians, has adopted a Christian model of marriage characterized by the legal registration of marriage, the predominance (but not requirement) of monogamy, and the free choice of a partner among young people. With the rapid schooling of young urbanites, the spread of Christianity (75 per cent declare themselves as such in Lomé), the rapid increase in transactions with Western countries, and the diffusion of means of communication (newspapers, radio, television), Western notions such as romantic love and emphasis on a monogamous couple as the core of a household have penetrated the cultural landscape of Lomé.

In certain ways, these imported models are contradictory to autochthonous/indigenous models and are perceived as such. The most flagrant contradiction is monogamy, but it is certainly not the only one. Contrary to the ideal of a closely linked nuclear pair, where each individual agrees to share freely with the other, is a strong tendency to separate holdings in the lineage model. And in contrast to the highly regulated form of Western marriage is the relative flexibility of the marriage rules in Ewe-Mina country, where unions are concluded by agreements between families.

Urban Variations, Starting with the Ewe-Mina Cultural Model

Entry into a Union: A Journey in Stages

Unions in urban Lomé often represent compromises between the private character of customary marriage, formalized only by the transfer of bride-price, and modern marriage, which is officially recorded, and in which the spouses choose each other. One anomalous result of this compromise is an inflation of bride-price, in spite of the official ceiling that has been set at a symbolic level of 240FF. This bride-price inflation reflects the increasing proclivity of the families to negotiate and, sometimes, of the woman to enter this negotiation. (Women themselves do not like the fact that bride-price could be fixed at a symbolic level.)

At ages 20–4, 35 per cent of the women of Lomé having at least one child declared themselves as having no formal union. This finding, however, does not necessarily indicate more permanently single women. Rather, it reflects the idea that there are intermediate stages between being unmarried and an 'official' union. As in rural areas, most first unions in Lomé are the final step in a series of successive stages. But increasingly it is not families who choose, but the partners themselves who do so and then consult their families. To marry against parental wishes is still frowned upon. Public opinion considers such unions as destined to fail.

Whether or not a woman declares herself as married, a first pregnancy is highly recommended, even by certain Catholic priests, to avoid the failure of a marriage through childlessness. In fact, a premarital pregnancy may precipitate a union by making the woman's family become anxious to conclude the union. Very often, bridewealth is a means to postpone entry into union; sometimes a pre-marital pregnancy serves as a check against extravagant bridewealth. Yet since there is sometimes a long period of 'frequenting' before marriage, it is also fairly common for a birth not to be followed by a union. In this case, the family of the young woman asks the father to help maintain his child or take the child into his family. The father rarely refuses to recognize his paternity. To have a child is almost always a source of pride.

If the union becomes stable, it may be legally concluded without the spouses moving in together even though bride-price has been paid. This is because in Lomé the notion of virilocality translates into the husband's obligation to have a roof where he can welcome his wife. When he is not yet well established, the wife may prefer to stay at her father's home for a time.

When the union is recorded, in conformity to the new family code, the husband may be asked whether he intends to be polygynous later on. If so, his partner, so forewarned, may not cite the arrival of co-wives as a reason for divorce, which she could do if her husband vowed monogamy. By including this condition in a modern family code (adopted in 1982; Pokanam, 1982), Togolese legislators sought to maintain cultural traditions and to avoid imposing prematurely the rule of monogamy.[2] According to preliminary statistics, in approximately one-half of recorded civil marriages in Lomé between 1981 and 1986, the husbands chose the option of polygyny, a pattern as common among illiterate as among educated men (Mignot, 1987). Ironically, however, fears of exposing themselves to lawsuit may have actually increased the incidence of 'outside' unions. A man who acquires further partners but never concludes the unions avoids legal incrimination, should his wife sue for divorce.

At the Time of the Union: Variations in Residence

If a union is concluded through bridewealth, whether registered or not, the constraints of both principal models—customary marriage based on an alliance between families, and Western-style marriage based on an intimately linked couple—will be stronger because substantial wealth is at stake. Yet concluding a union of either kind in these new circumstances may create new dilemmas or intensify inherent ones. Contrary to our assumptions, for example, the relative independence that women enjoy in traditional unions is often less in urban situations where husbands fear that the erosion of social control may allow their wives to go astray. And the logical tendency for couples in monogamous unions to share property may be offset by a growing mistrust of one's spouse and in-laws.

Most importantly, a durable monogamous union may calcify previous manœuvrability. Polygyny allows a man to advance his social mobility incrementally by successively entering unions that can reinforce it. Mobility is still important nowadays where many urban careers rapidly progress through a hierarchy of social success. Yet choosing legal monogamy means that spouses must effectively give up prospects of further advancement through marriage with this particular partner. Indeed, monogamy becomes all the more risky because first-time marriage occurs when the partners are young and their full potentials are not yet realized. If the man does not have ambitions to advance, albeit through polygyny, it may be the wife who regrets the loss of flexibility of customary marriage, where separations were relatively easy to carry out and allowed her to implement her individual strategies, whether matrimonial or not. She may also regret not having more autonomy to manage her own economic activity, to build her own social network and, in certain cases, to move to a subsequent partner.

We can understand now why polygynous unions without co-residence may come about. A number of people are deliberately minimizing the constraints of both customary polygyny and monogamy, giving the woman more autonomy and the man more liberty to accumulate concurrent partners. The irony is that both men and women are still seeking manœuvring room through conjugal links in contexts in which mobility through marriage is in theory precluded. Viewing residence not as an end result but as a means toward an end allows us to recognize the result as a modern version of the polygynous option.

The End of the Union: Discrete Exit rather than Abrupt Ending

It should not be surprising to learn that urban unions frequently end by divorce. In the APEL sample, 24 per cent of women aged 35–9 and 36 per cent of women aged 40–9 saw their first union broken by widowhood or divorce (see Table 10.4). What is less known is the role that multiple residences play in the process of separation or divorce. As we saw in Table 10.1, Census data from 1981 showed that the number of actually widowed or separated women remains quite small during the fertile period (ages 15–49). At older ages widowhood becomes a frequent situation (36–8 per cent of women aged 50 and over). This is not the case at these later ages for divorce and separation (10 per cent of Lomé women and 4 per cent of women in the surrounding rural area). Age differences between spouses and the existence of polygyny obviously increase the probabilities of widowhood. Remarriages are common, however, whether they be formal remarriages or consensual unions.

As for separations, they rarely entail divorce in the complete sense. Few couples have realistic recourse to the justice of modern law, and customary law is more often invoked to resolve disputes between families over the reimbursement of bride-price. As a result, unions often end by progressive slides into a separation. There again, women—since it is almost always they who take the initiative to separate—often take advantage of the multiple residence options available to them. By using the fact that she has family elsewhere, a woman can distance herself gradually from the conjugal household. She may leave on the pretext of family or religious duties (to this or that ancestor, to help a relative, etc.), or for professional reasons (to buy commodities in another city or even abroad).

An Overview

Lomé is notable for its significant number of female heads of households and of women who say they are in unions but do not live with their husbands. Yet instead of interpreting this situation as a trend towards the abandonment of customary forms of union for Western-style marriage, we should recognize it as a subtle manifestation of borrowing from different cultural models. Men and women appear to use the diversity of residence situations to bring about new forms of unions to minimize the tensions between their lineage and the urban culture, thus creating types of households that are particular to the new culture of African cities. The resulting diversity of forms of union has elements with roots in both Ewe-Mina culture and in Western culture. These recombinant forms are manifested in diverse marriage regimes and residential choices, especially at the beginning and at the end of a union. Consensual unions are not rare and if the urbanized élites have an official preference for monogamy, they have just as much unofficial penchant for a 'renovated' polygyny and a rigid separation of property between the spouses.

These new forms of union are far from being an anecdotal aspect of social change in Lomé society. Knowledge of their rationales may improve the interpretation of demographic data as well as development planning. Studies of social change, for example, tend to assume that family decisions (the number of children to have, the allocation of resources) are made in the heart of the nuclear family where the budget is common and where the husband is most often the head of the household. In Lomé this situation comprises the minority of cases. Indeed, the APEL survey revealed that women in monogamous unions living with their husbands represent about half of all ever-married women.

This finding has important consequences for the process of making household decisions: those concerning fertility as much as those concerning the allocation of resources, for example (Todaro and Fapohunda, 1988). The rigid maintenance of the separation of property and incomes between spouses has important consequences for caring for children, and thinking about parental responsibilities—especially paternal responsibility. Finally, the fact that there are numerous pathways of entry into a union leads one to question whether any future decline in fertility will result from increases in the age at first marriage. If it is true that this age is rising in certain African cities, there may be at the same time an increase in prenuptial births, in the births of children who spend much of their lives apart from fathers and the support of paternal kin.

In any case, the study of African marriage must move away from rigid, inappropriate frameworks based on classical definitions of marital states by linking what is qualitatively observed with statistical measurements.

Notes

1. For an extensive description of this analysis applied to Lomé marriage patterns, see Locoh, 1989. A succinct description of the method more generally and its application to a demographic survey can be found in Brouard (1980).
2. In Côte d'Ivoire, where legislation actually banned polygyny, the debates expressed by public opinion and the media are still lively. Many people criticize this legislation for penalizing polygynous people who were already married, especially because it would seem to disadvantage children born in unions that were officially 'illegitimate'.

Acknowledgements

The findings presented in this paper are the result of the APEL (Arrivée du Prochain Enfant à Lomé: The Arrival of the Next Birth in Lomé) Survey, conducted by the Demographic Research Unit (URD) of the University of Lomé, thanks to a grant from the Population Council which was part of AID contract number CP 82.93 A/CP 84.02 A. I would like to thank Sylvie Delcroix and Valerie Delaunay for their valuable assistance on the factor analysis.

References

Brouard, N. (1980), 'Analyse factorielle et typologique de l'enquête sur la fécondité et la contraception en Martinique', in Y. Charbit and H. Leridon (eds.), *Transition démographique et modernisation en Guadeloupe et en Martinique. Travaux et documents*, Cahier 89, INED, PUF.

Clignet, R. (1987), 'On dit que la polygamie est morte, vive la polygamie!', in D. Parkin and D. Nyamwaya (eds.), *Transformations of African Marriage* (Manchester: Manchester University Press).

Durand, C. (1979), *Le Mariage traditionnel au Togo: Traditions togolaises*, No. spécial, Annales de l'Université du Bénin, Série Lettres, Lomé.

Locoh, T. (1984), 'Fécondité et famille en Afrique de l'ouest: Le Togo méridional contemporain', *Travaux et documents*, Cahier 107, INED, PUF.

—— (1989), 'Changement social et situations matrimoniales: les nouvelles formes d'union à Lomé', *Dossiers et recherches*, Cahier 29 (Paris, INED).

Mignot, A. (1987), *Le Droit de la famille au Togo* (Togo: University of Benin).

Pokanam, V. (1982), 'Quelques aspects du Code Togolais de la famille', *Mariage et nuptialité*, i. University of Benin, Demographic Research Unit, Études Togolaises de Population, 4, Lomé.

Todaro, M. P., Fapohunda, E. (1987), *Family Structure, Implicit Contacts and the Demand for Children: A Consideration of Southern Nigerian Data*, Center for Population Studies, 136, The Population Council, New York.

11 Lineal Identities and Lateral Networks: The Logic of Polyandrous Motherhood

JANE I. GUYER

Introduction

The changes in African marriage patterns evidenced in both the cohort statistics and people's observations are difficult to contextualize and interpret by comparison with other parts of the world. This chapter explores some directions of thinking that arise out of juxtaposing focused quantitative data with ethnographic and historical data. The strengths of three kinds of logic are brought to bear on the problem: the inferential logic of statistical analysis, the systemic logic of social and cultural analysis, and the sequential logic in the known patterns of social and economic change.

The first part of the chapter examines the advantages and difficulties of bringing these literatures together, and some hypotheses one might pursue for Africa more broadly. The second part suggests that the later marriage pattern developing in the Yoruba areas of Western Nigeria over the past twenty years is associated with a growing informality of marriage, both first and subsequent marriages. The third part explores the history of marriage from the pre-colonial period to the present in order to place this informality in long term perspective. It seems that re-marriage was always informal, and that only under colonial conditions were measures developed to try to formalize bridewealth, divorce and repayment as part of a general policy to place a customary legal framework around the growing de facto existence of marital separation. The guidelines were never fully successfully institutionalized. One might see the increasing informality of first marriage, then, as an extension of informalities that already existed for remarriage. That said, however, parenthood has always been punctiliously recognized. Whereas marriage is a temporary contract, childbearing for both men and women creates life-long ties linking people into networks that are intrinsically important to Yoruba social and political life.

Under conditions of increasing economic unpredictability, the search for lateral ties of connection is intensified and added to the concern with the classic lineal ties of intergenerational interdependence. At its outside edge the logic of the 'primacy of parenthood over marriage' would generate a category of women practising what I call 'polyandrous motherhood', that is, cultivating co-parental ties with more than one father of their children.

Hakansson (1988) describes a situation for poorer women in a densely populated area of Kenya where such strategies seem their only viable option. One can imagine other situations where the incentives for women might be less desperate but nevertheless quite persuasive. Our empirical data are thin, but following out the intersecting logics of demographic patterns, family ethnography, and political-economic history suggests that parenthood and network-building by women deserve the kind of empirical attention that lineal identities and formal marriage alliances have commanded in the past.

Some Statistical Summaries of Changing Yoruba Marriage

The most recent published data for Western Nigeria is the Demographic and Health Survey for Ondo State, carried out in 1986 and published in 1989. Taken together, irrespective of age cohort, Ondo State women (of whom 86 per cent of the sample were Yoruba; DHS Ondo State 1989: 7) show two unusual characteristics by comparison with much of the rest of Africa. The median age at first marriage is now unusually high, age 20 for women now 25–49, even though 46 per cent of married women are in polygynous unions (ibid. p. xv). For most other countries the age of first marriage is in the teens.[1]

Yet more striking, are the recent trends in fertility in relation to marriage. Statistics from Ondo State suggest that there has been a dramatic reduction in both marriage and fertility for the cohorts 15–19 and 20–4. Of women now 15–19, 88.6 per cent are not married and 93.1 per cent have no children, and of those 20–4, 40.1 per cent are not married and 50.6 per cent have no children (ibid. 11, 23). If one uses cohort analysis to project life cycles, the peak period for marital and fertility inception will be between 25 and 30 for these youngest cohorts, whereas for the cohort now 45–9 marriage and first birth peaked in the late teens.

Can we, however, make a direct extrapolation from cohorts to directions of change, thereby implying that younger women are picking up *the same cycle* as their elders but at a later stage? From one perspective, this assumption makes sense; it is a change that seems to have occurred across the board, in urban and rural settings, in all religions, and to a more limited degree in all categories of education. And there is no question that childbearing is still highly valued, so that women may still bear the same number of children, simply at a later age.

On the other hand, the change seems great enough, and takes place in such a dynamic socio-economic context, that it seems likely to involve the entire matrix of kinship obligations and modes of economic support. Over the much longer run, as I will discuss later, the Yoruba divorce rate seems to have been rising, with the inevitable result that personal kinship networks become more complex. Classic anthropological theory has always viewed the reproduction of lineal identities and the optative creation of lateral ties as the two intrinsic features of all—and particularly African—kinship systems. What may be happening in modern Africa is a slow shift in the whole set of intergenerational and network links into new modes of marriage and childbearing. Marital processes are longer; there is less stability to some marriages. And more emphasis is placed on the value of children as links in lateral networks in addition to their value as resources in intergenerational relations: as labour in their youth and as devotees of the ancestors in old age.

A final alternative is that both old and new modes—if by that we mean central and normatively framed tendencies—are currently in question, and that the styles of life people are pursuing are in some sense 'on the cusp' of major transformative change, rather than a shift from lineal to lateral emphasis. As I will suggest below, the one critical aspect of current change that may fall into this category of major change is men's desire to recognize all their children. Old modes of intergenerational commitment and the value of a substantial following strongly support men's recognition of all their children. Africa has little developed in its cultural repertoire the idea of the utterly unknown and absent father familiar from the history of European illegitimacy. But under present economic circumstances support may attenuate to recognition alone, recognition may attenuate still further into selectivity, and selectivity into neglect, at least for some offspring. In this scenario, the lateral strategies of women may ultimately turn out to be illusory hopes based on the projected persistence of old patterns into a more radically new age than they imagined.

One needs, then, all three approaches: projections from cohort patterns, explorations of systemic logics, and entertainment of the possibility that current patterns have specific areas of indeterminacy and fundamental transformative change.

The Problems of Studying Change in African Marriage

Modern change in African institutions of kinship and marriage takes place in an unusual social context, one in which the authority of the church, the state, and the professions over the legal implications and the moral culture of family relationships is limited and inconclusive. The implications of sexual culture for marriage, marriage for fertility, fertility for legitimacy, and legitimacy for inheritance that were forged so punctiliously, consistently, and even brutally in Europe are unpredictable and negotiable in much of present-day Africa. By comparison with the involuted refinements of Western family law, mediated by specialists in everything from trusts and taxes to emotional compatibility, African marriage seems informal in its contractual specifications and indeterminate in its long-term implications. A series of new family codes has been put into effect during the post-Independence period, but implementation is left largely to the customary courts. For Nigeria, with its 250 different ethnic groups, two world religions, multiple denominations and independent churches, and many indigenous religions, the law for most citizens is exceptionally simple: 'As a general rule nothing contained in the statutes is to deprive any person of the benefit of customary law. Matters solely between Nigerians are to be decided under customary law' (Knapp 1979: N–60). The simplicity of local adjudication means very substantial indeterminacy at any aggregated level; although the statistics are regional and national, the institution of 'marriage' to which they refer is not.

Accordingly, there are three related problems in the study of change in marriage practice: the conceptual problem of deciding which unions fall under the heading of 'nuptiality', the logistical problem of having no long-term statistical series collected by state agencies to work from, and the interpretative problem of sensing that marriage may be more informal than in the past without having any clear 'data' or chronology to support such a conclusion. The fact of change is fairly obvious to observers and participants alike, but whether change has any particular direction is much more difficult to assess.

The potential historical sources to mobilize include works by anthropologists. These have to be used with great attention to detail because at a certain point in anthropological history there was a tendency to define typical or normative marital careers for each society, such that whole societies were described as 'bridewealth' or 'brideservice', 'patrilocal' or avunculocal'. The ethnography itself, however, often documented a much more eclectic set of unions and careers. African 'marriage' seems always to have taken varied forms. Before interpreting modern changes as breakdown or informalization it is worth considering the twin possibilities that (1) the anthropology of the 1920s to 1940s tended to homogenize social patterns, and/or (2) for a few decades marriage patterns actually were more homogeneous, thanks to colonial and mission diligence, but that current practices involve new recourse to a much older and more varied institutional repertoire.

The historical ethnography shows that pre-colonial Dahomey had at least thirteen named transactions which would fall under an anthropological definition of marriage (Bohannan, 1949: 273). A Beti ethnographer lists fourteen different kinds of transaction referred to as *aluk* or marriage, according to the form of the marriage, the social origin of the payments and the method of acquiring the woman (Ngoa, 1968: 195). Mary Douglas's (1963) classic ethnography of the Lele includes polyandry as well as polygyny, coexisting in the same society. The Jos Plateau peoples have practised highly formalized first marriage, followed by a series of unions referred to in the literature as secondary marriages. Comaroff and Roberts (1981) refer to the unfolding nature of Tswana unions as 'marriage as process', an idea which took firm hold not because the Tswana were idiosyncratic in this regard, but because of a theoretical shift in anthropology towards the study of process in general.

Looking back, the conceptual normalization of marriage in the comparative anthropology of the 1920s to 1940s seems related not only to the theory of that era, but to a dilemma of government. Many peoples lacked a criterion that could be unambiguously translated into colonial legal criteria for differentiating marriage from transactions involving slaves and debt pawns, both of which transferred rights in people against other items of value. Christian missions also worked towards narrowing down the range of transactions definable as marriage. In many areas they were militant against the very common practices of child betrothal and widow inheritance where consent of the parties was not a precondition of the union. By defining certain indigenous practices as indefensible, then, colonial governments and missions effectively narrowed down the acceptable forms of marriage, generally to some kind of transaction of goods between two families. At the same historical moment that governments and missions were paring down the range of unions they would recognize, anthropology excised the non-consensual and materially interested concept of 'bride-price', replacing it with 'bridewealth'. (The equivalent in French—*la dot*, 'dowry'—is technically inaccurate but like 'bridewealth' carries no connotation of 'sale'.)

State law has intervened throughout the century to contain and define certain strategic relationships and not others. The attempt to normalize indigenous marriage and to bring it into a clear relationship with a legal code was expressed in an exemplary way by Meek in his work in the Igbo areas of Nigeria: 'Parties applying to be married under the Marriage Ordinance are now required to show that their marriage would not be incestuous according to native law. . . . The normal system of contracting a marriage among the Ibo is by means of brideprice' (1937: 264–5). Other practices described in ethnographies and graphically explored by novelists, such as the acquisition of female slaves and marriage by capture, were conceptually distinguished from marriage and forbidden in the interests of social reform. In some places practices such as infant betrothal, that violated all notion of adult consent, were also abolished by law.

The literature to which we have recourse to understand change in African marriage over the past fifty years therefore reflects an administrative and intellectual need to define, distinguish, normalize, and reduce variety and ambiguity to simplicity and clarity, quite apart from any actual normative consensus on the basis of 'traditional' authorities or any actual transition to monogamy under the very considerable authority of the church. The devout faith of the church in the possibility of monogamous marriage in Africa, the colonial need to work within an unambiguous legal framework, modernization theory's belief in a 'transition', and an anthropology focused on 'norms' all tended to create a particular illusion of the recent past.

From a longer-term perspective, this regression towards a norm seems more like a passing phase than a benchmark state of affairs against which the present florescence (or fragmentation) of marital forms should be measured. The separability of the 'bundles of rights' people hold in one another has been a very long-term feature of social relations in Africa, and the very partiality and changeability of official interventions may have increased rather than decreased indeterminacy within local communities. From whatever angle, historical or contemporary, it is clear that 'marriage' simply does not have the symbolic and social clarity in Africa that it does in, for example, Asia.

Implications of Changes in Marriage for the Logics of Fertility

If we are concerned ultimately about the determinants of fertility it may be more fruitful, then, to start from parenthood than from marriage, asking how the shifting conventions and expectations of the various marital unions intersect with parenthood. An influential literature along these lines explores two different logics linking parenthood and marriage. The first focuses on the lifetime logic of intergenerational material interests. Caldwell (1976) has suggested that in Africa there are obvious material advantages to high fertility because wealth still flows between the generations from younger to elder. Cain (1984) has given similar explanatory prominence to intergenerational resource flows, but with a specific focus on the life-cycle needs of women in patriarchal systems of resource control; the more tenuous their rights, the more critically important their access to male kin, especially sons. 'Marriage', by implication, legitimates intergenerational rights and obligations in the eyes of the community, thereby providing collective and public insurance for the claims of the older generation. Marriage is—in part—a means of consolidating the rights of parenthood, in the parents' lifetime interests.

Esther Goody has pointed out another logic, one particularly characteristic of what she calls 'simple differentiated states' (1978). Building on the older observation that kinship involves network-building as well as lineage reproduction, she notes that 'opportunities for occupational, political and social mobility will be substantially increased' (1978: 246) where people can cast their networks of obligation as widely as possible. Through complex fosterage careers, for example, children establish their own economic and social futures from which their parents can benefit.

There are two different possible interpretative emphases in both logics, not necessarily mutually exclusive on the ground, but nevertheless interesting to separate. The first gives prominence to the intergenerational relationship between parent and child, with parents seeking 'more advantageous forms of training and sponsorship than they themselves can provide' (1978: 246), setting up their children with the conditions for success as a way of securing their own future. The implications for marriage are similar to those of the other intergenerational, or *lineal*, interpretations of fertility.

The other—and the version I want to pursue—is that to the parental generation, and particularly to women, children can be a means of creating new ties for oneself in a *lateral* strategy of network-building, giving 'opportunities for political and social mobility'. Children mediate relationships which can hardly be set up or maintained in any other way, including marriage itself. Given the comparative weakness of the legal framework for marriage, a child concretizes and expands the potential time horizon to a relationship which would otherwise be less significant. The time horizon of claims people can make through a sexual relationship alone is very short although the vistas are broad. Through marriage, claims are long but limiting; a persisting marriage in fact can be a restrictive commitment, especially for women, who can only be in one union at a time. Through a child there is a basis for claims over a fairly long time period along with some flexibility in the numbers of unions one can envisage.

The logic of a lateral strategy would seem, in the abstract, to differ in key ways from the lineal strategy. The critical difference seems to lie in the time frame over which people expect to make claims on each other. The lineal strategy involves predictions over very long periods, placing confidence in—and by marrying in a formal fashion, contributing to—the persistence of whatever legal or jural framework supports interpersonal claims. The lateral strategy works in a shorter time frame, with a greater premium on breadth and flexibility; children are the main means of giving some solidity to old relationships without foreclosing new ones. Each marriage, or union, becomes a stage of a social and reproductive career in which the children establish their parents' widening affinal networks.

There is a certain simple common sense to the lateral logic under present African conditions, especially for women. Only in the very upper reaches of the social scale can a married woman claim property from her deceased husband's estate if she has no children by him. Such is the inconclusive mix of a state law promoting nuclear family relationship and customary law sustaining the rights of broader kin categories and descent groups. But with a child, a woman may be able to make considerable claims even on a fairly casual relationship, and may manage to turn a good prospect into a longer-term lineal strategy through marriage. In Nigeria, stories and rumours circulate about fantastic amounts of money given by a particular eminent politician to women claiming to be pregnant as a result of passing liaisons with him at celebrations. Whether the stories bear any factual truth or not, in cultural terms they only make sense if we understand the child as the key to stabilizing an otherwise fleeting relationship. Without the child there is no basis to claim any longer-term assets.

Even in very modest circumstances in the rural areas of south-western Nigeria, a man does not expect a woman to stay with him unless she bears his child. Basically, she cannot afford to do so because her long-term claims on him simply as his wife are minimal. No woman will stay in a childless relationship unless she has given up conceiving. Once she has a child, she has a basis for claims, regardless of whether she stays married. One farmer known to me over the past twenty years has some sort of sexual or fertility problem, and has suffered the considerable grief of a long succession of 'wives', all of whom have left him regardless of any personal affection. Correspondingly, even in its earliest stages, fertility has a realizable value. A man can acquire a pregnancy caused by another, and may be eager to do so, but can only do so by paying for it.

These are simple examples of a lateral, shorter-term logic operating between the parents, in which the fact of the child growing to adulthood and supporting them in older age can almost be seen as a bonus which may or may not materialize. People clearly hope that it will but are prepared for the fact that it may not. Long before that, the parents will have benefited from the relationships the child has helped them to form between themselves and others, relationships that not only entail support for the maturation of the child itself, but result in ramifying connections through ceremonial, associational, and occupational networks. Any kind of ongoing linkage to a powerful personage may open up a whole world of possibilities. In the context of an unstable political, economic, and religious situation and without a powerful legal framework to enforce predictable resolutions, shortened time horizons in social life can shift the emphasis from lineal to lateral social ties, and revive experimentation in legal/jural relations between the sexes.

Caldwell, Orubuloye, and Caldwell have shown that sexuality for both men and women has always been expressed in networks rather than exclusive dyads, and they suggest that the present innovations are 'not in volume but in diffusion' (1991: 258). We cannot be sure, for the past, how clandestine but long term and tacitly acknowledged sexual relationships worked as socio-economic relations; but it would follow cultural logic very closely to assume that women always saw these relationships, not as an obligation to be endured within their marital lineage, but as an opportunity to be explored. Caldwell *et al.* do not suggest the motivations behind the current diffusion of sexual networks; they assume that wider networks simply result from the greater mobility of the population. But their finding is entirely consonant with my own suggestion here, that women may be building resource networks, no longer within the family and neighbourhood, but in a larger arena. The important issue that neither their study nor my own explores in sufficient empirical detail is the extent to which parenthood, as distinct from sexuality, is a purpose of these relationships, and the ways in which parenthood, as distinct from sexuality, affects women's longer-term material networks of support and resource access.

Rural Western Nigeria, 1968–1988

The study of changing interpersonal claims is one area of research where functional assumptions are essential. Each dyadic relationship is affected by shifts in others, composing a moving configuration rather than a set of static variables. The remainder of the chapter is devoted to exploring that configuration in Yoruba-speaking rural south-western Nigeria (Ibarapa District, Oyo State). I first worked in this area in 1968–9 and returned in 1988. The following suggestions stem from cases and dilemmas I came across during a study of changing production patterns in the agricultural economy.

I would suggest the following configuration of marriage and fertility. Although fertility has been very high during the oil-boom years, parents no longer expect much specific material support from sons and sons-in-law. The life-cycle rituals are differently prioritized than in the past: while marriage is less formal, naming ceremonies for babies are much more lavish and apprenticeship 'freedom' ceremonies are second only to funerals in expense. If these shifts can be seen as interrelated I suggest that they are consonant with a shift in emphasis from lineal to lateral strategies, from long-term to shorter-term logic by both men and women.

The argument I pursue is informed by Sara Berry's broader analysis of the Yoruba rural economy in the 1970s (1985). The title of her book, *Fathers Work for their Sons*, is an implicit critique of the idea that wealth still flows from junior to senior. Her statement about 'productive' assets expresses a similar direction of thinking: 'Low and uncertain returns to most forms of productive and commercial activity do not slow things down; rather they reinforce the impetus to keep moving, if only to avoid falling farther behind in the economic and political lottery of accumulating good connections' (1985: 192).

Focused as I was on changes in production, several aspects of intergenerational relationships were very striking to me in 1988. Most obviously, teenage and young adult sons no longer work on their fathers' farms, for three overlapping reasons. First of all, they did not particularly wish to in 1968, because their share in the farm's proceeds was totally subject to paternal discretion; but a young man had few personal needs in those days, except the bridewealth for a first marriage. Thirty pounds sterling, the cash component of bridewealth, accounted for over half a year's income from an average adult man's farm. The older man for whom a young man worked (not always his father but usually so) was committed to paying this sum eventually, but might take several years to do so. With the advent of better transport and a higher level of commercialization, a young man can earn quick money for himself.

More important, however, is the fact that cash bridewealth payments seem to be paid less often than in the past and fathers have given up trying to control even the first marriages of their daughters. In 1968, and still amongst some 'traditionalists' like the local bourgeoisie, the old formalities involve an elder kinsman of the groom mediating negotiations with the bride's father. But in several co-resident 'marriages' I knew of in farming villages, there had been no such mediation, and no payment made. In several cases of divorce and remarriage no bridewealth repayments were made as they used to be in the past. In 1968, the only circumstances I knew of where the former husband got nothing when his wife decided to leave him was when he voluntarily renounced his rights, either on the grounds of satisfaction with the number of children she had borne him or on the off-chance that she might change her mind. Non-payment and repayment of bridewealth is more frequent now, leaving young men with no great self-interest to be served by working for their fathers.

It is worth pausing to explore why bridewealth has gone into decline and to what extent this may affect the figures on marriage. The question 'Why?', no matter what its specific referent, does not tend to elicit illuminating answers from the people themselves. Explanations tend to be dismissive: 'We're not doing that now,' or 'We've abolished that.' While I would feel very hesitant to conclude anything definitive about bridewealth levels from the limited knowledge I have of court proceedings about divorce, this shift to greater informality would be consonant with a more general abandonment of traditional 'tribute' payments, including *ịṣákọ̀lẹ̀* ('rent' from tenants to land-owners). The latter has declined in frequency of observance and in the annual rate relative to income, in spite of increasing commercialization and some pressure on the land. Possibly the situation has been created in part by legal interventions, such that a father cannot, in fact, control his daughter's sexual and residential life, just as in the aftermath of the 1978 Land Use Decree a 'landowner' cannot easily eject a financially recalcitrant tenant. By contrast, debts incurred in explicitly contractual relationships are subject to Nigerian law. The Ọba puts pressure on people owing money to their associations or to migrant farm labourers by threatening them with the police. This is a very effective threat because the police can visit all kinds of unpredictable problems on the guilty party. People tend to be very acute about which debts can be collected, and do not worry at length about those which cannot, such as bridewealth payments and *ịṣákọ̀lẹ̀* for land. Certainly, to return to marriage, people say that a woman 'owns herself' and can come and go as she pleases. (Levels of reported bridewealth, *owó orí*, are referred to later in the paper.)

My final observation on intergenerational obligations is that work has been increasingly commercialized, meaning that fathers cannot count on their offspring to work from a sense of obligation. When asked, sons can be too busy elsewhere in the local economy or they can disappear into the urban labour market. As is well known, Yoruba women's work for kin and affines has been remunerated for a very long time. In the past, however, children used to work for their elders for no immediate return. In 1988 by contrast, the labour market for teenagers is very active. Schoolchildren hire themselves out after school, either in groups or individually. They tend to do lighter work such as weeding, planting, and fertilizer spraying. As a result I have seen even one of the most prestigious chiefs give cash payments to children from his own compound for their work on his farm, and I have heard of a

15-year-old girl bargaining for a daily wage rate with a man in a comparable position of seniority over her. Many of these teenagers and young adults are working to earn their own education costs, so their parents can hardly complain about their unavailability to work for nothing. In any case, a father has no strong sanctions to exert over an adult son. One farmer's three adult sons who were helping with farm work got tired of the conditions and went to Lagos. Asked several weeks later about where they were, what kind of work they were doing, and so on, their father did not know and was not trying to find out. In fact, many older farmers with adult sons away from home had no clear idea of what kind of work they were doing, hardly heard from them, and did not seem to spend energy worrying about it. Their adult children's activities were not a major topic of interest or conversation unless there was something particularly notable about them. Some parents do not know which city one or other of their children is currently living in. Many families of urban workers receive nothing in the way of remittances, except perhaps on a periodic visit for the festivals. The net flow of income can continue to go from senior to junior well into the middle age of the junior, involving very substantial expense and in one case I know of, the payment of a very large debt incurred by the junior in his home town and utterly neglected until heavy pressure was placed for his senior brother, a local farmer, to find the money. In a sense, the home residents take for granted that their children must eventually circulate back, but, even with a clearly intensified socio-political life that does draw many back on a regular basis, the parental generation is not necessarily expecting or demanding routine participation on the part of each and every youth.

Another index of a decline in parental access to youthful labour is the virtual disappearance of work by a son-in-law for his wife's father. In the past, a son-in-law was under profound obligations to his father-in-law. He could be asked to do more or less anything, even at short notice. At present, a man will give ceremonial support to his wife's father, but is rarely called on for farming or housebuilding. There may be a link between the shifts in marriage formality and this apparent attenuation of obligations between affines. To fill out this question one would need information on, for example, a divorced husband's relationship to his former wife's father, the maternal grandfather of at least some of his children. Certainly for the children the relationship with their grandfather is very important, but whether it continues to implicate their father, after divorce, remains a question to be explored. There are always ripples of ramifications which may themselves feed back as a direct influence: why should a father try to exert much control over a daughter's marriage if the life-time services of a son-in-law are uncertain or even inaccessible? In other words, it is not just intergenerational flows between parents and children that conditions changes in marriage; flows along kinship and affinal lines also affect relations between parents and children.

The picture of intergenerational relations presented here is entirely consonant with the fact that old people never stop making a living. In a sample of farmers studied in 1968 and re-studied in 1988, the farms of men now in their eighties were only marginally smaller than the farms of the same men in their sixties. For the men a decade younger, their farms were actually larger, in their seventies, than they had been in their fifties. The only people who don't farm are those too ill to plan, supervise, or carry out the work. The elderly also do much domestic work without recourse to the younger generation. Many an elderly Yoruba widower remarries an old woman 'for food', to cook for him rather than expecting his co-resident sons and daughters-in-law to support his daily needs.

Two major qualifications to this picture come to mind. There is still a very powerful expectation that children will make a funeral for their parents about one year after the death. Most people fulfil the obligation, partly through filial goodwill, partly for their own reputations, but partly, also, because there are public sanctions of criticism and attribution of misfortune to the neglect of this duty.

The second qualification concerns women. A woman is still expected to nurse her parents through their last illness; this is not the work of a daughter-in-law if a daughter is available. Women in their mid-years, bringing up their own children and generally running their own farming or trading enterprise, will be asked to move to their parents' house to take care of them. Marriage typically takes daughters away, but their help, even beyond moments of crisis, is by no means lost.

All this is circumstantial evidence of a loosening of intergenerational expectations of substantial routine support. Proximate generations of adults simply do not need each other as they used to, and they have few sanctions to exert even if they did. This is not to imply that fundamental notions of lineal identity which determine such critical issues as eligibility for chieftaincy and access to land have been altered. Membership in one's father's compound continued to be considered an absolute entitlement depending only on recognition, not on the 'legitimate' marriage of one's parents and not on one's own comportment. Neither am I suggesting that young people now neglect the older generation. Many of the older generation stay extraordinarily robust to very advanced ages, and certainly their children take care of their needs in sickness if they can. Members of the younger generation also still make the gestures of deference to their elders. It is rather the expectation of regular intergenerational transfers of tangible resources between adults which is at stake here: that range of transactions lying between the transmission of life-time identity and sustenance in a momentary crisis.

While those kinds of interdependence seem to be attenuating, certain life-cycle transitions are celebrated even more than before. Before looking at this in the familial context one needs to look at celebrations more broadly because the 'party' has become a major institution in Yoruba areas. Gradations of social importance—of occasions, of families, of individuals—are all given material meaning in the context of the party. It would be hard to exaggerate how much preparation and investment goes into these occasions: major roads blocked, music brought in, generators rented for reliable electricity, expensive gifts offered to guests, food provided, varieties of meat cooked, the latest design of traditional cloth worn, whole pages of announcement in the newspaper bought, and beer and schnapps consumed in legendary amounts by a crowd that necessarily encompasses not just invitees but, in principle, anyone who wants to come. Commercial banks that would balk at risky loans for productive business, loan out ceremonial money to their favoured customers with ease. On Friday there is a rush to withdraw cash in the right denominations for making donations to celebrants, dancers, drummers, and so on during the weekend, and on Monday it comes back again in other hands. Guests may travel very long distances to attend; relatives of the celebrants may travel far to purchase the right ingredients for the feast. The network-building and political work done at these gatherings is explicit and byzantine in its complexity. At the state level, major political alliances and status hierarchies are established and acted out. Anyone with political ambition has to participate when the celebrant is a power-broker of some kind.

The roster of occasions generating a party appears to be expanding from those embedded in the kinship nexus (birth, marriage, and death) and in traditional urban administration (chieftaincy) to include more achievements of a personal and professional sort (freedom from apprenticeship, book publication, another honorary chieftaincy title). Not only are there now more achievement-oriented ceremonies, but they seem relatively more expensive than in the past.

Except for the élite and the adherents of more conventional churches, marriage seldom provokes this scale of celebration or publicity in the newspapers. Announcements in the media are dominated by anniversaries: of death, accession to office, or headship of a company. In the rural community where I worked then, a naming ceremony for an infant, in which a man implicitly acknowledges paternity, is more important than it used to be, while a marriage is less so. A naming ceremony is a day-long affair, with food and drink for all guests, drummers and dancing in the evening. Guests are expected to make contributions to both parents, although people told me that with the downturn in the economy less is donated than in the recent past. Still, standards for the provision of food and drink do not seem to have declined. The child's parents are the immediate actors, with the older generation of grandparents doing the actual ritual of naming the child and receiving guests but not mobilizing their whole social network or spending the major amount of money.

A naming ceremony for a child offers both male and female parents in the young and middle years an opportunity to mobilize their kinship, affinal, occupational, and associational networks: to celebrate, invite guests, incur obligations, and repay others. Even an impecunious new parent is able to borrow money to finance a naming ceremony, and even a chronically unemployed guest will find a few naira to contribute. The kin, friends and associates of both parents will attend, affirming their links through the child in a manner which is hardly done any more at marriage. With chieftaincy a rare achievement limited mainly to the older generation, and the funeral of a parent being, obviously, equally rare and unpredictable, a naming ceremony for a child is the major social occasion on which the younger generation can become central participants in politico-ritual life.

The enormous growth in 'freedom' ceremonies can be thought of in the same framework. In 1968 there were hardly any apprentices or master-craftsmen in this small town. Now there is a freedom ceremony almost every weekend for graduation from training in a wide variety of skills, from electrical work to tractor driving, from motorcycle mechanics to battery charging. There seems to have been an increasing specialization in these apprenticeships, the conditions of which are controlled by district-wide associations. The costs are generally borne by the apprentice's parents and kinsmen who also host the freedom ceremony. The celebrant, however, also mobilizes his entire social network including the craft association, his social group *ẹgbẹ*, church membership and so on. Special cloth is bought, food prepared, and beer purchased. Music is often brought in for dancing which goes on most of the night. Attendance is open so the cost is substantial. Whatever interests are served they clearly include the accession of the celebrant to a position as an actor with a fairly permanent role in the local political economy, with not only mobilizable skills but mobilizable overlapping networks whose links his or her friends, associates, and kinsmen can activate for themselves.

Attendance itself—just being there—offers a stepping stone, a minimal common ground, for the creation of stronger ties between people. A friend who accompanied me to the chieftaincy installation of an eminent member of the national business élite was explicit about the possibilities opened up, even though he made no particular effort to circulate socially at the occasion itself. The importance of extending network links seems to make attendance at the kind of ceremonies that attract socially heterogeneous crowds far more important than those including more narrowly defined kin. Even in the past, marriage has tended to be a fairly private affair in rural communities; the spectacular weddings of the élite are patterned more on Western or national élite culture than on the marriage ceremonies from their own local backgrounds.

If marriage figures so little in rural social and ceremonial life cycles, the question remains of how people think, talk, and act when it comes to marriage. Although there seemed to be an expectation of mobility and informality in marital relationships, a group of farmers interviewed at a twenty-year interval had a core of more continuous relationships than I had expected, judging from the Yoruba literature on divorce (e.g. Lloyd, 1968).

Taking mobility and informality first, it is critical to note the stage at which a couple, and the larger kin group, will refer to the union as if it existed as a marriage. In present-day rural usage, a woman is referred to as *ìyàwó* (wife) by the man's kin as soon as the relationship has lasted some time and is no longer clandestine (even if it is not co-resident). The term is used automatically once she gives birth because it is the verbal expression of the claims the kin of the child's father make on the child. They will attend its naming, the senior members giving the child names of their own choice. That a woman is *ìyàwó* does not imply that she and her husband have gone through any marital ceremony or that he has paid bridewealth. These practices shed some light on why, statistically, marriage still precedes childbearing: for those for whom it logically might not, childbearing automatically confers the status of wife.

According to older women, some young women are beginning their sexual lives earlier than they used to, a trend they do not necessarily condemn but do find noteworthy. Young people say that the expanded 'party' life has offered more opportunity for forming liaisons than was available in the past. But passing relationships are quite different from regular relationships in which childbearing is an anticipated goal. Many young people in their twenties are not married in any formal sense, but expect to have children and if one miscarries or dies in infancy they try to conceive another. Several men were living with women for whom they had paid no bridewealth, and the relationships were more or less explicitly open-ended. A wife sometimes moved back home to her parents at her own wish, and it seems that she would be quite free to leave if she wanted to take up with another man. In other words, the authority that a married man had in the past over a younger wife's residence and activities seems to be the aspect of marriage that is attenuated in the present situation, but not the expectations and responsibilities surrounding fertility.

Young women themselves in the rural areas are quite keen to begin a childbearing career as soon as they have finished all training. Many daughters are sponsored by their parents for state education or for apprenticeship, and fertility usually seems to be delayed until this is over. But in an agricultural region, education has brought very few significant new occupational options which might encourage them to postpone childbearing further. In fact, from a brief interview with 222 women of all ages it seems that only 21 per cent of adult women have any education at all.

Throughout her life a Yoruba woman may leave one husband for another, either taking her small children with her or giving them to her mother to care for. It was never easy to ask questions or understand the answers relating to the reasons a woman might have to leave her husband. Men were inclined to think that it was either done on the spur of the moment or out of pure material interest, but this may reflect the fact that the final decision came suddenly from the husband's point of view. Apart from the obvious reason of infertility, many problems arise within domestic settings where the care of children of varying parentage has to be organized around potential jealousies within polygynous households. For example, a man can be responsible for a set of children with quite varying parentage: besides the children of his siblings, some will be his own children by previous wives, some his current wives' children by previous husbands, others jointly born to himself and one or more current wives, and even, these days, the child of an unmarried daughter. Or the husband's interest in another woman and his children by her may be the cause for his wife to leave, even if it takes several years for the break itself to occur. These are human situations of great complexity, giving rise to aggravations that may ultimately drive one wife to leave, or give the husband pretexts for the neglect that will make her leave. Women's interests in marital relations therefore draw structure from various underlying dynamics, not only the pure material self-interest which men tend to imply.

If she wants to, a woman can easily find a new husband until quite an advanced age, well past menopause, at which point domestic comfort and mutual affection ('for play', or 'entertainment') become important motivations. By this time, most women who remarry are widowed, rather than leaving one man for another, and one wonders whether these are simply the realizations of very long-term friendships. Certainly people indicated to me in 1968 that lifetime devotions which never resulted in 'marriage' or co-residence did exist. Caldwell *et al.*'s study (1991) gives support to the assumption that such relationships may have been frequent in the past. The discrete pursuit of non-marital relationships among older women has always passed without much comment.

When she is ready to leave her husband a wife now does so with even less formality than would have been the case twenty years ago. Many cases do still come to court, where the amount of repayment is really the central issue at stake. But once the initial marital relationship has been entered without payments, there is no reason to negotiate a formal separation. One man lost his third wife to another while I was there, but could claim nothing since she had been married to someone else before him and he himself had paid nothing at that time.

Women who leave their husbands are relatively unconcerned about the inheritance or other status implications of their own marital mobility for their children. As long as a child is recognized by its father, formal kinship status is not in question, and in any case all children have membership in their maternal as well as paternal compounds. Only in the case of succession to chieftaincy or the genealogical foundation of land claims does exactitude about parentage become the basis for exclusion.

Some women now simply give up altogether on relationships with men as a source of support. It has always been common for older women, past menopause, to take the option of returning to their home compounds or building their own houses. But in 1988 I met women who had taken this path at a younger age, housing and bringing up their own children from their own resources, sometimes helped by brothers.

In urban contexts the early informality and later independence may be gravitating towards one another so that an altogether shorter proportion of a woman's childbearing years is spent co-resident with a man. In the rural area I know, the middle stretch of a woman's marital career is more substantial. A certain stability—one is unsure whether due to inertia or contentment—does set in. For the thirty-three (out of sixty-six) male farmers I had interviewed at length in 1968 who were still full-time farmers in 1988, sixteen had the same wives—both number and individuals—in both years. A further nine had had wives die during that time, five had had wives leave them for reasons other than returning home after menopause, and three have more wives now than in 1968. For the remaining two the record is ambiguous, and indeed these data in their entirety should be treated only as indicative, since even the 'stable' husbands may have had evanescent relationships with no issue which they did not report. Older men are both reticent to enumerate their wives and children and somewhat unsure whom to report since the new relationships of old age are even more informal than those of youth. It was surprising to me, however, that almost half these men and their wives had sustained absolutely stable marriages over twenty years, six of them polygynous marriages. Apart from the sixteen, others had sustained at least one of their marriages. Of course, these are all men who were well established in their farming and marital careers in 1968 and therefore may represent the least mobile sector of the community.

These last cases suggest, in spite of the small numbers involved, that greater flexibility of choice does not necessarily produce generalized volatility. Different subsections of populations may be constructing differing trajectories of life, and a key area of enquiry would be the collective and congregational commentary that is bound to grow up when the life-cycle implications of differing patterns become discernible within communities.

Placing 1968–1988 in Longer Term Perspective

There are two interwoven themes in this brief section: first, the importance of re-reading the ethnography from the past, and not simply relying on summaries of distinctive features as providing 'benchmark' conditions for the analysis of change, and second, the reconstruction of certain themes in Yoruba change: the conditions of first marriage, the nature of divorce, and the nature of remarriage. The main conclusions are (1) that informality has long been a possibility for Yoruba first marriage, and it has been the *main* form of *re*marriage for the whole of recorded history, (2) that the nature of the formalities of first marriage was re-created under colonial conditions and poorly represented as 'bridewealth', largely for the purpose of creating legal conditions for divorce and remarriage. These conclusions leave open the possibility, then, that marital informality is perfectly familiar even to the oldest generations; if so, what needs research attention is potentially new practices of *parenthood* under new economic conditions.

Contextualizing bridewealth marriage is at the heart of re-examining marital practices. In his summary of Yoruba ethnography, Bascom refers to the various marriage payments for first marriage by the then-accepted shorthand term 'bridewealth', but he also makes very clear that there was a whole series of qualitatively different transactions that fell under that rubric (1969: 59–64). The earliest ethnographies mention quite small amounts of cash or cowrie shells changing hands (Johnson, 1921: 114; Fadipe, 1939: 68), in addition to precise numbers and kinds of kola nuts, yams, peppers, and so on.

Peel's history of Ilesha provides a historical context of the changes that resulted in rising bridewealth and greater monetization of marital transactions up to the 1950s. The basic condition was the demand of men to marry earlier, attendant on the greater economic independence of cocoa production and trade. Rates of reported seduction rose. In the first years of this century, the chiefs in charge of the colonial Native Court cut into a complex Gordian knot which included their own need for income from court fees and fines, by setting up a scale of monetary fees to be returned to a husband by a new suitor (1983: 105–6). According to the newspaper commentary in the *Lagos Standard*, it was 'a curious piece of legislation that has no parallel' (Peel, 1983: 105); in other words, it was an invention. Rates were set very high relative to current incomes, initially bore no relation to the bridewealth actually paid, and included one-fifth as court expenses. The measure was intended to be discouraging, in line with the chiefs' and elders' own interests in keeping the youth and their own wives under control, while providing themselves with a replacement income to the tributes, debt servitude, and so on that were now outlawed. Gradually the divorce settlement rates and marriage payments began to come into line, and the mean age of first marriage for men plummeted (Peel, 1983: 119). Clearly, the whole configuration of marriage and divorce payments is a recent, government-sponsored, and perhaps tenuous construction. Eades mentions even in the 1970s that 'The government publishes schedules of the maximum amount of each category of marriage prestation which is recoverable in the case of divorce' (1980: 57).

The information I have on recently paid divorce settlements indicates wide variation: in one instance, N80 for *owó orí* (bridewealth) and N120 for drinks; N120 for a woman who had already borne at least one child; N500 for a woman already pregnant, the amount to include N230 to the natural father to acquire the child as well as the mother; for a marriage three years old, no *owó orí* was paid but the husband financed his wife's apprenticeship freedom; N180 given to a first husband when marrying a divorcee. Certain compounds apparently do not ask for *owó orí* at all, just as—as mentioned above—certain landowning compounds seem fairly desultory about collecting *ìṣákọ́lẹ̀*.

In summary, the apparent indeterminacy of bridewealth payments is part of a long history that began, apparently, with a qualitative package, and developed a dominant monetary value in relation to changing divorce conditions. The overall aim was not to prevent divorce but to make it possible without unresolvable disputes.

According to the older ethnographies, divorce was rare in the past in the rural areas, and there were few institutions for dealing with it adequately. According to Johnson, writing at the end of the nineteenth century, 'Divorce is very rare . . . A woman divorced from her husband can never be married, or taken up legally by another man' (1921: 116). Yet the strength of such stated convictions simply attests to the need for interventions by the chiefs when the issue had to be faced. The same holds for the endorsement in 1937 of the formerly acceptable but not highly regarded 'marriage by consent'. Under the title of 'Other Recognized Forms of Marriage', Johnson describes one in which 'the woman's free consent, and the recognition of her by the members of a man's family, are all that is required for her to be regarded as the man's lawful wife' (1921: 116). Fadipe describes how, in 1937 when they faced an unravelling set of marital rubrics for their courts, the first Conference of Yoruba Chiefs decided on 'the popularisation of the practise hitherto largely confined to slaves and people without kin, of marriage by the mutual consent of both parties and without the preliminaries of consent of parents and the payment of bride-price' (1970: 92).

This overview history suggests that Yoruba marriage and divorce have been highly dynamic for at least a century, and that the authorities have intermittently reworked the conditions and possibilities in order to address social problems. In short, there has been marriage policy as well as marriage practice. Five specific themes need to be brought out from this brief overview of a long history. First, the present decline in bridewealth, seen in longer perspective, looks more like a reprise of old patterns than an innovation or disintegration. Second, the history of rising divorce rates is at least a hundred years long. Third, the informality of first marriage was an old but minor marital practice that was to some degree actively condoned in the twentieth century as a way of dealing with religious conversion and other *faits accomplis* of modern life. Fourth, there is the recurrent statement, for all historical periods, that actual marriage ceremonies were often 'simple, quiet affairs' (Bascom, 1969: 61, also Fadipe, 1970: 81). And finally, remarriage has never been a celebrational event in Yoruba social life. If one pieces together the historical development of all these elements, much of the present configuration can best be seen in terms of continuities with shifting emphasis, rather than sharp change or transformation. Yoruba history suggests that at key moments such moving configurations are grasped by the people or their leaders, and new possibilities are deliberated upon, opened up, and endorsed in public forums. We need to ask whether there is a comparative history and political intervention with respect to the parenthood of the children born to these changing kinds of marriage.

Conclusion

Always informal but never dominant forms of marrige in the pre-colonial period, remarriage and 'marriage by consent', took on much greater importance in this century. Divorce and remarriage to a man unrelated by descent to the previous husband became permitted in Yoruba customary law. By the 1980s practices of bridewealth payment and repayment that had been put in place as a framework for divorce had become very variable and optional. With repayment in doubt, even first marriage begins to have more informal qualities. For many couples marriage is now a far less important ceremony than the naming of their child. This means that although a woman cannot have concurrent husbands she can have concurrent recognized fathers to her children, men whose lineal identities are different from each other and whose children therefore tap into different kinship networks. In terms of resource access and daily needs for herself and her children a woman may be managing several men at once. Caldwell *et al.*'s (1991) study suggests that in a clandestine way—and with men of the same descent group—this may always have been the case. With divorce and remarriage, and then a growing informality to first marriage, the social networks can be cast more broadly now than in the past. If there is any brake to fertility in a lateral emphasis on creating networks, in contrast to a lineal emphasis on creating stable marriage, it lies in women's abilities to manage sequential relationships which are, by virtue of the children, cumulative relationships (see Karanja, Ch. 9).

As far as the historical ethnography can inform us accurately about the past, it seems that marriage has always been conceptually and organizationally distinct from parenthood in Africa (Guyer, 1989). The connection was formalized under colonial law but under fragile rubrics. Under present circumstances, marriage—in the sense of a legally recognized and exclusive relationship—may be becoming again more incidental to reproductive life than in the recent colonial past. The relative informality of first unions allows the younger generation a wider variety of options without, however, dramatically breaching the older generation's expectations about sexual relationships and the fixing of parenthood. In this long-term perspective, the present looks more like a shift of emphasis than a new departure altogether.

Reading the logics in this way focuses our attention on the need to understand how marital and parental relationships are envisaged and managed. How do children create links in kinship networks? How does a person cultivate the network, for example, to keep their child in touch with its grandparents and more distant kin on the other side? Is there a moral commentary on current marriage and parenthood from the religious authorities? How are claims being settled by the courts? Many of these questions have already been posed with respect to the old institution of polygynous fatherhood, but they are now equally important for an empirical approach to polyandrous motherhood. Present practices clearly derive from the past. But the centres of gravity may be shifting definitively enough—from lineal identity to lateral network, by women as well as men—that people are slowly creating a new configuration of kinship relations.

Notes

1. The main exception is Botswana DHS (Demographic and Health Survey, 1988), where women marry in their mid-20s; but the critical difference between the Tswana and the Yoruba is that Yoruba women's fertility is correspondingly delayed, whereas more than half of Tswana women bear children before the age of 20.

Acknowledgements

Field research in Ibarapa District of Oyo State, Nigeria, was carried out in 1968–9. It was financed by the National Institute of Mental Health, and in 1988 by the Anthropology Program, the National Science Foundation (BNS-8704188). I am grateful to these sources, and for affiliation in Nigeria with the Department of Sociology, University of Ibadan, in 1968–9, and with the Institute of African Studies of the University of Ibadan and the International Institute of Tropical Agriculture, in 1988. The author has benefited a great deal from the comments of Caroline Bledsoe.

References

Bascom, W. (1969), *The Yoruba of Southwestern Nigeria* (New York: Holt, Rinehart, & Winston).

Berry, S. S. (1985), *Fathers Work for their Sons: Accumulation, Mobility and Class Formation in an Extended Yoruba Community* (Berkeley Calif.: University of California Press).

Bohannan, L. (1949), 'Dahomean Marriage: A Revaluation', *Africa*, 19/4: 272–87.

Cain, M. (1984), *On Women's Status, Family Structure and Fertility in Developing Countries* (Washington, DC: The World Bank).

Caldwell, J. C. (1976), 'Towards a Restatement of Demographic Transition Theory', *Population and Development Review*, 2/3 and 4: 321–66.

—— and Caldwell, P. (1987), 'The Cultural Context of High Fertility in Sub-Saharan Africa', *Population and Development Review*, 13/3: 409–37.

—— Caldwell, P., and Quiggen, P. (1989), 'The Social Context of AIDS in Sub-Saharan Africa', *Population and Development Review*, 15/2: 185–234.

Caldwell, J. C., Orubuloye, I. O., and Caldwell, P. (1991), 'The Destabilization of the Traditional Yoruba Sexual System', *Population and Development Review*, 17/2: 229–62.

Comaroff, J. L., and Roberts, S. (1981), *Rules and Processes: The Cultural Logic of Dispute in an African Context* (Chicago: University of Chicago Press).

Demographic and Health Survey (1988), *Family Health Survey II, Botswana* (Columbia, Md.: DHS).

Demographic and Health Survey (1989), *Ondo State, Nigeria* (Columbia, Md.: DHS).

Douglas, M. T. (1963), *The Lele of the Kasai* (London: Oxford University Press).

Eades, J. (1980), *The Yoruba Today* (Cambridge: Cambridge University Press).

Evans-Prichard, E. E. (1951), *Kinship and Marriage among the Nuer* (Oxford: Clarendon Press).

Fadipe, N. A. (1939: repr. 1970), *The Sociology of the Yoruba* (Ibadan: Ibadan University Press).

Feyisetan, B., and Pebley, A. R. (1989), 'Premarital Sexuality in Urban Nigeria', *Studies in Family Planning*, 20/6: 34–54.

Goody, E. (1978), 'Some Theoretical and Empirical Aspects of Parenthood in West Africa', *Changing African Family Project, Monograph*, 41: 227–72.

Guyer, Jane I. (1989), Comment on S. J. Tambiah, 'Bridewealth and Dowry Revisited: The Position of Women in Sub-Saharan Africa and North India, Current Anthropology, 30/4: 428–9.

Hakansson, T. (1988), 'Bridewealth, Women and Land: Social Change among the Gusii of Kenya', *Uppsala Studies in Cultural Anthropology*, 10.

Hutchinson, Sharon (1990), 'Rising Divorce among the Nuer, 1936–1983', *Man*, NS 25: 393–411.

Johnson, S. (1899; repr. 1921), *The History of the Yorubas: From the Earliest Times to the beginning of the British Protectorate* (Lagos: CMS (Nigeria) Bookshops).

Knapp, V. (ed.) (1979), *International Encyclopedia of Comparative Law* (Tübingen: J. C. B. Mohr; and Alphen a/d Rijn: Sijthoff and Noordhoff).

Lloyd, P. (1968), 'Divorce among the Yoruba', *American Anthropologist*, 70/1: 67–81.

Meek, C. K. (1937), *Law and Authority in a Nigerian Tribe: A Study in Indirect Rule* (London: Oxford University Press).

Ngoa, H. (1968), 'Le Mariage chez les Ewondo: Étude sociologique', thesis, Sorbonne, Paris.

Peel, J. D. Y. (1983), *Ijeshas and Nigerians: The Incorporation of a Yoruba Kingdom, 1890s–1970s* (Cambridge: Cambridge University Press).

Tambiah, S. J. (1989), 'Bridewealth and Dowry Revisited: The Position of Women in Sub-Saharan Africa and North India', *Current Anthropology*, 30/4: 413–35.

Part V

Sexuality and Reproduction in Problematic Unions

12 The Impact of Marriage Change on the Risks of Exposure to Sexually Transmitted Diseases in Africa

MICHEL CARAËL

Introduction

There are now thousands of references on sexually transmitted diseases (STDs) in Africa and on the relationship between sexuality and health (Barton, 1989). Most works on these topics focus on the clinical aspects of STDs, laboratory techniques for studying them, and the difficulties in diagnosing and treating them. Moreover, many of the studies describe the incidence of STDs among patients who have come to specialized services or family planning clinics: people who are not at all representative of the general population. The data in such studies must be interpreted with great caution in any attempts to generalize more broadly. Another serious drawback to this apparent richness of materials is that most are not concerned with the behavioural or socio-cultural underpinnings of STDs, especially those associations between STDs and forms of nuptiality. Even STD studies that do address behaviour seldom touch on nuptiality. Such oversights clearly reflect a lack of knowledge among clinicians and epidemiologists about the topic of nuptiality, though it is obviously important to their work.

Different kinds of unions, with their associated obligations and expectations for partners and their families, affect the range of sexual behaviours that occur before and outside the union: elements that are important for understanding sexual behaviour overall and the prevalence of STDs. Sociological variables used most often in STD studies such as age, sex, income, education, occupation, and marital status appear to be useful for overall characterization of differences in prevalence. But they have proven to be weak correlates of STDs because they are used as substitutes or proxies for more precise risk factors that are rarely incorporated directly: age at the start of sexual activity as an indicator of the duration of exposure, number and type of sexual partners for determined durations, use of condoms, and the timeliness and efficacy of treatment.

The complexities of conjugal partnerships are crucial to our present concerns for understanding STD transmission. Unfortunately, both anthropological and demographic studies often view marital status categorically:

people may be married, divorced, widowed, or single. Few studies have considered other forms of partnership or the importance of concubines, mistresses, 'deuxième bureau' ('second office') arrangements, 'outside wives', etc. Standard survey procedures often fail to capture the complexity and diversity of unions, particularly in urban settings: a man who has a customary marriage in a rural setting, cohabits in the city with a regular partner, and sometimes stays with more casual partners will be classified simply as 'married'. The 'legal' union classification masks semi-regular or occasional sexual relations which are crucial for the epidemiology of STDs and the human immunodeficiency virus (HIV).

Whereas social correlates of STDs in Africa are rarely examined, sociocultural perceptions of STDs are even less studied. People with little schooling or low incomes rarely use what Westerners would regard as sound health methods for prevention or modern treatment once symptoms develop. They are similarly less likely to seek treatment for their sexual partners. Why they fail to seek treatment is unclear. Studies sometimes mention, often anecdotally, that STDs are considered banal or mundane sorts of illnesses and are accorded little importance (Griffith, 1963; Bennett, 1962); hence, several months may go by before someone seeks treatment for symptoms considered minor. However, the health, social, and economic consequences of STDs, knowledge about the sources of the diseases and how to treat them, and whether sexual partners are informed are topics that remain largely unexplored.

After describing the epidemic of STDs in Africa and its implications for public health, I examine the association of STDs and nuptiality, emphasizing the importance of casual sexual relations for the diffusion of STDs and HIV infection. The epidemic of HIV and other STDs in Rwanda provides a case study to illustrate the importance of commercial sexual relations and to examine the relationship of the epidemic to nuptiality.

The Epidemiology of STDs and their Sequelae in Africa

The following data shed light on the epidemiology of STDs in Africa. Prevalence levels for these diseases were obtained from among pregnant women who sought prenatal care.

For *neisseria gonorrhoeae*, the prevalence rates among pregnant women, generally from urban areas, vary from 1 to 15 per cent, with a mean of 5–10 per cent (Schulz *et al.*, 1987). For *chlamydia trachomatis* other studies have suggested that the prevalence is about 5–8 per cent, but may reach 15–20 per cent in Nairobi, Kenya, and in Libreville, Gabon, where more refined diagnostic techniques are used. The incidence rate of *chlamydia* infections, the most common cause of urethritis, could be similar or even higher than that of gonorrhoea (Meheus *et al.*, 1986). Infection from this disease is often asymptomatic among women but is asymptomatic among only about 25 per cent of men. The prevalence rates for *t. pallidum*, which is responsible for syphilis, are on the order of 3–19 per cent (Meheus *et al.*, 1986) but the techniques that have been used to calculate these rates have serious limitations and interpreting them is difficult. Syphilis is responsible for from 12 to 25 per cent of genital ulcers (Mabey, 1986). The sexually transmitted *h. ducreyi*, which is responsible for chancroid, and for 30–50 per cent of genital ulcerations, has an incidence about the same as that of syphilis (*The Lancet*, 1982). The relative frequency of genital ulcers is high due to chancroid, to venereal lymphogranulomes, and to inguinal granulomes. Other STD vectors such as *herpes simplex* virus, *trichomonas vaginalis*, and *condylomata acuminata* are equally widespread, but their prevalence is not well known.

The best-known effects of sexually transmitted diseases are their consequences for primary and secondary sterility among men and women (Retel-Laurentin, 1974; Cates *et al.*, 1985). It is estimated that secondary sterility may be two or three times more common than primary sterility (WHO, 1990). In cases of sterility among women, 40–85 per cent can be traced to pelvic inflammatory disease (PID) (Meheus *et al.*, 1986). The annual incidence of PID reaches 3–4 per 1,000 in certain regions of Africa, compared to 1 per 1,000 in other continents (Muir and Belsey, 1980). The difference can be explained by the high prevalence of *chlamydiae*, *gonoccocies*, and genital infections at the time of childbirth. Another important consequence of pelvic inflammatory disease is the risk of pregnancies that begin outside the uterus: it has been estimated that 1 out of 25 pregnancies to women with PID were ectopic, compared with 1 out of 147 among members of a control group (Meheus *et al.*, 1986).

With a prevalence of syphilis of about 10 per cent among pregnant women, 5–7 per cent of all pregnancies end either by spontaneous abortion/miscarriage in the first months, or with perinatal or infant mortality (Schultz *et al.*, 1987; Ratnam *et al.*, 1982). *Gonoccocal ophtalmia* of the newborn is 50 times higher in Africa than in industrialized countries (Galega *et al.*, 1984).

In sum, the presence of undetected and untreated STDs among pregnant women can lead to miscarriage or false pregnancy, premature onset of labour, placental infection of the foetus, premature birth, infection of the newborn at the time of childbirth, or increased risk of maternal mortality.

In Africa, STDs are classed as the sixth most important public health problem. The epidemiology of STDs in Africa, however, differs considerably from that in other regions of the world in several respects:

- the number of STD cases is generally higher in Africa. Moreover, it is more prevalent among men than among women. While this difference could imply a difference in sexual behaviour, it could stem from the fact that the symptoms are more readily recognized in men, and thus reported.
- The complications of STDs and their consequences are more common in Africa, particularly because of the lack of material and personnel resources needed to diagnose them. Adequate treatments, moreover, are often expensive and are inaccessible to most people.
- Simultaneous infections are more common, particularly among women among whom STD symptoms are generally less evident. Moreover, those

bacterial strains of gonorrhoea that are resistant to the classic treatments—20–50 per cent of gonorrhoea infections—are more common in Africa than elsewhere. The widespread practice of self-medication with antibiotics often indirectly produces complications and secondary infections because people usually stop treating themselves as soon as the symptoms diminish or disappear, although the infection is not yet cured.

• Finally, the role of prostitutes and their clients in the diffusion of STDs is much more important in Africa than in Western countries; we will return to this point later.

Urban *v.* Rural Risks of STDs

The rates of incidence (that is, the frequency of new cases during a given time period) of STDs are generally much higher in the city than in rural areas; but because rural areas lack laboratories and qualified health personnel, it is sometimes difficult to determine the importance of these diseases there (Arya *et al.*, 1977). Sexually transmitted diseases in rural settings are less common and less easily treated than in the city. Their prevalence (number of cases in a defined population at a specific point in time) can thus be higher there even though the incidence may be low. In the city the situation is often the inverse—a low prevalence accompanied by a higher incidence—because there is more contagion, but diseases are treated more readily. In Cameroon, for example, the prevalence rate of gonorrhoea among women in their fertile years was estimated at 14 per cent in Yaounde and at 22 per cent in rural areas (Nasah *et al.*, 1980). When the interchange between city and countryside is extensive, or when circular migrations often take people far from their home communities, the rural setting can itself constitute a vast reservoir of STDs (Arya *et al.*, 1980).

Like those of migration, the mechanisms of infection may be circulating and complex: young adults—men, and less commonly, women—migrate to the city to look for jobs through which they can amass money for bridewealth or other marriage expenses, then return to their rural homes to marry. In these cases, STDs are contracted in the city through casual sexual relations or relations with prostitutes and are introduced into the rural populations. In other cases, the circulation involves only rural locales: young Fulani men of northern Cameroon, for example, who travel about in order to solidify a social network, often have sexual partners among divorced or married women; but these partners come from within their own groups rather than from prostitutes in the cities (David and Voas, 1978).

STDs and Nuptiality

Patterns of STD transmission have been linked to marriage patterns. For example, multiple marriages have been linked to greater risk than marriages with one partner, whether simultaneously or in sequence. A recent study in Cameroon, Kenya, and the Sudan showed that women married more than once and women in polygamous unions were more frequently sterile than monogamous women in their first unions (Larsen, 1989). In some instances, these differences in conjugal and STD patterns may be linked to ethnic affiliation. A rural study in Uganda (Arya *et al.*, 1980) compared the adults in the Teso region where gonorrhoea was widespread and fertility was low with the Ankole region where gonorrhoea was rare and fertility was high. Teso women were more likely to have had more than one marriage than the women from Ankole (33 versus 10 per cent). But it is not clear whether marriage was the cause or the effect. The greater number of marriages among Teso women could have stemmed from a greater proportion of sterile women (20 per cent of Teso women versus 2 per cent of Ankole women), with sterility rapidly leading to remarriage.

While there has been some attention to ethnic patterns of STD transmission, urban correlates have drawn the most attention. In the study in Cameroon, Kenya, and the Sudan, the education of the woman and her husband and the husband's profession were not significantly associated with the level of sterility, in contrast to place of residence; levels were higher in the city than in rural settings, especially in the capitals. According to Larsen (1989), the higher prevalence of venereal diseases in the cities probably reflects more prostitution.

But quite apart from prostitution *per se*, the duration of urban residence appears to be an important correlate of STDs. A study in Kenya examined the socio-demographic predictors of contracting gonorrhoea by comparing men who cohabited regularly (had contacts with their partners at least once a week) to those who were married (Verhaeghen and Gemert, 1972). Among 444 male patients in an urban setting who came to a clinic for treatment of gonorrhoea, a multivariate analysis that controlled for age indicated that one of the most important correlates of gonorrhoea was a long period of residence in the city and the lack of cohabitation with a wife or regular partner. The clinic population was extremely young: 80 per cent were under 30 years of age. Marital status was not significant (40 per cent in both groups were married), but cohabitation did appear to lower the risk of contracting gonorrhoea. Only 9 per cent of those with gonorrhoea lived regularly with their wives, compared to 19 per cent of those without gonorrhoea.

The urban environment has been described as a place where behaviour is more individualistic, customary rules governing marriage are disappearing, and acculturation is intensifying especially because of schooling and the diffusion of Western models of behaviour. Yet stereotypical dichotomies of 'urbanization, individualism, and modernity' versus 'village, community, and tradition' obscure the fact that, far from losing their ethnic or cultural identities, urban residents often adapt to the demands or constraints of the new environment. Cultural practices can continue to define sexual and marital options (Parkin, 1966; and other chapters in this volume). Sexual freedom in the cities is thus not necessarily brought about by the abandonment of custom, but may represent the adaptation of customary rules to a new environment. For example, respect for a long-lasting period of post-partum sexual abstinence between spouses, which is supported in rural settings by polygamy, may induce urban husbands to contract sexual relations outside marriage with 'free' (unmarried) women. One study found that the longer the duration of the post-partum abstinence in Ibadan, Nigeria, the more common were STDs (Caldwell and Caldwell, 1983).

Although part of the rise in STDs can result from adaptations to new environments, the notion of a 'crisis', of disaster or social disintegration, can be applied to critical events that sharply intensify the spread of STDs. Wars, migration flows, and economic crises—events that rupture social structures and cause breaches of norms—are often associated with an increase in sexually transmitted diseases (Hart, 1974). The vast sexual market in certain African cities, therefore, may be less a response that is culturally determined by traditions among particular groups than the result of ruptures brought about by a series of society-wide misfortunes and subsequent efforts by individuals to cope. More generally, widespread underemployment, the build-up of rural poverty resulting in an increase of migrants to cities, and the dearth of opportunities that women face when divorced or abandoned when pregnant by occasional partners (Robertson and Berger, 1986) can also be viewed as crisis situations leading to the spread of STDs.

Lesotho illustrates an extreme situation where massive male emigration to South Africa led to the absence of a large proportion of the adult male population; this long-term imbalance has as corollaries late marriage, the decline of polygamy, an increase in marital dissolution, and frequent sexual relations outside marriage (Timaeus and Graham, 1989). At the time of a national survey based on a representative sample of 1,582 persons (Lesotho, 1989), 45 per cent of the men and 17 per cent of the women who were sexually active reported casual and commercial sexual relations during the preceding 12 months. The incidence of clinical symptoms of STDs in the 12 preceding months—vaginal/urethral discharge and pain for several days, as indicators of gonorrhoea—was around 25 per cent. So common, apparently, are sexual exchanges among married adults because of the male partner's absence, that married people or people who had been married said that they had been treated in the last 12 months for these symptoms more often than single people who were sexually active (20 per cent and 14 per cent, respectively).

STDs and Prostitution

For the epidemiology of STDs in Africa, it is the prostitutes—persons exchanging sexual relations for an immediate payment in monetary form—their clients, and the secondary contacts of their clients who run the greatest risks of infection in the absence of condom use. Little (1973) had proposed as a definition of prostitute, 'a person whose means of subsistence during a period of time depends principally on the sale of sexual services and whose relationship with a client does not go beyond that of the sexual act'. This definition of the prostitute–client relationship is not entirely satisfactory; it covers only one limited aspect of an often complex, ongoing relationship, although an important component is always the financial aspect.

The definitions of prostitution that we commonly employ obscure numerous problems in the African context. Under this rubric are grouped specific practices which do not necessarily bear the same meaning from one culture to another. And the concepts of sexuality, sexual relations, and sexual partners are themselves relative and culturally specific, as are the exchange networks, the relationships between clients and prostitutes, and the place that they occupy in the sum total of sexual exchange. The data which follow often suffer from a lack of clarity of the concepts used but they reflect, none the less, a specific aspect of the epidemiology of STDs.

Sexual relations with prostitutes are cited as a probable source of infection by 50–90 per cent of male STD patients: 50 per cent in Khartoum (Taha *et al.*, 1979), 62 per cent in Ethiopia (Plorde, 1981), 84 per cent of the students in the University of Kampala (Arya and Bennet, 1974), 90 per cent of the patients in Butare (Meheus *et al.*, 1974). Of male patients with gonoccocal gonorrhoea in Nairobi, 90 per cent declared that they had acquired their infections at the time of sexual contact with prostitutes (Plummer, 1988). By comparison, less than 30 per cent of male patients in Europe and in the United States cited prostitutes as the probable source of their infections (Turner and Morton, 1976; Potterat *et al.*, 1980).

These disparate data suggest that in Africa, relationships with prostitutes play a central role in the spread of STDs in urban areas: 30–50 per cent of prostitutes are infected with a sexually transmitted disease at any given time, with clients infecting other prostitutes and other sexual partners. Thus 56 per cent of women seeking treatment for an STD in a health centre in Nairobi said that they had been infected by their husbands (Plummer, 1988).

Prostitutes often represent an extremely mobile group; in Mombasa, Kenya, they came from forty different ethnic groups and seven countries of West Africa; 40 per cent of them had lived in the city for less than one year (Verhaeghen and Gemert, 1972). Several studies have indicated that prostitutes and their clients play a considerable role in the propagation of STDs and the spread of *n. gonorrhoeae* bacteria that are resistant to penicillin (D'Costa *et al.*, 1985; Kreiss *et al.*, 1988). In Nairobi a study among three groups of prostitutes (Kreiss *et al.*, 1986) of high, medium, and low socio-economic status showed that their numbers of clients per day were respectively 0.3, 5.0 and 5.6. The prevalence of gonorrhoea among them was respectively 16 per cent, 28 per cent, and 46 per cent. Of the prostitutes studied, 70 per cent had at least one sexually transmitted disease. The mean time for reinfection by *n. gonorrhoeae* was 12 days.

HIV/AIDS and Sexually Transmitted Diseases

Because of its transmission through sexual means, the long period of incubation before symptoms appear, and its incurability, HIV infection has drawn epidemiological publicity to prostitutes and their clients. It has quickly become clear that the STDs that prostitutes and their clients carry have probably facilitated the transmission of HIV (van de Perre *et al.*, 1988; Caraël *et al.*, 1985; Kreiss *et al.*, 1988). Seropositivity for HIV has been associated in Africa with genital ulcers, *t. pallidum*, and *c. trachomatis* (Kreiss *et al.*, 1986; Greenblatt *et al.*, 1988; and Simonsen *et al.*, 1988). A prospective study among 422 clients of prostitutes has shown that men with genital ulcers more frequently seroconverted to HIV than clients with urethritis (Cameron *et al.*, 1989). This seroconversion was independently associated with the numbers of contacts with prostitutes, with genital ulcers, and with lack of circumcision. Although men ordinarily are more effective transmitters of HIV to women than vice versa, these data confirmed a high rate of HIV transmission from women to men with active STDs. Genital ulcers clearly contribute to the infectiousness of HIV in a seropositive person of either sex and/or the susceptibility of the person who is not already infected with it.

The Case of Rwanda: HIV and Nuptiality

The HIV epidemic in Rwanda is probably older than in other regions. A retrospective study showed that already in 1982, 12 per cent of blood donors from Kigali were infected with HIV (Sontag, 1986). In 1984, 18 per cent were infected (van de Perre *et al.*, 1987*c*). In 1987 a national study (Rwandan HIV Seroprevalence Study Group, 1989) established that HIV seroprevalence was 18 per cent in urban areas for the general population and 30 per cent among people aged 26–40. In rural settings, the prevalence was 1 per cent for the general population and 3 per cent among people aged 26–40. In Burundi in 1989, a national survey (Burundi, 1990) established HIV seroprevalence in the population aged 25–44 to be 19 per cent in urban areas, 18 per cent in semi-urban areas, and 1 per cent in rural areas. The risk factors for men were contacts with prostitutes, a history of STDs, and, for rural men, visits to the city during the 5 preceding years (van de Perre *et al.*, 1987*c*).

In the national HIV seroprevalence survey in Rwanda, women in urban areas were significantly more infected than men at all ages (20.7 per cent v. 15.5 per cent; $p < .05$). In Burundi, women both in urban and rural areas were significantly more infected than men. This difference in the rates of infection by sex could be due to an HIV transmission rate that is higher from men to women and the fact that infected men have greater numbers of partners than infected women. Other possible factors include differences in exposures as a result of the age difference in infection rates (older men having sex with

TABLE 12.1. *HIV seropositivity in women age 14 and over, by marital status and rural/urban residence: Rwanda 1987*

Marital status	% Sero +	
	Rural	Urban
Single	6.3	25.4
Married	0.4	13.0
Common law union	3.3	27.4
Widowed, divorced, separated	4.5	29.5
All	2.0	21.8
No. of cases	441	1,529

younger women) and similarly differences in age and gender-specific prevalence of STD. The national survey in Rwanda also revealed that divorced, separated, or widowed people, and people living in informal unions were more likely to be infected by HIV than those living within marriage (Table 12.1).

Divorce and separation are relatively common in Rwanda, compared to several other African countries (ONAPO, 1985), and formerly married people are likely to have had several partners. Most separated and divorced people were between 20 and 30 years old, and lived alone but had had more than one sexual partner within the last year. On the matrimonial market, the low status of separated and divorced people, especially women, may incite them to initiate several relationships before finding a stable partner. Even urban women with some economic independence who are uninterested in remarrying still need sexual partners for social support. Conjugal instability itself is often attributed to the lack of faithfulness of one partner (usually the man) or both, a pattern which (if true) puts separated and divorced persons at further risk of HIV infection (Caraël *et al.*, 1988).

A cohort study in Kigali in 1988, based on a representative study of 1,469 sexually active women aged 17–45 (Lindan *et al.*, 1991), showed an even clearer difference in terms of marital status and HIV infection. While only 22 per cent of married women were seropositive, 54 per cent of widows, divorcees, and separated women were seropositive, as were 48 per cent of single people, and 35 per cent of women living in informal marriages. Similar results were shown in Sassandra, Côte d'Ivoire, where women in free unions showed a relative risk of seropositivity of 7.1 compared to married women (Rey and Guillame, 1990).

In Burundi as in Rwanda, civil, religious, and customary marriage have been declining for more than a decade although we cannot clearly measure the change because we lack precise data that separate marriages from free unions; the latter, for instance, were included in the category of

TABLE 12.2. *HIV seropositivity in women aged 17–45 by marital status, length of union, and partner's income (± 10,000 FrR): Kigali, 1988*

	% Sero +	
	Union < 6 yrs	Union > 6 yrs
Married (N = 613)		
Lower income	24	16
Higher income	35	17
Common-law union		
Lower income	38	16
Higher income	52	39

Source: adapted from Lindan *et al*., 1991.

non-legitimate unions, as were polygamous unions. Civil or customary marriage implies the payment of the bridewealth that links the families of the bride and groom and gives the patrilineage certain rights over the children. In former times, a couple who were married under customary procedures settled in a separate dwelling near the husband's parents with a parcel of land. Bridewealth allowed the family to establish an equilibrium in the marriage exchange; the wealth received for girls permitted young men to marry. The diminution of agricultural lands in Rwanda, the increase in population density, and the general poverty have considerably transformed the marriage system and have led to informal unions (Wils *et al*., 1987).

Free unions or 'common law' marriages with regular or irregular cohabitation, without payment of the bridewealth, have become widely accepted, yet they have a low status in Rwandan society, and they exert fewer constraints on sexual behaviour than do formal legal marriages. Many free unions are transformed into formal marriages with the passage of time, the birth of children, and the accumulation of a bit of financial capital, but others end in dissolution. In Rwanda in 1984, free unions represented 41 per cent of first unions, civil marriages 44 per cent, and customary marriages 15 per cent (ONAPO, 1985).

Lindan *et al*. (1991) found that women in free unions were generally more infected than married women, after age was controlled (Table 12.2). Furthermore, for a cohabitation lasting less than six years, women whose husbands/partners had higher monetary incomes were at greater risk of HIV seropositivity, compared to women with very low incomes whatever the nature of the union. Informal unions were, on average, shorter in duration than legal marriages and they were more often associated with lower income (56 v. 29 per cent; $p < .001$). Table 12.3 shows two additional factors that were associated with higher HIV seropositivity: more than one lifetime sexual partner and a relationship that is currently non-monogamous. Women in informal

TABLE 12.3. *Proportion of HIV seropositive women, by number of lifetime sexual partners and perceived types of sexual partnerships: Rwanda, 1988 (N = 992)**

	% Sero +	
	Monogamous	Non-monogamous†
One lifetime sexual partner	21 (N = 408)	36 (N = 246)
More than one lifetime sexual partner	38 (N = 85)	51 (N = 253)

* Women who were unsure whether their partners were faithful (21%) are not included in this table (HIV seropositivity was 26% in this group).

† Non-monogamous relationships include those cohabiting with non-monogamous men (76%), those cohabiting but themselves having other partners (3%), and those not cohabiting but sexually active (21%).

Source: adapted from Lindan *et al.*, 1991.

marriages declared that their unions were non-monogamous more often than married women (38 per cent v. 24 per cent; $p < .001$). They were also more likely to report more than one regular union in their lives. Both of these factors are associated with higher HIV seropositivity.

Informal unions reflect several marriage strategies. For a couple that is experiencing financial difficulty, informal unions represent a period of temporary cohabitation or a definitive union. For men with high incomes, on the other hand, they constitute temporary liaisons. These men either 'accumulate' women by issuing multiple marriage promises and by continuing to seek other partners for a marriage that is socially more advantageous, or practise a discrete form of polygyny in a society that is predominantly Christian, without taking on the obligations that are traditionally attached to polygynous marriage. For women, free unions, though far from ideal, are preferable to divorce or separation; marriage options for women are reduced after the first union, and are often responses to economic necessity. What resources women obtain from these unions allow them to become less dependent on their parents. The evolution towards a greater range of forms of marriage that result from different economic constraints has been described in other contexts (Cherlin and Chamratrithirong, 1988; Comaroff and Roberts, 1977). The increase in informal unions seems to be associated with a general rise in the age at first marriage/union. Between 1976 and 1983, the mean age of entry into union for women went from 18.6 to 22.0 years (ONAPO, 1985).

Data from the national CACP survey confirm this trend (Rwanda, PNLS/OMS 1988). Figures 12.1–12.4 show the proportion of men and women aged 15–30 by three mutually exclusive categories: no sexual experience, currently sexually active (before a regular union), and in union or having already been in a regular union. The median age at first sexual relations for women was 19–20 in rural areas and 18–19 in the city; the figures for men were 20–1 and

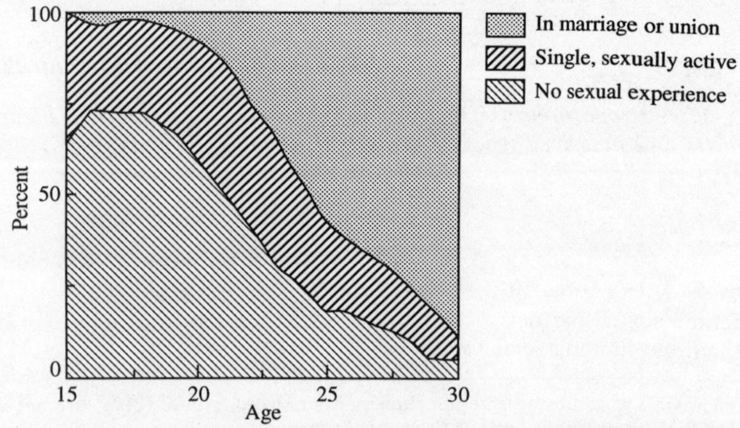

FIG. 12.1. Sexual and conjugal status of male respondents: rural Rwanda

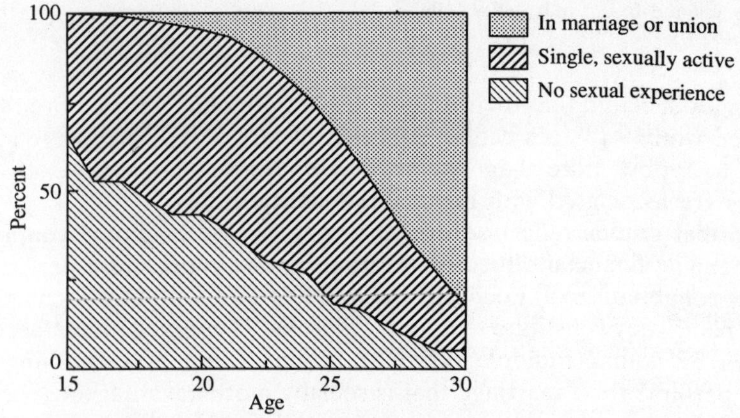

FIG. 12.2. Sexual and conjugal status of male respondents: urban Rwanda

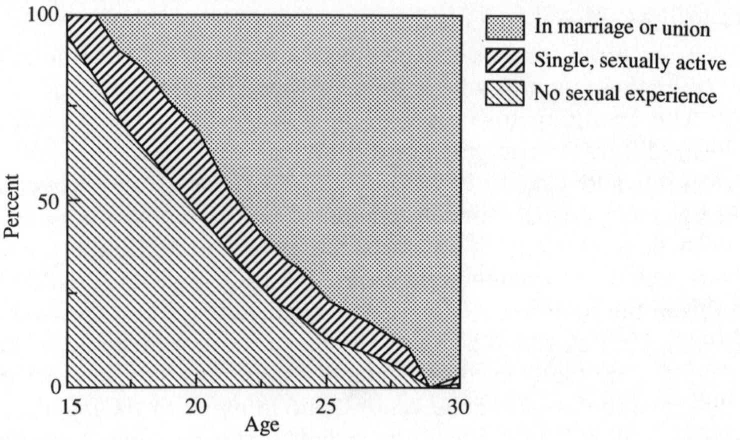

FIG. 12.3. Sexual and conjugal status of female respondents: rural Rwanda

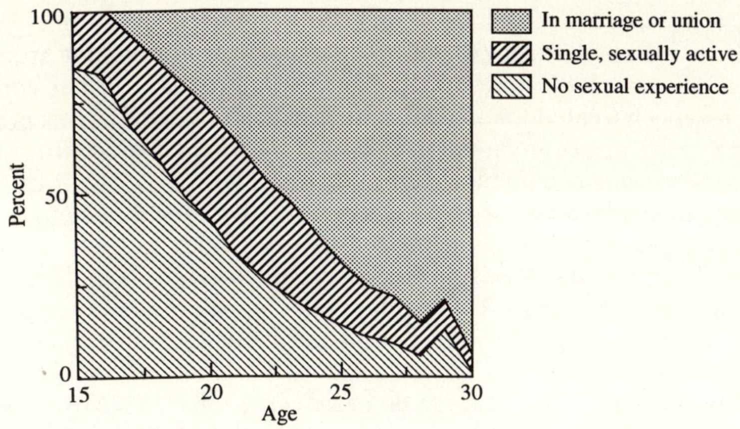

FIG. 12.4. Sexual and conjugal status of female respondents: urban Rwanda

17–18 respectively. The median age at first marriage was 21–2 years for women in rural settings and 22–3 for women in urban areas. These figures were 23–4 and 26–7 for men, respectively.

The proportion of women and men who were sexually active and had not yet had a first union was 10.3 and 16.8 per cent in rural areas, and 15.7 and 31.6 per cent in urban areas. Men in urban areas may put off marrying because of higher levels of schooling, a lack of financial resources, and greater access to sexual partners outside marriage.

The apparent increase in age at entry into the first union for women accompanied a surprising decline in the mean age at first pregnancy (from 21.3 to 19.5 years) between 1978 and 1988. In addition, pregnancies to single women represented about 10 per cent of annual births in the prefecture of Butare in 1988 (ONAPO, 1989), a fact which supports the thesis that age at which sexual activity begins has been independent of the increase in the age at entry into the first union.

In urban areas, the high proportion of single men who are sexually active—both immigrants and residents—has important consequences in terms of sexual behaviour and sexually transmitted diseases. Two groups in particular, blood donors and military personnel from Kigali, declared that they had had between nine and eleven different sexual partners during the previous year, the majority being unmarried (Caraël and Piot, 1989). A multivariate analysis indicated that the men most likely to use prostitutes were single and slightly older (25–9 years v. 20–4 years). Having a money income that was higher than average was also associated with more contact with prostitutes. These data suggest that for men who can afford these services, going to prostitutes represents an important element in men's sexual transactions.

The role of prostitutes and their clients in the dynamics of an epidemic of sexually transmitted diseases was illustrated by a retrospective analysis of the medical records of annual cohorts of the military contingent from Kigali between 1981 and 1986 (Caraël and Piot, 1989). In 1981, 20 per cent of the recruits had been treated for an STD after two years of service; this proportion increased to 60 per cent in 1986. The soldiers who enlisted after 1981 acquired similar rates of infection more quickly. Thus 60 per cent of the 1984 recruits had been treated for an STD after only two years of service and the per capita incidence of STDs increased as well. Since there is no evidence that sexual behaviours in this particular group changed in the space of five years, it was very clear that STD infection rates among the group of prostitutes near the military camp had risen because of increased infection among the soldiers.

A study in Butare, Rwanda (van de Perre *et al.*, 1985, 1987*b*) analysed risks of HIV infection among prostitutes and their clients. The prostitutes were aged 24 on average; they had a mean of two children and 82 per cent among them were single. They had practised their profession for an average of four years. They declared a monthly median of 44 partners, and only 5 per cent regularly used condoms. The results showed a positive serology for syphilis among 58 per cent of the prostitutes and 40 per cent of the clients (6 and 4 per cent, respectively, among the women and men of a control group); an immunofluorescence test indicated a positive reaction to *c. trachomatis* among 94 per cent of prostitutes and among 25 per cent of clients. In this context, an outbreak of HIV became an immediate epidemic; 88 per cent of prostitutes and 28 per cent of clients were seropositive. Among clients, the likelihood of being seropositive increased significantly with the number of contacts with the prostitutes. The median number of prostitutes that these clients saw each year was 31, compared to 3 sexual partners per year among members of the control group.

A study among 124 couples (Caraël *et al.*, 1988) of whom one or both partners were infected by HIV (Table 12.4) showed that in nearly half the cases, one of the two partners had already been in at least one union previously. Compared to 150 seronegative couples, the male partners of infected couples were more likely to have (1) experienced long periods of sexual activity before entering their first union (7.2 v. 4.6 years); (2) had contact with prostitutes (81.5 per cent v. 27.3); (3) a higher frequency of contacts (2.1 v. 1.5 per month); and (4) a history of STDs in the two preceding years (64.5 v. 15.3 per cent). Furthermore, they were more likely to have travelled frequently to the interior of Rwanda (25 v. 6 per cent with more that one week in the interior per month), events that may have created opportunities for contact with casual sexual partners.

The women in the infected couples were more often in free unions (65 v. 48 per cent). But the seropositive women, like seropositive men, were more likely to have experienced more than one regular union (31.4 versus 5.3 per cent). Compared to women in seronegative couples, these women were more

TABLE 12.4. *HIV seropositive and seronegative couples, by sociodemographic variables and risk behaviour: Kigali, Rwanda, 1986*

	Sero + couples N = 124	Sero – couples N = 150
Age median, years		
Men	32.2	31.9
Women	26.2	26.5
Age at first sexual intercourse, median, years		
Men	18.7	19.7
Women	19.4	19.0
Duration of premarital sexual activity, median, years		
Men	7.2	4.6*
Women	0.6	0.3
Union's duration, median, years	5.6	5.3
Monogamous union, %	79.8	96.7†
First union, %		
Men	73.4	82.0
Women	68.6	94.7†
Parity, median, n	2.2	2.7
History of STDs in last 2 yrs, %		
Men	64.5	15.3†
Women	47.6	11.3†
Male sexual contacts with prostitutes, %	77.9	27.3†

* $p < 0.01$;
† $p < 0.001$

Source: adapted from Caraël *et al.*, 1989.

often in polygamous unions (20 v. 3 per cent). Polygamy apparently intensified two risk factors: the man's accumulation of several sexual partners and the women's likelihood of having been in other unions. But although the women in the infected couples were more likely to report episodes of STDs (47.6 versus 11.3 per cent), the study concluded that the risk factors for HIV infection in the unions were principally those related to male sexual behaviour.

Discussion

Based on these observations made in urban areas of Rwanda and other cities of Central and East Africa, we can suggest the following generalization (Caraël,

1987): the greater the imbalance in sexual freedom between men and women, the more rapid the progress of the HIV epidemic. When men are not constrained yet the virginity of young girls and the faithfulness of married women remain norms, single men with little access to women through marriage commonly seek the services of prostitutes. Once they are married, moreover, a large percentage may continue to frequent prostitutes.

The rapid spread of HIV infection among prostitutes is intensified by their numerous sexual partners—between 500 and 1,500 clients per year—and by the high prevalence of STDs. This high prevalence in turn increases the rates at which prostitutes infect a variety of clients of different ages and social networks and results eventually in the contamination of the clients' regular female partners. These particular characteristics of its transmission may help explain why HIV infection spreads at an exponential pace, as it has in Kigali, Lusaka, and Nairobi (Caraël and Piot, 1989).

These situations of rapid evolution of HIV infection should be compared with other cultural contexts where, paradoxically, sexual liberty for *both* sexes actually limits the spread of HIV infection. Here men turn to commercial sexual relations less often, meaning that there is little concentration of numerous sexual partners around a small number of women. The fluidity of sexual relations and the diversity of possible unions—'deuxième bureau', mistresses, etc.—slows the development of both the HIV and other STD epidemics and confines them to limited networks. The HIV epidemic in Kinshasa, Zaire, which has stabilized at the level of 5–6 per cent of the adult population for the past five years, corresponds to this latter model.

The important point to stress is that urban situations themselves do not necessarily intensify transmission of STDs. Rather, inequalities in male and female sexual behaviour—particularly the double standard in urban areas where men may accumulate sexual partners as a sign of status or most notably as a transgression of traditional control on their sexuality in rural areas, while women are expected to be faithful to one partner—appear to comprise a social structure that is particularly susceptible to the AIDS epidemic. In fact, urban societies which allow women to adopt a variety of marital strategies and which facilitate, through the flexibility of forms of unions, their re-entry into regular unions, whether monogamous or polygamous, may even keep the spread of HIV from becoming an epidemic.

Whereas this study has examined the effects of forms of union on STD and HIV epidemics, the possible effect of HIV worries on conjugal behaviour is relevant as well. No specific study so far has attempted to evaluate this question. None the less, a shift toward monogamous unions and a greater stability of unions as a response to the threat of the epidemic seems unlikely in urban contexts of economic decline where conjugal strategies become elements of survival for individuals. The change that seems more likely—and, indeed, has appeared already in several surveys on sexual behaviour—is the shift towards more protective behaviour (the use of condoms) within relations that are strictly casual (Caraël *et al.*, 1992). This could reduce the incidence of HIV and STDs without eliminating them altogether. Should that happen, infection by HIV would become another item on the list of endemic tropical diseases.

References

Arya, O. P., and Bennett, F. J. (1974), 'VD Control: A Case Study of University Students in Uganda', *International Journal of Health Education*, 17/1: 53–65.

—— and Lawson, J. B. (1977), 'Sexually Transmitted Diseases in the Tropics: Epidemiological, Diagnostic, Therapeutic, and Control Aspects', *Tropical Doctor*, 7/2: 51–6.

Arya, O. P., Taber, S. R., and Nzanze, H. (1980), 'Gonorrhea and Female Infertility in Rural Uganda', *American Journal of Obstetrics and Gynecology*, 138/7: 929–32.

Barton, T. (1989), 'Sexually-Related Illness in Eastern and Central Africa: A Selected Bibliography', in N. Miller and R. Rockwell (eds.), *AIDS in Africa: The Social and Policy Impact* (Lewiston, NY: The Edwin Mellen Press), 269–91.

Burundi (1990), PLNS/OMS, Enquête nationale de séroprévalence au VIH, NAP/ WHO.

Bennett, F. J. (1962), 'The Social Determinants of Gonorrhoea in an East African Town', *East African Medical Journal*, 39/6: 332–42.

Caldwell, J. C., and Caldwell, P. (1983), 'The Demographic Evidence for the Incidence and Cause of Abnormally Low Fertility in Tropical Africa', *World Health Statistics Quarterly*, 36: 2–22.

Cameron, W., D'Costa, J. L., and Maitha, M. G. (1989), 'Female to Male Transmission of HIV-1: Risk Factors for Seroconversion in Men', *The Lancet*, 2: 403–7.

Caraël, M. (1987), 'Le SIDA en Afrique', in E. Hirsch (ed.), *Le SIDA, rumeurs et faits* (Paris: Les éditions du Cerf).

—— and Piot, P. (1989), 'Le SIDA in Afrique: Aspects épidémiologiques et sociaux', in G. Pison, É. van de Walle, and M. Sala-Diakanda (eds.), *Mortalité et société en Afrique*, Cahier 124 (Paris: PUF and INED).

—— Van de Perre, Ph., and Akygeneye, E. (1985), 'Socio-Cultural Factors in Relation to HTLV-III/LAV Transmission in Urban Areas in Central Africa', Abstract 05.1, First International Symposium on AIDS in Africa, Brussels.

—— van de Perre, Ph., and Lepage, Ph. (1988), 'HIV Transmission among Heterosexual Couples in Central Africa', *AIDS*, 2: 201–5.

—— Cleland, J., Deheneffe, J. C., and Adeokun, L. (1992), 'Research on Sexual Behaviour that transmits HIV: The GPA/WHO Collaborative Surveys, Preliminary Findings', in T. Dyson (ed.), *Sexual Behaviour and Networking: Anthropological and Socio-cultural Studies on the Transmission of HIV* (Liège: Ordina Press), 65–87.

Cates, W., Farley, T. M. M., and Rowe, P. J. (1985), 'Worldwide Patterns of Infertility: Is Africa Different?', *The Lancet*, 2: 596–8.

Cherlin, A., and Chamratrithirong, A. (1988), 'Variations in Marriage Patterns in Central Thailand', *Demography*, 25/3: 337–53.

Comaroff, J. P., and Roberts, S. (1977), 'Marriage and Extra-marital Sexuality', *Journal of African Law*, 21/1: 97–123.

David, N., and Voas, D. (1978), 'Societal Causes of Infertility and Population Decline among the Settled Fulani of North Cameroon', *Man*, 16: 644–64.

D'Costa, L. J., Plummer, F., and Bowmer, J. (1985), 'Prostitutes are a Major Reservoir of Sexually Transmitted Diseases in Nairobi, Kenya', *Sexually Transmitted Diseases*, 12/2: 64–7.

Galega, F. P., Heymann, D. L., and Nasah, B. T. (1984), 'Gonococcal Ophthalmia Neo-natorium: The Case for Prophylaxis in Tropical Africa', *Bulletin WHO*, 62: 95–8.

Greenblatt, R. M., Lukehart, S. A., and Plummer, F. A. (1988), 'Genital Ulcerations as a Risk Factor for HIV Infection', *AIDS*, 2: 47–50.

Griffith, H. B. (1963), 'Gonorrhoea and Fertility in Uganda', *The Eugenics Review*, 55/2: 103–8.

Hart, G. (1974), 'Factors Influencing Venereal Infection in a War Environment', *British Journal of Venereal Diseases*, 50: 50–68.

Kreiss, J. K., Koech, D., and Plummer, F. A. (1986), 'AIDS Virus Infection in Nairobi Prostitutes', *New England Journal of Medicine*, 314: 414–18.

—— Caraël, M., Meheus, A. (1988), 'Role of STD in Transmitting HIV', Editorial, *Genito-urinary Medicine*, 64: 1–2.

The Lancet (1982), Editorial, 'Chancroid', 2: 747–8.

Larsen, J. (1989), 'Levels and Differentials of Sterility', in R. J. Lesthaeghe (ed.), *Reproduction and Social Organization in Sub-Saharan Africa* (Berkeley, Calif.: University of California Press), 167–211.

Lesotho (1989), *KABP about AIDS*, National Survey, NAP/WHO.

Lindan, C., Allen, S., and Caraël, M. (1991), 'Knowledge, Attitudes, and Perceived Risk of AIDS among Urban Rwandan Women: Relationship to HIV Infection and Behaviour Change', *AIDS*, 5: 993–1002.

Little, K. (1973), *African Women in Towns: An Aspect of Africa's Social Revolution* (Cambridge: Cambridge University Press).

Mabey, D. C. W. (1986), 'Syphilis in Sub-Saharan Africa', *African Journal of STD*, 2/2: 61–6.

Meheus, A., De Clercq, A., and Prat, R. (1974), 'Prevalence of Gonorrhoea in a Central African Town', *British Journal of Venereal Diseases*, 50: 50–2.

—— Reniers, J., and Collet, M. (1986), 'Determinants of Infertility in Africa', *African Journal of STD*, 2/2: 31–6.

Muir, D. G., Belsey, M. A. (1980), 'Pelvic Inflammatory Disease and its Consequence in the Developing World', *American Journal of Obstetrics and Gynecology*, 138: 913–28.

Nasah, B. T., Nguematcha, R., and Eyong, M. (1980), 'Gonorrhea, Trichomonas and Candida among Gravid and Non-Gravid Women in Cameroon', *International Journal of Gynecology and Obstetrics*, 14: 48–52.

ONAPO (1985), *Enquête nationale sur la fécondité, 1983*, Rwanda.

—— (1989), *Aspects socio-culturels des grossesses non désirées des jeunes filles rwandaises de la préfecture de Butare*, Rwanda.

Parkin, D. (1966), 'Types of Urban African Marriage in Kampala', *Africa*, 36/3: 269–84.

Plorde, S. (1981), 'STD in Ethiopia: Social Factors Contributing to their Spread and Implications for Developing Countries', *British Journal of Venereal Diseases*, 57: 357–62.

Plummer, F. (1988), unpublished data.

Potterat, J. J., Phillips, L., and Rothenberg, R. B. (1980), 'Gonococcal Pelvis Inflammatory Disease: Case-Finding Observations', *American Journal of Obstetrics and Gynecology*, 138: 1101–4.

Ratnam, A. V., Din, S. N., and Hira, S, K. (1982), 'Syphilis in Pregnant Women in Zambia', *British Journal of Venereal Diseases*, 58: 355–8.

Retel-Laurentin, A. (1974), *Infécondité en Afrique noire: Maladies et conséquences sociales* (Paris: Masson).

Rey, J. L., Guillaume, A. (1990), personal communication.

Robertson, C., and Berger, I. (eds.) (1986), *Women and Class in Africa* (New York: Africana).

Rwanda (1988), PLNS/OMS, *Enquête nationale connaissances, attitudes et croyances à l'égard du Sida.*

Rwandan HIV Seroprevalence Study Group (1989), 'Nationwide Community-Based Serological Survey of HIV-1 and Other Human Retrovirus Infections in a Central African Country', *The Lancet*, 1: 941–3.

Schulz, K. F., Cates, W., and O'Mara, P. R. (1987), 'Pregnancy Loss, Infant Death and Suffering: Legacy of Syphilis and Gonorrhoea in Africa', *Genitourin Med*, 63: 320–5.

Simonsen, J. N., Cameron, W., and Gakinya, M. N. (1988), 'HIV Infection in Men with Sexually Transmitted Diseases', *New England Journal of Medicine*, 319: 274–8.

Sontag, D. (1986), unpublished data.

Taha, O. M., Ali, M. H., and Omer, E. E. (1979), 'Study of Sexually Transmitted Diseases in Patients Attending Venereal Disease Clinics in Khartoum, Sudan', *British Journal of Venereal Diseases*, 55: 313–15.

Timaeus, I., Graham, W. (1989), 'Labor Circulation, Marriage and Fertility in Southern Africa', in R. J. Lesthaeghe (ed.), *Reproduction and Social Organization in Sub-Saharan Africa* (Berkeley, Calif.: University of California Press), 365–400.

Turner, E. B., and Morton, R. S. (1976), 'Prostitution in Sheffield', *British Journal of Venereal Diseases*, 52: 197–203.

van de Perre, Ph., Clumeck, N., and Caraël, M. (1985), 'Female Prostitutes: A Risk Group for Infection with HTLV-III', *The Lancet*, 2: 524–6.

—— Caraël, M., and Nzaramba, D. (1987*a*), 'Risk Factors for HIV Seropositivity in Selected Urban-Based Rwandese Adults', *AIDS*, 1/4: 207–11.

—— Clumeck, N., and Steens, M. (1987*b*), 'Seroepidemiological Study on Sexually Transmitted Diseases and Hepatitis B in African Promiscuous Heterosexuals in Relation to HTLV-III Infection', *European Journal of Epidemiology*, 3: 14–18.

—— LePolain, B., and Caraël, M. (1987*c*), 'HIV Antibodies in a Remote Rural Area in Rwanda in Central Africa: An Analysis of Potential Risk Factors for HIV Seropositivity', *AIDS*, 1/4: 213–15.

—— de Clercq, A., and Cogniaux-Leclerc, J. A. (1988), 'HIV p 17 Antigen in Lymphocites but not Epithelial Cells from Cervicovaginal Secretions of Women Seropositive for HIV: Implications for Heterosexual Transmission of the Virus', *Genito-urinary Medicine*, 64: 30–3.

Verhaeghen, A. R., and Gemert, W. (1972), 'Social and Epidemiological Determinants of Gonorrhoea in an East African Country', *British Journal of Venereal Diseases*, 48: 277–86.

WHO (1990), unpublished data.

Wils, W., Caraël, M., and Tondeur, G. (1987), *Le Kivu montagneux: surpopulation, malnutrition et érosion des sols* (Brussels: Académie Royale des Sciences d'Outremer).

13 Marital Status and Abortion in Sub-Saharan Africa

JOHN CALDWELL AND PAT CALDWELL

Abortion is not openly disclosed in many societies, and hence statistics and information about it are often scanty and poor. Among the regions with the poorest information is sub-Saharan Africa. The reason is that fertility was the central tenet of pre-modern religion and culture (Caldwell and Caldwell, 1987). The Yoruba of Nigeria were a typical case. Barrenness was associated with wickedness because it was believed that childless women had connived with evil forces to avoid bearing children. The fact that they most frequently kept to this pact by willing the death of the foetuses within them, rather than merely failing to conceive, meant that both miscarriages and induced abortions were regarded with the deepest suspicion. In addition, inducing an abortion meant refusing to bear a child already conceived and ready to augment the members of the lineage, the proper woman's noblest duty. These sentiments have survived the conversion of most people to either Islam or Christianity. Indeed, Catholicism has reinforced attitudes to abortion. Such attitudes have not prevented abortion, but they have made it difficult to study.

This chapter discusses abortion among women who report themselves as single as well as married, partly because the incidence of abortion is higher among the former and partly because its practice is not so carefully hidden. In addition, the abortion practices of the unmarried may also condition the women involved to be more receptive to the idea of abortion when they do marry. We argue that despite its minimal impact on overall fertility, abortion can be an important means for individual women to achieve their wider life and marital goals.

There are problems in defining, and hence studying, marital status in Africa, as there are in the contemporary West. For many purposes there is an advantage in both societies in defining as marriage any kind of reasonably stable conjugal union, as we did in the Nigerian Segment of the Changing African Family Project (Okediji *et al.*, 1976). However, this solution is inadequate when studying abortion because different forms of marriage give rise to different degrees of guaranteed stability which in turn determine the likelihood of abortion. Marriages which have the sanction of neither the full customary, Christian, or Islamic rites, nor registration under statutory law, can be viewed as binding to different degrees according to the standpoints of each partner, or their relatives, and of society. In his studies of Ghana, Bleek (1978: 113) has described marital status as a 'treacherous variable' with the descriptions of the respondents often not approximating reality. He decided that there was a group of marriages which were really not very different, particularly in their impact on abortion, from the unmarried state (Bleek, 1978: 119).

The dearth of adequate studies of abortion is not a problem unique to Africa. Most surveys suffer from massive understatement in abortion reporting. In rural Ethiopia a random response technique, which adjusted the level of response to a question with many missing answers in accordance with levels of response to other questions where the answers were known, convinced one set of investigators that their survey had identified less than one abortion in every 300 that had actually taken place (Chow *et al.*, 1979: 275). The actual numbers here are probably irrelevant and probably reflect the unsuitability of the survey technique for this kind of investigation in this kind of society. Nevertheless, the study does carry conviction that the incidence of abortion is far greater than the level reported and that it is a matter of such sensitivity that few other subjects are as likely to be underreported to the same extent. In 1962, we omitted questions on the respondents' practice of abortion in Ghana (Caldwell, 1968: 161) after preliminary field tests had shown that their inclusion would jeopardize the whole study because of fears arising as a result of government announcements that the death penalty was being imposed for the offence (although it appears that no legislation was actually enacted, cf. Dugbaza, 1984: 46–9). Ladipo (1978: 112) concluded that there was no possibility of social scientists getting adequate information in Nigeria on abortion because of social attitudes and legal sanctions. Attitudes may amount to more than aversion to discussing a sensitive topic, for women's secrets of pregnancy and birth were until recently protected by a kind of religious sanction; Rattray (1927: 55–6) reported that, until the 1920s at least, Ashanti women revealing such matters to men were thought likely to die.

The most thoroughgoing attempt to overcome these problems was made by Bleek (1978) in Ghana during his 1971–3 field work, subsequent to the announcement in 1969 of the planned death penalty for abortion but in a situation of continuing but unenforced illegality. He employed six Ghanaian nurses who interviewed solely in a health clinic, claiming that the information was needed for medical purposes and assuring complete confidentiality. These unusually intensive methods allowed him to estimate that 4 per cent of the women had ever had an abortion, a figure which his subsequent research indicated might be less than one-tenth of the real level. He later worked with a single lineage which had great confidence in him, and even there much of his data came from one informant with whom he had particularly close relations. Because of the intensity of the approach required, he was reduced to collecting data on nineteen women of whom six had certainly had abortions and four had probably aborted. Later (p. 117) he encountered another problem which might well invalidate some of his respondents' testimony as well as that found in other surveys. He was able to obtain information on the type of abortion for only half the cases, and, where he obtained that information, he judged it improbable that the method had actually induced a miscarriage on half of all occasions. Indeed, in a region where there appears to be a good deal of menstrual irregularity, a substantial proportion of all abortion attempts that were reported to have been successful may not in fact have ended a pregnancy, although this should not obscure the fact that those involved were willing to procure an abortion and did make the attempt. He employed information about abortion both from women and from their male partners. With regard to male partners, Harrell-Bond (1975: 258) reported of Sierra Leone that the men often do not know about an abortion that has taken place, and, when they do know, are less likely to report the information than are the women. In Accra, Dinan (1983) also obtained abortion and other sensitive information in the mid-1970s by concentrating intensively on a very small sample of twenty-five unmarried women aged in their twenties and in white-collar occupations, but even in these circumstances she almost certainly failed to get information on a considerable number of abortions. It should be noted that, in the first survey of the Nigerian Segment of the Changing African Family Project (henceforth referred to as CAFN1), we obtained lower levels of reported abortions among respondents than they reported among their friends (Caldwell and Caldwell, 1987: 242).

Much of the evidence in this chapter on abortion in contemporary transitional society, especially among married women, draws on Bleek's (1974, 1977, 1978) work among the Akan Kwahu in Ghana; Dinan's (1983) study in Accra, Ghana; Schuster's (1979) work in Lusaka, Zambia; Harrell-Bond's (1975) study in Freetown, Sierra Leone; as well as our own experience in the Changing African Family Project in Ibadan, Nigeria, which helps us to interpret the other studies, and Ampofo's (1971) investigation of patients in Korle Bu Hospital, Accra. These and other cited studies all include work with an anthropological approach. In addition, two surveys of the literature (Liskin, 1980; Frank, 1987) proved valuable.

Abortion in Pre-Modern Society

There is near consensus among anthropologists that means of aborting were widely known in pre-modern society—or at least persons who could provide information were readily identifiable. There is almost as much agreement that the level of practice was very low. The only recent research in a remote rural area that contradicts this is the use of the randomized response technique in Ethiopia by Chow and colleagues (1979). Still, their direct collection of data could not obtain a figure as high as one woman in a thousand having induced an abortion.

With regard to East Africa it is striking how few of the anthropologists surveyed by Molnos (1973) had learnt anything of abortion while in the field. Raum (1973) reported evidence from doctors that Chaga women in north-east Tanzania did not want indefinitely high fertility even early in the present century and cites Ittameier's work, published in German in 1923, as concluding that there were approximately equal numbers of premarital and marital abortions but without any estimate of levels. Mwambia (1973), reporting on the Meru of Central Kenya, and Kettel and Kettel (1973), on the Tuken of the western Kenya Highlands, both report a low level of traditional abortion, but only Kettel and Kettel ascribe any of the abortions to married women.

In West Africa, Gessain (1963: 36) described a significant, but non-quantified, level of abortion among single girls among the Coniagui, a matrilineal group in Guinea; but no abortion among married women. Harrell-Bond (1975: 256–7) described abortion in traditional Sierra Leone as being 'practically unknown' and 'terribly frowned upon and a serious crime'. In Kwahu, Ghana, Bleek (1978) worked in a changing society but cross-questioned the old about the situation between the Wars: 'Old people in Kwahu . . . confirm that abortion was never induced in their time' (p. 104). This is possibly an idealized picture because cocoa-growing was already revolutionizing society at that time (cf. Field, 1960) although, as we shall see, economic and general social change may be relatively unimportant factors in raising the abortion rate when compared with educational changes. Pellow (1977: 50) also reported of the Akan, 'abortifacients may be used only if the mother's life is in danger'.

Among the Nupe of Nigeria's Middle Belt, Nadel (1942: 7, 8) reported unexpectedly low birth rates and observed that 'The Bida women know how to bring about abortion, and even know the use of contraceptives—or rather of certain practices and drugs which are believed to work as contraceptives'. He did not report abortion levels or clearly relate them to low fertility. Undertaking field work in the same area forty years later, Adeniyi (1978: 448) stated: 'Personal discussions with the Sagi (chief of women) at Bida, Mokwa, Doko, Katcha and Agaie and other Nupe women confirm the widespread practice of contraception among Nupe women, but do not confirm the practice of abortion', and again (p. 449), 'All the Sagi I interviewed were of the opinion that abortion is rarely practised among Nupe even though it is easily procured from boka medicine men. It is said that abortion is unheard of after marriage, and it may be accepted that, even if common before marriage it is probably negligible in its effect on Nupe fertility.'

Pre-modern society abhorred abortion throughout the region. Nevertheless, the practice was known because some circumstances were so serious that they overrode the objections. Certainly, abortion was practised on such a small scale that it had little impact on most families and none on societal fertility levels. Nevertheless, its existence provided an armoury of abortifacients which are still known and used.

Indigenous abortions have been attempted by four methods. The first is by spells or witchcraft (Bleek, 1978: 114; Schuster, 1979: 164). The second is by means of vegetable extracts, usually from leaves or bark (Rattray, 1927: 55; Nadel, 1942: 8; Raum, 1973: 28–9; Kettel and Kettel, 1973: 415–16; Bleek, 1978: 115–16). Bleek provided a list of twenty-eight supposedly abortifacient herbs used in Kwahu, and Schuster (1979: 164) refers to the large collection of such herbs in the Botany Museum in the University of Zambia. The third method, that of massaging, squeezing, or beating the abdomen, usually by old women specialists, is reported among the Meru of central Kenya by Mwambia (1973: 60–1). We have discussed similar methods in Ghana where the borderline between contraception and abortion is not always clear. The fourth method is the introduction of a foreign body into the cervix, often a twig from a specific bush or a mid-rib from a specific leaf.

One problem is why abortifacients were widely known in the past if they were not used on any great scale. Bleek (1978: 114) suggests that herbal abortifacients were probably known for two reasons: (1) because they are identical with the medicines employed for cleaning out the uterus after a birth; (2) because they were used in witchcraft to attempt to make one's enemies abort. Such enemies, in our experience in Nigeria, are most likely to be co-wives, according to women who miscarry. The connection between sorcery and abortion receives support from Colson (1958: 156), who writes of a Plateau Tonga woman of Zambia who

> sought an older woman to ask for medicine to terminate her pregnancy. The older woman had her summoned to court on a charge of slander, as this was tantamount to saying that she had medicines for sorcery. The court and spectators agreed with her, and the young wife was made to pay her damages. Thus, abortion, at least by the use of medicines, is classified with sorcery and suffers under the same moral censure.

In much of the region, clitoridectomy was practised and this was believed to be essential before either sexual relations or a birth could take place. Indeed, a baby born of an uncircumcised mother was very ill-omened indeed. In this case, a Coniagui mother in Guinea might be as active as her pregnant daughter in arranging an abortion (Gessain, 1963: 36). Fortes (1954: 265) reported among the Ashanti (who have nubility ceremonies but do not practise clitoridectomy) in the 1930s and 1940s that 'It was, and, in rural areas, still is, regarded as a very serious crime and sin for a girl to become pregnant before her nubility ceremony.' Although he added that he had been told that abortion was not practised in such cases, his subsequent description of 'the religious horror of pregnancy before the nubility ceremony', and his attestation that, in such circumstances, 'the shame for the girl's parents is so great that, as the Ashanti say, they would rather be dead than endure it', must cast some doubt on the matter. We have heard reports of abortions occurring as a result of other unnatural sexual relations, particularly incest.

Among those peoples where births to single girls are likely to make marriage more difficult, for instance the Meru of Kenya (Mwambia, 1973: 60–1) and the Tuken of Western Kenya (Kettel and Kettel, 1973: 415–16), abortion is likely to be induced to prevent the birth of an illegitimate child. This is reported elsewhere in East Africa but in situations where it is unclear whether this impediment to marriage arises from indigenous attitudes or from Christianity and the more Westernized life in urban areas. This impediment to marriage is little reported in West Africa, where it is commonly said that the concept of illegitimacy does not exist, and where children are easily fostered among relatives (however, cf. on the Nupe of Nigeria, Katcha, 1978: 53 and our analysis below of the Nigerian Segment of the Changing African Family Project). A similar picture is drawn for Zambia by Schuster (1979: 46–7). Gessain (1963: 36) also reports abortions by single Coniagui women in Guinea in order to delay the time when they must marry and move away from the natal home, a situation which may partly arise from the fact that this is a matrilineal society.

What is clear is that abortion was rare among married women (cf. Busia, 1954: 349). Almost the only reason was a pregnancy where the conception had clearly taken place during the husband's absence, a relatively uncommon circumstance in settled agrarian societies except during periods of warfare; cf. Baker and Bird (1959: 101) on a husband's right to his wife's exclusive sexual services. Rattray (1927: 55 ff.) recorded that married Ashanti women were encouraged to abort if it were disclosed, except by their prior unforced confession to their husbands, that their pregnancies were adulterously conceived; if they were to go the full term, the women would almost certainly die in childbirth. Fortes (1954: 265) heard of such abortions twenty years later.

The reason that abortion was probably very limited in the past is that it was so strongly abhorred, which is still widely true in rural areas. In a society where high fertility and the reproduction of the lineage were the central aim, abortion was regarded as unnecessarily removing a recruit to the lineage and hence as murder. It was also recognized as very dangerous and as having a potentially suicidal element. Abortion, like contraception, was also thought to imperil the reproduction of the lineage in that it was believed to jeopardize the future fecundity of the woman. In societies where even secondary sterility was regarded with horror, the woman as well as the lineage had reasons for apprehension (Caldwell and Caldwell, 1985, 1987).

Abortion in Contemporary Society

Except on very limited medical grounds, induced abortion is illegal nearly everywhere in sub-Saharan Africa: indeed, the region includes half the countries in the world where there are legally absolutely no grounds for induction (Liskin, 1980: F-107; Frank, 1987: 191). Only Burundi and Zambia allow abortions on social grounds, the latter having enacted legislation in 1972 similar to the 1967 British abortion act (Mati, 1977: 77). In practice, the situation is much more complex, as the case of Ghana demonstrates. Abortion has long been punishable by heavy penalties but Bleek (1978: 105–7) notes that its occurrence is *rarely* reported to the police and action is hardly ever taken. Yet state-enforced customary law assumes in Ghana, as it does in much of the region, that extramarital sexual relations take place and permit fines to be imposed (in the case for which Bleek provided details, on a schoolteacher who had impregnated a pupil) on the evidence of an abortion having taken place. Harrell-Bond (1975: 258) records that no doctor in Sierra Leone has ever been charged with inducing an abortion. The liberalization of Zambia's abortion law was accompanied by controversy for years both before and after legislation with assertions that abortion was 'killing' and that it was unnecessary because 'Zambia is a huge rich country' (Schuster, 1979: 164). It has, in fact, 'been implemented on a very limited scale, at least in part due to lack of facilities and trained staff' (Mati, 1977: 75). The South African position is rendered complex by governmental and racial attitudes making medical abortion almost inaccessible to African women and by the women themselves often being guilt-ridden about abortion because of the strong stand against abortion taken by the Christian churches, Calvinist as well as Catholic.

Whatever its actual prevalence among certain sectors of the population, abortion is undoubtedly understated. Reports in surveys do not reach the level of 5 per cent anywhere (Liskin, 1980: F-142). CAFN1 recorded a level of 2.3 per cent of all females 15–59 years of age in Ibadan in 1973 and 2.5 per cent of those claiming to have had sexual relations, but intensive assessment of individuals certainly suggested a higher figure. Retrospective reporting of fertility control in Ibadan showed that abortion, while doubling in incidence between 1960 and 1973, fell as a proportion of all fertility control from 9 to 2 per cent (Caldwell and Caldwell, 1987: 234–5) because of a doubling in contraception every four years (Caldwell and Ware, 1977: 491). Our only other measures of the incidence of abortion are indirect or intensive studies of very small numbers. Chow *et al.* (1979) estimated by the randomized response technique that 35 per cent of women of one rural area of Ethiopia had resorted to abortion, in contrast to less than 0.1 per cent who were willing to state that they had done so, but it is doubtful if the experiment was meaningful.

Bleek (1978: 171) concluded that 11 out of 42 ever-pregnant women in a lineage which he investigated intensively (26 per cent) had certainly, and a further 10 per cent had probably, procured abortion. But his later assessment suggests that the real proportions, as determined by the numbers actually pregnant, may have been only half as great (i.e. 13–18 per cent). Furthermore, there remains some possibility that the lineage to which he was closest was unusually non-traditional. Levels of this type, or higher, are supported by the study of persons 14–25 years of age carried out by Nichols *et al.* (1986) in Ibadan during 1982. Of 454 female respondents, 54 per cent had been pregnant and 40 per cent had aborted (i.e. 74 per cent of all pregnancies). These results may be inflated because the sample was largely self-selected. Even so, one-third of the respondents did not supply information for this question, suggesting that many of these responses would also have been affirmative.

Because of fears of punishment, women are likely to admit attempting abortion only when emergency threatens. Hence, most abortion *information* for the region comes from hospital records (cf. Liskin, 1980: F-112–13 and F-141; and also Mati, 1977: 75; Adekunle, 1978: 107–8; Binkin *et al.*, 1984: 9; Moto, 1984/5; Njuki, 1987: 88). Except for the demonstration that abortion is not unknown and is often badly performed, little can be done with these statistics. The induced abortions are not distinguished from other miscarriages, and estimates suggest that the former might contribute anything from 20 to 50 per cent of the total. They cannot be compared with the total number of births in the hospitals because women with septic conditions come to the hospital from a much larger catchment area than do women seeking normal institutionalized childbirth. In situations where the great majority of births, and probably the majority of abortion deaths, do not take place in hospitals, there are problems about interpreting total miscarriage figures which usually amount to less than 1 per cent of the estimated births in the country. Hospitals frequently regard abortion complications cases as taking up too much of their limited resources (cf. Binkin *et al.*, 1984; Njuki, 1987: 89) and imply that the situation is worsening. Where Liskin (1980: F-141) was able to obtain hospital data over a number of years, most, but not all, hospitals revealed an increase in the number of cases, but it was far from clear whether this represented a rise in the abortion rate or greater recourse by the infected to hospitalization. Finally, in terms of our focus in this chapter, it might be noted that the data are *rarely divided by marital status*.

Levels. We had recorded that in 1962, half of urban élite males and one-third of élite females in Ghana believed abortion to be common (Caldwell, 1968: 162). Certainly, abortion is widely known. Bleek (1978: 114) described the level of knowledge as 'staggering' and reported that most of the members of the lineage he examined knew more about abortion than contraception. In CAFN1 in Ibadan in 1973, 93 per cent of respondents reported knowing of abortions performed by doctors, 92 per cent by women themselves, and 90 per cent by other persons.

Costs. Everywhere medical abortions can be expensive, ranging in Ghana in 1971 from US$6 to $48, or almost one-fifth the annual per capita income at that time. In Ibadan, Ladipo (1978: 112) claimed that the majority of medical abortions are done by persons without medical qualifications (who, if this is really the case, may have been regarded as doctors by most of the CAFN1 respondents). By the beginning of the 1990s we obtained testimony that nearly all doctors in Ibadan, and reportedly in southern Nigeria, performed abortions, usually for around US$4, a reduction of the earlier price only in terms of international currency and reflecting massive devaluation of the currency. It is significant that this is one-tenth of the price usually charged for a tubectomy. The evidence generally supports the CAFN1 picture of most partners being consulted about abortions, being supportive and often helping to meet the cost and make arrangements.

Methods. The means of abortion are everywhere surprisingly similar. Apart from doctor-induced abortion, the important method of abortion in urban Nigeria and Ghana (cf. Ampofo, 1971: 93) is proprietary pills, usually containing ergot, apiol, ferrous sulphate, and sometimes pennyroyal. These are mostly manufactured in Britain and are often described as being for 'ladies' hygiene'. They are usually obtained from either chemists or nurses. Pellow (1977: 174) reported abortionists in Accra who were not doctors but nevertheless used an 'instrument that looks like scissors'. Around Johannesburg, Longmore (1959: 136–7) cited nurses selling 'steel' (ferrous sulphate) pills, M and B tablets, and quinine, as well as women using both herbs and such modern substances as permanganate of potash, glycerine, ergot, Epsom salts, washing blue, and laxatives.

The aborters. What, then, are the socio-economic characteristics with which abortion in contemporary societies is associated? Abortion is thought to be common in towns. In CAFN1 the highest rates found were among women who had been brought up in Lagos. It is also most common among the educated as Harrell-Bond (1975: 257) has reported for Freetown, and Schuster (1979) for Lusaka. All studies report higher levels of education for those who have aborted, partly reflecting the dominant role that education itself plays in encouraging abortion (CAFN1; Bleek, 1978: 112; Binkin *et al.,* 1984: 9). In fact, most studies report a number of pregnancies caused by schoolteachers and others in the training process.

Abortions are predominantly to *childless* women because it is the fact of bearing any children which may change life courses by forcing marriage or the cessation of schooling or training. Frank (1987: 193) found 50–80 per cent of all abortions among this group; Bleek (1978: 112) found 65 per cent; and Binkin *et al*. (1984: 10) found 61 per cent. Few women had practised contraception before abortion: 12–19 per cent according to Frank (1978: 193), although CAFN1 showed that they were more likely to do so in the course of their subsequent life with 22 per cent doing so compared with 16 per cent of the total population. In terms of *religion*, two studies of Ibadan (CAFN1; Adekunle, 1978: 107–8) showed Christians more likely than Moslems to practise abortion, but the differential largely, but not wholly, disappeared when education was controlled.

Abortion is also most common among the *young*. Sexual relations begin relatively early, as Dinan (1983: 436–47) reported in Accra and WFS surveys revealed in the region. WFS and DHS (two international demographic survey programmes, the World Fertility Survey and the Demographic and Health Surveys) results show the median age at first sexual relations for Ondo State, Nigeria, as 15 years, Cameroon and Côte d'Ivoire as 16 years, and Benin as 17 years (Ofosu, 1989). Analysis by age cohort in Benin and Côte d'Ivoire showed a stable situation, with some apparent tendency for the youngest cohort (controlled for truncation effects) to begin sexual activity at a slightly earlier age. Bleek (1977: 71) observed that 'Premarital sex is almost universal among the present generation of Kwahu youngsters.' The situations reported by Schuster (1979) in Zambia and Harrell-Bond (1975) in Sierra Leone are similar. Harrell-Bond wrote of a fear of contraception among schoolgirls because they thought it was sterilizing (p. 259).

Given these young ages at first sexual relations it is not surprising that all studies report a higher level of abortion among young women: a majority under 25 years (Liskin, 1980: F-142); 75–92 per cent under 25 (Frank, 1987: 193); 64 per cent under 21 (Njuki, 1987: 88); 56 per cent under 21 years and 82 per cent under 26 years with the modal ages 17–20 (CAFN1); and a median age for induced abortion of 22.1 years compared with 26.6 years for spontaneous abortions (Binkin *et al.*, 1984: 9). Pointing out that, 'In Ghana public opinion suggests that induced abortion is the most common method of fertility control', Bleek (1978: 103, 104) added: 'Abortion, together with sexual laxity and laziness, is one of the main items on the list when the elderly condemn present-day society.' He reported (p. 118) that nowadays, 'When a young woman dies, rumours immediately attribute this to abortion, more so than to witchcraft which is the usual explanation for sudden death.'

The limited data from studies such as those by Bleek (1978) and Nichols *et al.* (1986) suggest that as many as 20 per cent of the women of the towns of southern Ghana and Nigeria will have abortions, even at present levels of abortion, during a lifetime. This does not suggest that the rates are as high among married women, or that the impact on fertility levels is particularly great, as the following analysis of CAFN1 data will show. *Marital status*, when it is assessed, appears to be an important determinant of the level as well as the correlates of abortion. Abortion is clearly less common among *married* women. Only 18 per cent of abortions in the Malian study were procured by married women, in contrast to 86 per cent of spontaneous abortions (Binkin *et al.*, 1984: 9) and 8 per cent in a 1963–70 study of University College Hospital, Ibadan (Adekunle, 1978: 107–8). Dinan (1983: 347–8) concluded that a majority of the twenty-five single young women in white-collar occupations she studied in Accra had procured abortions, totalling 32 abortions in all, but this was a very select group whose single marital status in many cases was explained by abortion, for this had been the mechanism which had allowed them to finish their education rather than marrying and so secure their present jobs and afford residence in the studied area.

For married women, abortion appears to be used extensively for spacing children. Liskin (1980: F-143) reports this practice among women in widely different parts of Africa. But although he later associates abortion with the erosion of the post-partum abstinence period, he does not specifically relate these actions to fears for children's survival or fears that relatives and neighbours will learn of the irresponsible and indecent early resumption of sexual relations. In Lusaka, the strict period of abstinence is only three months, but this is followed by the substantial period when sexual relations are permitted only if withdrawal is used, resulting at times in abortion because it has been employed unsuccessfully (Schuster, 1979: 35). There may be similar problems among the Akan of a tension between short post-partum abstinence periods and the expectation of considerable birth intervals, for Bleek (1978: 108) cites a short interval as the cause of abortion. Because some of the other notable intensive studies have been done in societies which never emphasized long post-partum sexual abstinence (i.e. Bleek on the matrilineal Akan of Ghana) or do not now have a long period (i.e. Schuster on Lusaka), there has been less emphasis on this aspect of marital abortion than may be warranted by the total sub-Saharan African experience.

Everywhere the problems of adulterous conception and unstable marriages are important. Schuster (1979: 46–7) found in Lusaka that some men in loose unions felt their virility to have been proved once they fathered a child and took the pregnancy to show that it was time to take another woman, thus encouraging their first partners to seek to abort pregnancies. Nowhere was family limitation an important cause of marital abortion.

The reasons given for abortions by the *unmarried* are everywhere similar. Bleek (1978: 112) wrote of a country town in southern Ghana,

> The first pregnancies are often not wanted. They come too early because often the person does not want to marry yet, cannot support a family, wants to complete schooling, or simply does not yet want to be burdened with a child. The presence of the matrilineage, which could take care of the child, does not always convince the expectant mother to have the child. Men in particular are reluctant to bear responsibility for their unwanted pregnancies. They may attempt to circumvent this responsibility either by abortion or by desertion of the mother and child.

Bleek (1978: 119) explains why the lineage's willingness to accept the child does not change the situation:

> The unwillingness to bear a pregnancy to term is related to, and partly caused by, the reluctance of the present young generation to settle in an early customary union. The ambitions of the young lie outside the traditional sphere of the village where marriage and children used to provide social status. At the present time they are considered rather a hindrance to success.

In Zambia, as in Ghana, families do not reject pregnant girls although they may negotiate either for marriage or damages (Schuster, 1979: 47–8). Bleek (1978: 119) agrees that even in provincial Ghana there is in practice, if not in theory, some concept of illegitimacy and of disadvantage to the mother, while Harrell-Bond (1975: 257) reports that in Freetown premarital pregnancy might well spoil a girl's chance of a good marriage. There is evidence across the region, in fact, that men react ambivalently to their girlfriends' pregnancies, some helping to get an abortion and others opposing abortion in their pride of fatherhood although they may be unable or unwilling to offer marriage (cf. Dinan, 1983: 346–7).

Being young and single are often associated with *economic hardship* which, in turn, may increase the likelihood of abortion. There is evidence from Ghana that many mothers, far from being opposed to their daughters having sexual relations while at school, encourage such relations with males who can provide the girls with some economic support (Akuffo, 1987: 161). As a result, over one-third of the girls surveyed had dropped out of school because of pregnancy, while others had been allowed to continue by means of an abortion; the latter could not be quantified but it was thought to happen 'often'. Of seventy-five girls aged over 15 who were surveyed, seventy had boyfriends: most said that the main reason was financial need, and over half said that without boyfriends they would have nowhere to sleep (ibid.).

So important is economics that the line between married and unmarried women's motivations for abortion often become blurred when considering the potential impact of a pregnancy on how women support themselves. Ladipo (1978: 113) reported the maintenance of a career to have been a significant response in answers given by women in Ibadan's teaching hospital. In Lusaka, Schuster (1979: 20 ff.) found both new élite and career women avoiding large families because they derived prestige from their jobs rather than from the lineage. She also found 'toughened worldly-wise women' in the shanty towns and elsewhere who danced and drank with men in the bars and who would have been disadvantaged by pregnancies and infants. Both groups of women may be formally married, have more tenuous relationships, or be single, as was the case among the women studied by Dinan (1983) in Accra. Schuster (1979: 20) blamed labour laws and the ban on women's employment in towns and mining areas during the colonial period for making southern African towns and their women tougher. Certainly, such causes and effects are still found in South Africa where Longmore (1959) studied the Johannesburg townships and found abortions among married women pregnant by men other than their husbands, career women, and 'tough women about town' who drank and socialized with men. She reported:

> My informant stated that professional women were largely to blame for starting abortion practices. It was stated that European opinion frowns on pre-marital pregnancy as an unforgivable sin, so that professional unmarried women such as teachers, nurses, social workers and so on, must procure abortion, otherwise they will lose their jobs. Unmarried school teachers, for instance, never get their jobs back if they are retrenched on account of becoming pregnant while unmarried (Longmore, 1959: 137).

Abortion in Ibadan City: In-Depth Information from the Nigerian Segment of the Changing African Family Project, 1973 (CAFN1)[1]

Change in the level of abortions can only be adduced by the link between abortion and modern society in contrast to the pressures felt in former times. Accordingly, we now examine the demand for abortion in the context of a society where female age at first sexual activity has probably fallen, long post-partum sexual abstinence is under pressure, women are getting jobs in the modern sector of the economy, the nature of marriage is changing, and above all, levels of urbanization and education are rising. We focus on the effects, or implications, of marital status or relationships with men on abortion.

The 1973 CAFN1 survey was based on a 1 in 24 probability sample of Ibadan city with interviews with 6,606 Yoruba females 15–59 years of age. It yielded some statistical data on 143 females who reported having had abortions. Of these, there are in-depth investigations of 89. We found abortions most common in higher occupational groups (as did Binkin *et al.*, 1984: 9). In CAFN1 respondents who worked in professional, managerial, or white-collar jobs or who were students accounted for almost half of all those in the survey reporting abortions, compared with only one-tenth of all respondents being in these categories. Of women who had had an abortion, 4 out of 5 reported that they had experienced qualms about going ahead with the abortion, but this is not significantly different from reactions reported for the first use of contraception. Of women who had aborted, 9 out of 10 said that they had had no trouble in finding out how to do it or in meeting the cost, although cost clearly helped to determine the type of abortions, especially among schoolgirls.

Marital status. Of the eighty-nine women on whom we have intensive studies, 74 per cent procured abortions when single while 26 per cent did so while married. The proportion of all abortions occurring among married women was 25 per cent in CAFN1, where marital status was defined very broadly. There is a cohort effect here because of increasing levels of abortion and a truncation effect because some women have not yet married and most still had many years of marriage to experience. But analysis by cohort still shows that most females are twice as likely to have abortions before their first marriage as after it, while the age-specific rate among the single is many times higher than among the married. There is doubtless some understatement of abortion, especially marital abortion, in spite of our very intensive approach. Nevertheless, the relative understatement of marital abortion supports the conclusion that abortion is more greatly feared by married women and is probably less common among married than single women.

Marital status appeared to affect the method of abortion which, in turn, may affect prospects for the woman's safe recovery. By 1973 Ibadan was a city with many doctors and a range of hospitals. Half of all abortions to the single and three-quarters to the married were carried out by persons said by the respondents to be doctors (but see below), usually by surgery. The difference is accounted for by the greater resources available to the married, their greater self-confidence, and a wider network of contacts. Of the remaining married women, three used pills (discussed below), one herbal decoctions from a native doctor, and one charms and spells. Among the single females, sixteen resorted to pills, one to herbal remedies, two to codeine and APC, two to mixtures of washing blue, starch, rock salt, lemon, stout, etc., two to twigs, one to massage, and one to what she described as contraceptive jelly.

In Ibadan, like much of the rest of Africa, no dire consequences flow from early sexuality, and female sexual relations in Ibadan typically begin around 15 or 16 years of age (Caldwell and Caldwell, 1987: 239–40). But, among single women, we found ignorance, difficulties in obtaining contraception (with some of the fear arising from a confusion between contraception and abortion; ibid., 1987: 239–41), and sex forced unexpectedly by boyfriends on unprepared girls, all played roles leading to abortion. The strong desire of adolescent girls in urban areas to form close relationships with boyfriends even though they fear the possibility of both pregnancy and abortion seems to arise from a lack of close relationships within the family, particularly with fathers (ibid., 1987: 239–40; see also Schuster, 1979: 36–7, on Lusaka).

The predominance of abortions among the single is explained by the causes given by the women. *Schooling* has certainly been the main instrument of explosive change in abortion as Busia (1950: 38 ff., 60, 109) found in the Social Survey of Sekondi-Takoradi. Among the single, 60 per cent of all abortions (i.e. almost half the total abortions in the sample) were attributed to the need to finish schooling, university, or training in nursing, teaching, or other technical colleges. These are the routes into the modern world and nearly all CAFN1 respondents felt that a pregnancy carried to full term would cancel that journey. By 1973 almost half of all the women of Ibadan had some education, and almost one-fifth some secondary schooling, both proportions being much higher among the young.

There were also reasons for abortion that are not connected with education and explain abortions among those not in training. In spite of a good deal of testimony that there is no such thing as an *illegitimate* child in West Africa, fully 25 per cent of all single females securing abortions did so because of the disadvantage suffered by unmarried women in bearing a child in terms of strains with relatives or employers or because of the economic strain or possible impact on a future marriage. Some of this group were further pressured to abort because their boyfriends denied *paternity*, and hence they feared ridicule and being charged with either promiscuity or an inability to identify the father. A further 7 per cent did not want a marriage to be prematurely dictated. In contrast to the historical situation in the West, only 2 per cent of unmarried couples aborting cited the need for abortion in terms of the fact that they were not in a financial position to marry.

The situation was very different among *married* women. On the whole, abortion had little to do with the problems of raising large families. The single most important cause was attempting to maintain the expected minimum interval between births, followed by efforts to suppress evidence of adultery, and finally to prevent births which might strain (or undesirably strengthen) a weak relationship. Limiting family size was of little importance, as was limiting the damage to a woman's career, again in contrast to the situation that has obtained in the West. Pressures for early resumption of sexual relations among urbanized, and often educated and middle-class, couples in Ibadan, living away from the watchful eyes of relatives, now constitute a major form of marriage crisis. Husbands claim that urban, educated wives are more sexually attractive and some of these husbands do not want to obtain other women or do not know how to go about it. For their part the wives are afraid that their refusal to resume sexual relations may force their husbands to find other women and even other wives. At the same time both husbands and wives remain terrified that an early pregnancy will reveal to their more conservative relatives that they prematurely resumed marital sexual activity (Caldwell and Caldwell, 1987: 243–5). Thus, among married women, abortion had much to do with pressures with regard to expectations about proper forms of sexual relations and childbirth. Fully 35 per cent of abortions were obtained to prevent babies being born too close together, sometimes because of the fear of the ill-effects upon the child but commonly to hide the fact from relatives, neighbours, and friends that sexual relations had been resumed without observing a decent period of post-partum sexual abstinence. Such an expectation has long existed because of the belief, probably well-founded, that short birth intervals increase infant and child mortality; grandparents could remove children from parents who did not observe the full post-partum abstinence period on the grounds of wilful endangerment of life (cf. Caldwell and Caldwell, 1981: 85–6). The fact that the pregnancy originated in an adulterous relationship accounted for another 20 per cent of abortions, and an equal number were explained by the instability of the union.

Among married women, only 5 per cent of abortions arose from the need to keep a job, testimony both to the fact that West African employers understand high fertility and that children can be looked after by relatives and siblings, as well as to the fact that relatively few women are employees in the modern sector of the economy. Very few petty traders or market women felt that their activities necessitated the termination of pregnancy or the limitation of births, an observation also made in Accra by Ampofo (1971: 89), who concluded, however, that such women must be unusually reluctant to admit to abortion. Only 20 per cent of marital abortions were occasioned by economic problems or hardship, or for reasons of health, and only a fraction of this proportion regarded the condition as permanent in the sense that the abortion was an attempt to terminate childbearing. This situation is similar to the one regarding the use of contraceptives (cf. Caldwell and Caldwell, 1987). The reasons include the respect for high fertility, the benefits accruing from children, the economic dependency that older women often have upon their children, and the ease of fostering children among relatives.

The following more statistical information on who plays a role in abortion decisions is derived from 143 abortions detected in CAFN1. Most of the analysis is not subdivided by marital status. Still, it is significant to note that relationships to men figure prominently in women's abortion decisions. Of those women who aborted, only 8 per cent were not at the time in any kind of stable relationship. Of the remainder, over a quarter did not attempt to discuss the situation, usually because they thought the man would oppose an abortion and sometimes because they were not prepared for him to insist upon marriage. Where discussion did take place, 89 per cent of men agreed on an abortion, 4 per cent opposed it, and 7 per cent were indifferent or said that that kind of thing was a woman's business. Those women who said that their male partners opposed their abortion claimed that this opposition was almost entirely based on wanting the child and not because they felt abortions presented some danger to the life or health of the women involved. In almost 60 per cent of abortions only the couple knew what had occurred, but this is not significantly different from the situation with regard to the use of contraception. What is significantly different, especially with regard to single women, is that female friends are likely to play a larger role in the decision to adopt contraception while mothers are equally important when it comes to abortion. Mothers are also the most likely to oppose the abortion, usually on the grounds of its danger to their daughters.

If women in contemporary society see abortion as a means to a higher-status way of life or a more advantageous marital situation, one can examine the characteristics of the aborters in the CAFN1 study at the time of marriage to see who they were by this time, and, to some extent, whether the abortion succeeded in its aim. In terms of occupation, 17 per cent of those procuring abortions while single were still students or trainees; of the rest, 50 per cent had gone on to professional or white-collar employment, compared with 8 per cent of the general population. By contrast, of those who aborted while married, only 24 per cent were in professional or white-collar employment. Among the rest, those aborting were overrepresented in every occupational category and underrepresented among housewives not otherwise employed. Thus, although they rarely explained abortion in these terms, pregnancy, childbirth, or young children must have presented some inconvenience to at least some working women. Those who aborted while single were twice as likely as the general population to have come from middle-class backgrounds (perhaps only a measure of the likelihood that they would reach secondary school), while this was not true of married women who aborted. Married women who procured abortion had only a slightly greater chance than the general population of having a husband with a middle-class occupation: proof that the matter of significance in forcing the abortion was their own job rather than the socio-economic class in which they were found; while

70 per cent of the subsequent husbands of single aborters were middle class: evidence that their persistence with the route to success had worked out. Educational differentials were striking: those with at least some secondary education formed 19 per cent of the general population, 37 per cent of women aborting while married, and 77 per cent of those aborting while single. Moslems made up 55 per cent of the general population, 41 per cent of married aborters, and 30 per cent of single aborters, very largely a reflection of educational differentials. Islam in West Africa probably emphasizes marriage and motherhood more and worldly success less than Christianity, perhaps because the latter has been channelled by Western influences and is laden with Western values. What is clear is that those who aborted were not adopting a pattern of successful fertility control: by the time they were in their late thirties, Ibadan women had given birth to five children and so had those who aborted when single (their fertility pattern at younger ages does not suggest that the pattern is changing); those who aborted when married had given birth to six, evidence, perhaps, that some of them were trying to avoid excess fertility.

Likely Developments

Like other studies, the CAFN1 research suggests that people in African societies are pragmatic, and attitudes toward behaviour can change quickly as circumstances alter. The main changes in circumstance have been associated with education, urbanization, and new patterns of conjugal relationships. Among single women, abortions are important for three reasons. First, they undoubtedly still provide a very substantial majority of all abortions within the region, because the disadvantages imposed by terminated schooling or training are greater than the burdens imposed upon married women by additional pregnancies or children. Second, abortions among the single may pose more dangers because single women lack the confidence and social connections to procure abortions by safer available methods. Third, abortions to single women have provided individual and social familiarity with the practice and have undoubtedly been the single most important influence promoting marital abortions. Yet we stress that because of the instability and ambiguity of many urban marriages, the borderline between the problems leading to abortion among the single and among many of the married is by no means sharp. A woman who believes she is in a stable union or even married may find when she becomes pregnant that her partner does not take that view. Indeed, she may abort to save the existing relationship.

What is surprising is that the rise of urban abortion has been accepted very pragmatically by those involved and by much of the broader urban, educated society. In Ibadan we found that, although most women who procured abortions were frightened of the danger to their health and future fecundity, few regarded it as a moral issue or something at odds with societal norms.

Similarly, Dinan (1983: 348) reported of her sample of single women in Accra that 'Termination of pregnancy was undertaken naturally and guiltlessly by the women.' Schuster (1979: 91 ff.) said that her tough, older women of the Lusaka bars could cope with their lives and that many were clearly prepared to have abortions. Kumekpor (1970: 23) found in southern Togo a significant proportion of women who would be prepared to resort to abortion if a very short birth interval threatened and whose sole apprehension of abortion was its danger. Bleek (1978: 109–11) reported that, while 61 per cent of those he interviewed thought abortion bad on the grounds of danger, only 9 per cent said that it amounted to murder. Indeed, he concluded that most members of the lineage he studied intensively would not oppose an abortion by a reputable doctor. Harrell-Bond (1975: 257–8) reported that, among the Freetown élite, one-third of the men and one-quarter of the women were not opposed to abortion, and that, in any case, 'Most of the objections centred on the possible ill-effects of abortion on the future fertility of the woman.' Bleek (1978: 119) went further in discussing the possible legalization of abortion, and wrote: 'It is not likely that there are fundamental objections to it from the Akan world view.' Taken together, these attitudes are astonishing, given the older religious views towards the termination of foetal or infant life, and the disapproval of lineages to any restriction on the continuing supply of descendants. These changes may reflect the impact of pragmatism, urban secularism, or of Christianity and Islam. The truth is probably a mixture of all these forces.

If attitudes appear to be changing rapidly, why is abortion more widely illegal in sub-Saharan Africa than in most other parts of the world? One reason is that there is little demand for it among the rural population or the urban married, and hence political considerations still weigh against change. Politicians realize that, in spite of the matter-of-fact way that many women now seek abortion, a larger proportion of the politicians' supporters are still influenced by the pro-fertility element in African religion and culture, and especially by the associated horror in eliminating, whether by witchcraft or modern means, a life once it is conceived. These considerations have been sufficient to prevent most countries following the leadership in legislative change of either Britain or France: the societies which, because of their colonial links, have had great influence in other legal domains. Even those parents who would help their daughters have abortions while still at school might well oppose the legalization of abortion, just as many have opposed the provision of contraceptive services, on the grounds that this would legitimize non-marital sexual activity by young women. On the other hand, governments have made little attempt to take action against abortionists. The real problem is that so many abortions are dangerous, both because of a lack of doctors and because the illegality of abortion means that most doctors and hospitals do not practise it on a significant scale.

In a world region where little contraception is practised, it is possible that at least until the 1970s, abortion was as widely practised as contraception in most areas. Certainly abortion may have been more widely known in Africa than contraception. Like contraception, abortion has so far acted almost entirely to prevent rises in unwanted pregnancies as post-partum periods shorten and as schoolgirl pregnancies rise—tendencies that the availability itself of abortion and contraception may have encouraged. Yet even if we adopt what may be extravagantly high estimates of abortion (from the few very intensive, but perhaps atypical studies) that 20 per cent of urban women and 5 per cent of rural women may have on average 1.5 abortions in a lifetime, we still find that abortion has little effect on overall fertility. Marital abortion probably reduces annual fertility by no more than 1 per cent, and marital abortion aimed at limiting family size by no more than 0.1 per cent. All told, abortion may reduce African lifetime fertility by 2 per cent and annual fertility (given the age structure of society and abortion) by 3 per cent. Perhaps modern contraception as yet does no more.

Family planning programmes are likely to be a more important feature of Africa's future than of its past. Governments will increasingly establish them, partly in the hope that slower population growth will improve economic performance. There will be much more publicity encouraging fertility limitation and discouraging premarital pregnancies. In some cases the additional contraceptive services will prevent pregnancies and hence abortions. But, on balance, the antinatalist messages will almost certainly raise the abortion rate as well as the practice of contraception for decades to come. Both will help to drive the future African fertility decline.

Note

1. The Changing African Family Project was a joint programme of the Department of Sociology, University of Ibadan, Nigeria, and Department of Demography, Australian National University, Canberra, carried out during the 1970s under the direction of Francis O. Okediji and John C. Caldwell and with funding from the Population Council in twelve African countries. The Nigerian segment consisted of three major surveys, with associated in-depth research, carried out in 1973. The first of these three surveys was confined to Ibadan city and is the major source of data for this chapter.

Acknowledgements

This chapter has benefited from research by Pat Quiggin, assistance by Wendy Cosford, and word-processing by Pat Goodall.

References

Adekunle, L. V. (1978), 'Some Aspects of the Sociology and Epidemiology of Abortion', *Nigerian Behavioural Sciences Journal*, 1: 104–11.

Adeniyi, E. O. (1978), 'Population Growth and Fertility among the Nupe of Nigeria', in C. Oppong *et al.* (eds.), *Marriage, Fertility and Parenthood in West Africa*, Part 2 (Canberra: Australian National University), 435–51.

Akuffo, F. O. (1987), 'Teenage Pregnancies and School Drop-Outs', in C. Oppong (ed.), *Sex Roles, Population and Development in West Africa* (London, NH: Heinemann), 154–64.

Ampofo, D. A. (1971), 'Abortion in the Accra Korle Bu Hospital: The Demographic, Social and Medical Perspectives', in N. O. Addo (ed.), *Symposium on Implications of Population Trends for Policy Measures in West Africa* (Legon, Accra: University of Ghana), 79–105.

Baker, T., and Bird, M. (1959), 'Urbanisation and the Position of Women', *Sociological Review*, NS 7: 99–122.

Binkin, N. J., Burton, N. N., Touré, A. H., Traoré, M. L., and Rochan, R. W. (1984), 'Women Hospitalized for Abortion Complications in Mali', *International Family Planning Perspectives*, 10: 8–12.

Bleek, W. (1974), 'Perception and Practice of Birth-Control in a Ghanaian Matrilineage', in D. G. Jongmans and H. J. M. Claessen (eds.), *The Neglected Factor: Family Planning, Perception and Reaction at the Base, Van Gorcum, Assen*, 138–61.

—— (1977), 'Family Planning or Birth-Control: The Ghanaian Contradiction', *Cultures et developpement*, 9: 64–81.

—— (1978), 'Induced Abortion in a Ghanaian Family', *African Studies Review*, 21: 103–20.

Busia, K. A. (1950), *Report on a Social Survey of Sekondi-Takoradi* (London: Crown Agent for the Colonies on behalf of the Government of the Gold Coast).

—— (1954), 'Some Aspects of the Relation of Social Conditions to Human Fertility on the Gold Coast', in F. Lorimer (ed.), *Culture and Human Fertility* (Paris: UNESCO), 341–50.

Caldwell, J. C. (1968), *Population Growth and Family Change in Africa: The New Urban Élite in Ghana* (Canberra: Australian National University).

—— and Caldwell, P. (1985), 'Cultural Forces Tending to Sustain High Fertility in Tropical Africa', *World Bank PHN Technical Note*, 85–16.

—— and Ware, H. (1977), 'The Evolution of Family Planning in an African City: Ibadan, Nigeria', *Population Studies*, 31: 487–507.

Caldwell, P., and Caldwell, J. C. (1981), 'The Function of Child-Spacing in Traditional Societies and the Direction of Change', in H. J. Page and R. Lesthaeghe (eds.), *Child-Spacing in Tropical Africa: Traditions and Change* (London: Academic Press), 73–92.

—— (1987), 'Fertility Control as Innovation: A Report on In-Depth interviews in Ibadan, Nigeria', in J. A. Ebigbola and É. van de Walle (ed.), *The Cultural Roots of African Fertility Regimes: Proceedings of the Ife Conference, 25 Feb.–1 Mar. 1987* (Ile-Ife, Nigeria: Obafemi Awolowo University, and Philadelphia: University of Pennsylvania), 223–52.

Changing African Family Project: Nigerian Segment (CAFN1) (1973), Data files of the Department of Sociology, University of Ibadan, and Department of Demography, Australian National University.

Chow, L. P., Gruhn, W., and Chang, W. P. (1979), 'Feasibility of the Randomized Response Technique in Rural Ethiopia', *American Journal of Public Health*, 69: 273–6.

Colson, E. (1958), *Marriage and the Family among the Plateau Tonga of Northern Rhodesia* (Manchester: Manchester University Press).

Dinan, C. (1983), 'Sugar Daddies and Gold-Diggers: The White-Collar Single Women in Accra', in C. Oppong (ed.), *Female and Male in West Africa* (London: George Allen & Unwin), 344–66.

Dugbaza, T. (1984), 'Population Policies and Programmes in Ghana from 1966 to 1984', United Nations Regional Institute for Population Studies at the University of Ghana, Legon, Accra, mimeograph.

Field, M. (1960), *Search for Security: An Ethno-Psychiatric Study of Rural Ghana* (London: Faber & Faber).

Fortes, M. (1954), 'A Demographic Field Study in Ashanti', in F. Lorimer (ed.), *Culture and Human Fertility* (Paris: UNESCO), 253–350.

Frank, O. (1987), 'The Demand for Fertility Control in Sub-Saharan Africa', *Studies in Family Planning*, 18: 181–201.

Gessain, M. (1963), 'Coniagui Women', in D. Paulme (ed.), *Women of Tropical Africa* (London: Routledge & Kegan Paul), 17–46.

Harrell-Bond, B. E. (1975), *Modern Marriage in Sierra Leone: A Study of the Professional Group* (The Hague: Mouton).

Ittameier, C. (1923), *Die Erhaltung und Vermehrung der Eingeborenen Bevölkerung* (Hamburg: L. Friederichsen).

Katcha, A. (1978), *An Exploratory Demographic Study of the Nupe of Niger State: The Case of Sakpe Village* (Canberra: Australian National University).

Kettel, B. L., and Kettel, D. W. (1973), 'The Tuken of the Western Kenya Highlands', in A. Molnos (ed.), *Cultural Source Materials for Population Planning in East Africa*, iii. *Beliefs and Practices* (Nairobi: East African Publishing House), 415–27.

Kumekpor, T. K. (1970), 'Rural Women and Attitudes to Family Planning, Contraceptive Practice and Abortion in Southern Togo', Socio-Demographic Survey of the Republic of Togo, Report No. 1, Department of Sociology, University of Ghana, Legon.

Ladipo, O. A. (1978), 'Abortion among Nigerian Youth', *Nigerian Behavioural Sciences Journal*, 1: 112–16.

Liskin, S. (1980), 'Complications of Abortion in Developing Countries', *Population Reports*, series F.7.

Longmore, L. (1959), *The Dispossessed: A Study of the Sex-Life of Bantu Women in Urban Areas in and around Johannesburg* (London: Jonathan Cape).

Mati, J. K. G. (1977), 'Abortion in Africa', in F. T. Sai (ed.), *Family Welfare and Development in Africa* (London: International Planned Parenthood Federation), 74–9.

Molnos, A. (1973), *Cultural Source Materials for Population Planning in East Africa*, iii. *Beliefs and Practices* (Nairobi: East African Publishing House).

Moto (1984/5), 'Why abortion will not be wished away', 30: 10.

Mwambia, S. P. K. (1973), 'The Meru of Central Kenya', in A. Molnos (ed.), *Cultural Source Materials for Population Planning in East Africa*, iii. *Beliefs and Practices* (Nairobi: East African Publishing House), 60–7.

Nadel, S. F. (1942), *A Black Byzantium* (London: Oxford University Press).

Nichols, D., Ladipo, O. A., Paxman, J. M., and Otolorin, E. O. (1986), 'Sexual Behavior, Contraceptive Practice, and Reproductive Health among Nigerian Adolescents', *Studies in Family Planning*, 17: 100–6.

Njuki, S. K. (1987), 'Abortion: Is It a Problem at Nakuru General (Provincial) Hospital?', in K. D. Rogo (ed.), *Adolescent Fertility: Proceedings of a Workshop Held at Kwale, Coast Province, Kenya, August 1986*, Kenya Medical Association and Ministry of Health, Nairobi, 88–90.

Ofosu, D. Y. (1989), 'Hazard Models Analysis of Birth Intervals: A Study Based on West African Data', Ph.D. thesis, Australian National University.

Okediji, F. O., Caldwell, J. C., Caldwell, P., and Ware, H. (1976), 'The Changing African Family Project: A Report with Special Reference to the Nigerian Segment', *Studies in Family Planning*, 7: 126–36.

Pellow, D. (1977), *Women in Accra: Options for Autonomy* (Algonac, Mich.: Reference Publications).

Rattray, R. S. (1927), *Religion and Art in Ashanti* (London: Oxford University Press).

Raum, O. F. (1973), 'The Chaga of North-eastern Tanzania', in A. Molnos (ed.), *Cultural Source Materials for Population Planning in East Africa*, iii. *Beliefs and Practices* (Nairobi: East African Publishing House), 27–46.

Schuster, I. M. G. (1979), *New Women of Lusaka* (Palo Alto, Calif.: Mayfield Publishing Company).

14 The Implications of Premarital Childbearing for Infant Mortality: The Case of Côte d'Ivoire

DOMINIQUE MEEKERS

> Marriage is a duty, a requirement from corporate society, a rhythm of life in which everyone must participate . . . Failure to get married under normal circumstances means that the person concerned has rejected society and society rejects him in return.
>
> John Mbiti, quoted in Kuria (1987)

Introduction

Previous studies suggest that mortality levels of children born before marriage are higher than those of children born within marriage (Adegbola, 1987*a*, 1987*b*; Gueye and van de Walle, 1988). However, these studies have failed to take into account the fact that cohabitation and marriage do not necessarily coincide. I suggest that it may be less the marital status of the parents *per se* that is important for the survival chances of the child than whether the mother and the child are cohabiting with the father—or even with another man. This paper addresses these issues using data from the 1980–1 Côte d'Ivoire Fertility Survey (Ministère de l'Économie et des Finances, 1984). This survey is ideal for our purposes because it separates some of the key events of conjugal unions: the date of first cohabitation, the marriage ceremony, and first sexual relations with the first husband.

If we define the consummation of the first marriage as first sexual relations with the man who will become the first husband, then all children born before the consummation are biologically fathered by a man other than the first husband. If these children face active discrimination during the course of the subsequent marriage, rather than being merely disadvantaged because they were born before marriage, then we expect them to have higher mortality levels than those children fathered by the husband—even those born before the marriage ceremony or before the onset of cohabitation.

The Concept of Illegitimacy

Many scholars have argued that there is no such thing as illegitimacy in African societies (Adegbola, 1987*a*, 1987*b*; Gueye and van de Walle, 1988), because children born to unwed mothers are believed to be, as are all children, 'gifts of God'. Consequently, all children are said to be wanted. If an unmarried woman bears a child, acknowledgement of paternity as well as some form of compensation by a man who may or may not be the genitor is generally sufficient to give the child a proper position in society (Adegbola, 1987*a*). Given that paternity over the child of an unwed mother is indeed virtually always claimed, the concept of illegitimacy is seen as inappropriate to African societies.

I argue the opposite: the very fact that some man is supposed to claim paternity over premarital children suggests strongly that these children need a socially approved father, whether or not he is the biological father (for a similar argument, see Malinowski, 1927; 1930). In a similar vein, Mair (1971: 91) stressed that it is not the loss of virginity that is disapproved in Africa, but rather the bearing of children in circumstances not socially authorized. This lack of social approval is precisely what distinguishes legitimate from illegitimate births: 'an illegitimate child is a child whose conception and birth did not conform to the institutional rules which, in its parents' community, govern reproduction' (Teichman, 1982: 80). If this definition of illegitimacy is accepted, rather than a definition based simply on the legal status of children born to unwed mothers, then it is clear that the notion of illegitimacy also applies to African societies.

Although it is important to have a father, the roles attributed to the father vary in importance depending on the type of filiation and residence (Paulme, 1960: 21). In societies with patrilineal descent and patrilocal residence the role of the father is critical. Though less so in matrilineal and matrilocal societies, the social role of the father still remains important: 'even in matrilineal systems, there are rules to control the mating of the sexes, and with very few exceptions there is stressing of the role of some man as father. A woman should have an accredited husband to be the father of her children: Malinowski called this "the principle of legitimacy"' (Gluckman, 1969: 68).

In patrilineal societies children clearly need a father since they take his name. Among the Ga the naming ceremony of a child, *kpodziemo*, necessitates a 'father', although this father need not be the child's progenitor. In the case of a husbandless mother the child usually belongs socially to her own father (Fields, 1940: 24). Lorimer (1954: 70) points out that among the Ashanti a child born before marriage suffers no disgrace, provided that it is conceived after the woman's nubility ceremony and that the father has acknowledged and named the child.

Acceptance of paternity not only confers rights on the child; it also implies that a man accepts the responsibilities associated with fatherhood. If a man is not available to take these responsibilities by marrying the mother or by acknowledging his responsibilities over the child, then other 'back-up' institutions are available to appoint an alternative custodian, often the mother's brother, or her father, or even a new conjugal partner. However, at least in patrilineal societies, these alternatives are the last resort. Moreover, the existence itself of such alternatives suggests that children born out of wedlock lack something important: not the parents' marriage, but a 'father'.

In matrilineal societies, the situation is quite different. Children have automatic membership in the lineage of their mother and it is often the mother's brother who gets formal custody over them. But even in these societies there seems to be a need for a socially approved father or custodian for children. Manoukian (1950: 27–33) reports that in Akan society the father, although in a weak legal position, is responsible for the care and welfare of the children until they reach puberty. Even in the case of a divorce, the children remain under his custody. Similarly, among the Alladian of Côte d'Ivoire, although the mother's brother retains legal authority over the son, it is the father who gives him permission to build his own hut, marry, and start fishing on his own account (Augé, 1971). The example of the Baoulé is even more striking. A Baoulé woman enjoys considerable sexual freedom, and when she is young and unmarried she may have several lovers simultaneously (Étienne and Étienne, 1971). However, if she becomes pregnant she must indicate a genitor, who, once identified, must accept paternity. The man concerned cannot escape responsibility on the grounds of ambiguous biological fatherhood. People will tell him: 'You have slept with the girl; if it is not your child, then put your hand in her belly and remove it' (ibid. 168 ff.). The man can, however, respond that the child is a gift for the woman's parents, in which case her brother will take responsibility for the child. Still, the parents of the girl (her father or his brother) will call the child *gwa su ba* ('the child of the public domain'), because 'we do not know the name of the father' (ibid.).

Thus, although matrilineal societies usually identify the mother's brother as a child's custodian, the child still seems to need a socially approved father. Teichman (1982) remarks that although children born in a matrilineal community have automatic lineage membership, these societies still distinguish between legitimacy and illegitimacy. In fact, relationships in which childbearing is not sanctioned are known in all societies, if only because of the universal prohibition against incest.

Even though virtually all African societies have institutions to regulate paternity over children of unwed mothers, it should be no surprise that some children born outside wedlock suffer disadvantages, even when paternity is claimed. Social disadvantage does not necessarily imply discrimination, although this possibility cannot be ruled out. It is possible that children born out of wedlock are born in unfavourable conditions (e.g. their mother may be poor), but this does not necessarily mean that these children are treated differently from children born within wedlock. Yet occasionally laws may, indirectly or unintentionally, negatively affect the status of certain children. The introduction of laws in Côte d'Ivoire setting the conditions for a valid marriage—including the banning of polygyny—led ironically to 'the logical necessity to specify a different legal status for the children born of lawful wedlock and for those not so born, with the former having rights that the latter should not share' (Levasseur, 1971: 164). In many countries such as Nigeria, the legal system attempts to ensure that no child will suffer disabilities because of the circumstances of its birth (Karanja, this volume). But despite such legislation, fathers may give more support to the children of their legally married wives. Gage and Bledsoe (this volume) and Harrell-Bond (1975) argue that fathers in Sierra Leone are inclined to give more support to children of their legal wives than to children of their 'outside' wives, and Locoh (1988) points out that this also appears to be the case in Togo.

One disadvantage suffered by out-of-wedlock children is that the father may not reside with the mother, and may be unavailable in case of an emergency. Additionally, if the union has broken up, a prime source of the woman's income may be cut off, and if her behaviour is considered reprehensible, she can expect little support from her own lineage. Oppong (1988: 425) observes as well that when a woman enters a new union, her new partner often refuses responsibility for her children from a previous union.

Measuring Premarital Childbearing

Whether a child was conceived or born before or after marriage obviously depends on how marriage is defined. Most anthropological definitions of marriage have been based on Malinowski's 'principle of legitimacy', which states that a legal marriage gives a woman a socially recognized husband, and her children a socially recognized father. In his classic introduction to 'African Systems of Kinship and Marriage', Radcliffe-Brown (1950: 5) defines marriage as 'a social arrangement by which a child is given a legitimate position in the society, determined by parenthood in the social sense'.

Although the above definitions of marriage are conducive to demographic applications because they link marriage and childbearing, demographers have generally opted for more easily quantifiable definitions, such as a self-reported marital status. However, in the African context where marriage is a process stretched out over time, the dichotomy between 'single' and 'married' is an extreme oversimplification (van de Walle, 1968). In the World Fertility Surveys marriage has been defined as 'cohabitation', in an attempt to include less stable and informal unions. But in several cultures marriage does not imply the start of cohabitation. For example, among the Baoulé and Guéré of Côte d'Ivoire a woman often continues to live with her parents after her marriage and starts cohabiting with her husband after she has borne several children (Étienne and Étienne, 1971; Schwartz, 1975). Other examples of unions without cohabitation exist, including some new forms of polygyny known as 'deuxième bureau' or 'outside marriage', where a married man has a wife 'on the side' with whom he does not cohabit (Karanja, 1987, and this volume; Lacombe, 1986). Even when multiple wives are formally recognized, partners may live separately, a pattern found increasingly in urban contexts (see Locoh, this volume). Because 'cohabitation' does not apply to all types of union, it is not a satisfactory criterion to measure the onset of a union.

Whereas all the disparate components of marriage do not necessarily coincide, Étienne van de Walle (1968) also pointed out that the simple census dichotomy between 'single' and 'married' is unsatisfactory because of the processual nature of African marriage. What are needed are indicators for various stages in the union-formation process. The 1980–1 Côte d'Ivoire Fertility Survey made such an attempt (Ministère de l'Économie et des Finances, 1984; Sombo, 1985). It expanded the battery of questions relating to marital status, thus providing information unavailable in any of the other World Fertility Surveys. The main indicators pertaining to the first union are the type of union, the date of cohabitation, the date of the marriage ceremony, and the date of consummation. A full explanation of how these dates were derived is presented elsewhere (Meekers, 1990; van de Walle and Meekers, this volume). In this study, the definition of marriage is not the main concern; rather I focus on the three separate events in the union-formation process.

The Côte d'Ivoire questionnaire does not contain a question on the legitimacy of the births. Yet because it breaks the marriage process into three principal events, we can define 'legitimacy' in three ways: by comparing the date of first birth with the three indicators of the first union, i.e. the dates of cohabitation, ceremony, and consummation.[1] I have defined consummation of the first marriage as the first sexual relations with the man who is to become the woman's first husband. Hence, as all children born before the consummation are fathered biologically by a man other than the first husband, it is not possible to identify the social father of such a child. Whereas the WFS data do not tell us whether a man eventually accepted social fatherhood for such a child, it is likely that when the woman enters her first union her partner takes some responsibility for it. Even if the husband has not formally accepted social fatherhood for such a child, it is reasonable to assume that it will benefit indirectly from any economic support the mother gets from her husband. For example, even in the case where the husband does not directly support the child, the woman can use any economic support he provides for her to buy food and clothing for the child, or to take the child to a clinic when it is ill.

Prevalence and Correlates of Premarital Childbearing

Premarital childbearing is likely to be more prevalent in societies that lack strong sanctions against premarital sexuality. It is also likely to be common in urban areas where lineage control over female sexuality is diminished. By contrast, Islamic societies will probably have less childbearing before marriage since premarital sexuality is strongly disapproved and is precluded by a low age at marriage for females (Murdock, 1967; Lesthaeghe, Kaufmann, and Meekers, 1989).

TABLE 14.1. *Correlates of premarital childbearing*

	Women with a birth before					
	Cohabitation		Ceremony		Consummation[1]	
	%	No.	%	No.	%	No.
Ethnic Group						
Baoulé	39.1	642	69.6	632	17.4	642
Other Akan	18.4	478	52.8	470	11.7	478
Krou	15.4	518	38.7	512	12.7	518
Mandé Nord	7.3	482	10.8	481	2.5	482
Mandé Sud	11.0	400	22.5	395	8.0	400
Voltaic	7.4	461	28.8	459	4.8	461
Other African	6.3	1,008	14.5	998	4.4	1,008
Literacy						
Literate	26.7	491	63.1	477	17.3	491
Illiterate	13.2	3,510	28.8	3,481	7.4	3,510
Residence						
Abidjan	16.4	676	35.8	670	9.9	676
Other	14.6	3,325	32.4	3,288	8.3	3,325
Type of First Union						
Ceremonial	4.0	2,268	3.4	2,268	3.5	2,268
Free union[2]	27.5	867	100.0[3]	824	20.8	867
Visiting union	71.2	156	34.6	156	10.3	156
Legitimated union	22.0	710	49.4	710	9.6	710
TOTAL	14.9	4,001	33.0	3,958	8.6	4,001

[1] Children born before consummation of the first marriage are by definition fathered by a man other than the first husband.

[2] Excludes women in free unions who had their first birth less than eight months before the survey.

[3] By definition, all births to women in free unions occur prior to the marriage ceremony.

Notes: Data pertain to women who reported that they have married or have lived with a man. The subsample for 'ceremony' is smaller than for the other two events because not all respondents had a marriage ceremony; the number of cases by ethnic group does not add up to the total because some women did not specify their ethnic group.

Table 14.1 shows the proportion of ever-married women who had their first births before each of the three separate events that potentially comprise a first union: cohabitation, ceremony, and consummation. As expected, the matrilineal ethnic groups that ethnographies characterize as having considerable sexual freedom for females have high proportions of premarital first births, as defined by all three of the indicators. These include groups such as the Akan and, particularly, the Baoulé. Among the Islamic groups, including the Mandé Nord, the Voltaic, and the immigrants, premarital births are much less common. Literacy is strongly and positively associated with premarital childbearing. Urban residence is also associated with a higher proportion of women with a premarital first birth, but the difference is relatively small.

Although these patterns hold for each of the indicators of marriage, the proportion of women with a birth before the consummation of the first union is much lower than the proportion of women with a birth prior to cohabitation or to the ceremony. This finding was expected because children born before the consummation are, by definition, fathered by a man other than the first husband.

Additional information we can compile on the sequencing of the marital events—shown under 'Type of first union' in Table 14.1—allows us to ask how childbearing relates to the marriage process itself. This analysis necessitates identifying different types of union sequences (Meekers, 1990, 1992). One type consists of union formation sequences that started with the marriage ceremony. Such women have very low proportions of premarital births, for each of the indicators. Another type, 'visiting union', consists of women who did not start cohabiting with their husbands at the time of the marriage ceremony; rather, their husband would visit them in their own residences. Not surprisingly, the large majority of these women (71 per cent) had their first birth before they began cohabiting. These findings are consistent with literature on the Baoulé and the Guéré, which suggests that women often do not start co-residing with their husband until they have borne one or more children (Étienne and Étienne, 1971; Schwartz, 1975). The third type of union is the so-called 'free union', characterized by the absence of a marriage ceremony. More than one out of every four women whose first union was a free union had a child before first cohabitation, and one out of every five women in free unions had a child before the consummation of that particular union. Finally, some women started their first union as a free union, but later legitimated it with a marriage ceremony. As is the case for free unions, this type of legitimated union is associated with high levels of premarital childbearing.

The next sections examine the extent to which the survival chances of first-born children are affected by the marital statuses of their mothers at the time of birth.

Premarital Childbearing and Child Support Systems

A substantial body of literature suggests that premarital children have higher mortality levels than children born after marriage (e.g. Adegbola, 1987*b*; Gueye and van de Walle, 1988; Berkov and Sklar, 1976; Teichman, 1982). This has been observed in both historical and contemporary Western societies, and may also be true for African societies. Gueye and van de Walle (1988) remark, however, that for Africa, because of the lengthy nature of the marriage process and because of the existence of options other than marriage to claim paternity over a child, this mortality differential should not be discussed in terms of the 'legitimacy' of children. Rather, it should be interpreted in terms of 'the social acceptance of premarital births and of the support systems from which the various categories of children benefit' (Gueye and van de Walle, 1988: 5).

In other words, although these children are legally accepted, they do not necessarily have the same support as children born within marriage. Most African societies have a broad spectrum of institutions to deal with paternity and responsibility for premarital births.[2] Of course, these backup institutions should begin when the child is very young (or perhaps unborn) and therefore most vulnerable. Therefore, any delays or problems with the issue of paternity, and especially responsibility, may affect the child's survival chances.

The fact that the brother of an unwed mother can claim or is attributed paternity of her child does not imply that there was no problem initially. Rather it means that the problem is dealt with in the best possible way given the circumstances. Indeed, by having a child an unmarried woman may put her brother in a difficult position. A brother often depends on bridewealth from his sister's marriage to contract a marriage of his own. If the sister bears a child and still does not marry, not only may he have to postpone his own marriage, but he may also have to take responsibility for her child. As Mair (1971: 165) expresses the problem for the Nuer, 'one cannot suppose that such a brother bears his disappointment with beautiful meekness'. Though this may not comprise a reason for discrimination, it is reasonable to assume that the child will not get preferential treatment either. Even in matrilineal societies where the brother anticipates custody over his sister's children, it is said that a man commonly loves his own children more than his sister's children (Mair, 1971: 134). And in patrilineal societies a man may have limited resources to provide for his sister's children since he may be trying to accumulate the bridewealth for his own marriage or to provide for his own children as well.

However, in most cases the genitor will claim paternity, and a premarital pregnancy will often be followed by the initiation or acceleration of the marriage proceedings. Although this will generally benefit the child, it may also result in an unsatisfactory marriage which breaks up quickly. In such case, no father is around while the child is an infant.

These early pregnancies may have undesirable implications for both parents and child. Even when the pregnancy is followed by marriage the situation is often far from ideal. Generally, men are eager to claim paternity over a child because it increases their social status (Paige and Paige, 1981), but the costs associated with raising a child imply a substantial financial obligation early in married life. The heavy financial burden or the increased likelihood of a conjugal mismatch, both of which are associated with hastily conducted marriages, is likely to have negative implications, not only for the parents but also for the child.

In many indigenous African societies, marriage was an economic union in which the spouses jointly provided for the subsistence needs of their children (e.g. Étienne, 1977). Changes in economic relationships including male labour migration have forced women in many African societies to assume increasing responsibility for the daily household costs and the survival of their children. Nevertheless, most women are still economically dependent on husbands, who contribute for specific expenditures (Locoh, 1988). The type of the union affects both the economic support and the social status that a woman obtains by entering such a union (Karanja, this volume; Mann, this volume). In the case of informal or 'free' unions, which constitute an increasing proportion of marriages among younger women, economic support from the husband is not guaranteed (Lystad, 1975; Nelson, 1988).

> The [Agni] woman who is partner to an irregular union remains in her mother's home, and her children remain with her. Although her mate is expected to provide for their support, he may do as he wishes, and the woman has no recourse if he does not. Father–child ties are hence tenuous, and the primary burden of rearing the children falls upon the mother and her matrilineal family (Lystad, 1975: 198).

The current economic crisis reinforces this problem. Single mothers and women who are living alone because their husbands are absent or their marriage process has not been completed are confronting increasing problems in supporting their children and themselves (Locoh, 1988: 60). The support that both the woman and the child get depends on the man's commitment to the union.

Mortality of Premarital Children

Using data for Bobo-Dioulasso (Burkina Faso), Gueye and van de Walle (1988) confirmed that the higher mortality levels of premarital children largely stem from their lower birth order. But the effect of the parents' marital status nevertheless remains significant after controlling for birth order and maternal education. In support of this conclusion, other research has shown that premaritally born first births also have an excess mortality over first births following marriage (Adegbola, 1987*b*; Lesthaeghe, Kaufmann, and Meekers, 1989). This was the case in Benin, Cameroon, Côte d'Ivoire, Ghana, Lesotho, and southern Nigeria, while there was little difference for Senegal, northern Nigeria, and Kenya. Controlling for age at first birth, mother's education, and rural/urban residence did not eliminate these mortality differentials. These findings support the claims that even if they are legally equal to other children, premaritally born children are socially disadvantaged. Yet I argue that the main contributor to the higher mortality of premarital children is probably not discrimination but simply the fact that unmarried mothers themselves are generally in disadvantaged social positions.

TABLE 14.2. *Zero-order effects of indicators of the timing of mother's first marriage on infants' likelihood of dying in the first year of life: hazard model estimates*[1]

	Model 1	Model 2	Model 3
Timing of cohabitation[2]			
Months after cohabitation	–.091*	—	—
	(.023)		
Months before cohabitation	.091	—	—
Timing of the ceremony[2]			
Months after ceremony	—	–.018	—
		(.018)	
Months before ceremony	—	.018	—
Fathered by first husband[3]			
No	—	—	.049
			(.028)
Yes	—	—	–.049
Constant	–1.875	–1.934	–1.901
	(.023)	(.018)	(.028)

*Significant at 5% level.

1 Excludes children born in the month of the survey, and children born to never-married women. N = 43,143 person-months. Standard errors in parentheses.

2 Time-varying covariate.

3 Children born before the consummation of the union are, by definition, fathered by a man who did not become the first husband.

Let us ask now whether each of the three indicators of marriage in the Côte d'Ivoire WFS is associated with differential infant mortality. In other words, do infants born before the onset of cohabitation have a higher probability of dying than those born afterward; do infants born before their mother had a marriage ceremony have higher mortality levels than those born afterwards; and, do infants who are not fathered by the first husband (i.e. who are born before the consummation of the first union) have a greater chance of dying than those infants fathered by the first husband? Table 14.2 shows hazard model results of the effects of the various indicators of the timing of mother's first marriage on the likelihood that her child will die in the first year of life. The dependent variable in each model is the log odds of a child dying before reaching its first birthday. The coefficients for each variable show the deviation from the likelihood of dying for the reference group. A positive coefficient indicates that infants in this category are more likely to die than those in the reference group; a negative coefficient means that infants in this category are less likely to die than those in the reference group. Model 1 estimates the likelihood of an infant dying during months before the mother has started cohabiting, relative to the probability of dying in months after the onset of cohabitation (i.e. cohabitation status is a time-varying covariate); Model 2 compares infant mortality during months before the marriage ceremony and during months after the marriage ceremony has taken place. Finally, Model 3 estimates the likelihood of dying for infants fathered by the first husband relative to infants who are not fathered by the first husband.

The results for Model 1 indicate that the likelihood of dying before age 1 is significantly higher in months before the onset of cohabitation, as opposed to afterward. If children were discriminated against simply because they were fathered by a man other than the first husband, then they would likely have higher mortality levels than children by the first husband. Yet Model 3 shows that whether a child was fathered by the first husband of the woman or by another man does not significantly alter an infant's survival chances. The crucial aspect of the marriage process for infant mortality appears to be whether the mother is cohabiting presently with a partner. (Infant mortality is also higher in the months before a marriage ceremony has taken place, but this effect is small and not significant, see Model 2.)

Of course, children of unmarried mothers can be expected to show higher mortality levels because in most cases these are first births and their mothers tend to be very young. Both of these factors are known to reduce children's survival chances. On the positive side, though, unmarried mothers, perhaps because they are young, are more likely to live in an urban environment and to have an above-average education—both of which are associated with increased survival chances of the children (Akoto and Hill, 1988; Cantrelle, 1980; Mosley and Chen, 1984). In Table 14.3, I examine the impact of the three indicators of marriage on infant mortality, after controlling for age at first birth, literacy, type of place of residence, union type, and ethnic group.

While Table 14.2 shows that infant mortality is significantly higher in the months before the mother has started cohabiting with her first husband, Table 14.3 confirms that this is independent of whether these children were fathered by the man who eventually became the first husband or by a different man. After controlling for the other variables, including whether the husband is the father of the child, the likelihood of dying in the first year of life remains significantly higher during the months before the onset of cohabitation.[3] The effect of the timing of cohabitation is of the same order of magnitude as the effect of mother's literacy, and it is stronger than the effect of urban residence.

The timing of the marriage ceremony has no significant effect on the child's likelihood of dying in the first year of life, and whether the first husband is the father of the child has no significant effect on mortality either. If anything, children fathered by a man other than the first husband have slightly lower mortality. This result suggests that discrimination by husbands against their non-biological offspring is not responsible for the mortality differential observed in other studies (e.g. Adegbola, 1987*b*). Rather it appears that the higher mortality levels of children born before cohabitation is a result of the absence

TABLE 14.3. *Hazard model estimates of the effect of background variables on the likelihood of dying in the first year of life*[1]

	Coefficient	SE	t
Constant	−2.035	.046	−44.47
Timing of cohabitation[2]			
Months after cohabitation	−.113	.038	−2.94
Months before cohabitation	.113	—	—
Timing of the ceremony[2]			
Months after ceremony	−.018	.038	−.48
Months before ceremony	.018	—	—
Fathered by first husband[3]			
No	−.035	.040	−.88
Yes	.035	—	—
Ethnic group			
Baoulé	−.042	.045	−.94
Other Akan	−.059	.049	−1.18
Krou	−.004	.044	−.10
Mandé Nord	.088	.043	2.04
Mandé Sud	.015	.049	.31
Voltaic	.100	.043	2.31
Other African	−.098	—	—
Type of first union			
Ceremonial	.020	.043	.46
Free union	.005	.056	.08
Visiting union	.025	.073	.35
Legitimated union	−.050	—	—
Literacy			
Can read	−.116	.032	−3.58
Cannot read	.116	—	—
Place of residence			
Abidjan	−.095	.027	−3.50
Other	.095	—	—
Age at first birth			
< 18	.086	.017	4.93
> = 18	−.086	—	—

[1] Excludes children born in the month of the survey, and children born to never-married women. N = 43,143 person-months.

[2] Time-varying covariates.

[3] Children born before the consummation of the union are, by definition, fathered by a man who did not become the first husband.

of a husband *per se*. These findings support the argument that single mothers and mothers not living with their husband may be more likely to live in poverty (Locoh, 1988; Oppong, 1988).

Conclusions

In Africa, as in most other parts of the world, marriage is the preferred context for childbearing. However, as a result of both the stepwise nature of African marriages and the sexual freedom of unmarried females, a substantial proportion of first births occurs before a marriage has been concluded.

It has often been argued that these children are not stigmatized, because they can be legitimated by institutions other than marriage. It is true that these children are 'legitimate' with respect to the strictly legal aspects (although in Côte d'Ivoire children from a second wife in polygynous marriages are 'illegitimate' according to the law). However, the concept of 'legitimacy' goes much further than that. Teichman (1982) points out that illegitimacy refers to socially unauthorized conception and childbearing, which would include childbearing without the payment of the bridewealth.

The fact that alternative institutions exist to compensate for deviations from the rule does not imply that the problem does not exist. Ethnographic studies suggest that a child still needs a social father. Even in matrilineal societies where lineage membership is transferred through the mother and where the mother's brother has authority over her children, the 'father' still has important responsibilities for the children. It is impossible to determine the social father of a child with the WFS data, but it seems reasonable to assume that the man who cohabits with the mother has at least some responsibility for the child. Even in the case where this man has not accepted social fatherhood for a child of whom he is not the biological father, his economic support for his wife will indirectly also benefit her child.

It should be no surprise that the marital status of the mother at the time of birth can have important implications for the child. Other research had already shown that premarital births have higher mortality levels. Yet this analysis shows that marital status itself consists of discrete components, some of which are much more important for child welfare than others. The analysis has confirmed higher levels of mortality for infants born prior to cohabitation, but not for those born prior to the marriage ceremony or prior to the consummation of the marriage, the latter implying that the child was fathered by a different man. Furthermore, children fathered by a different man do not have higher mortality levels than those fathered by the woman's first husband, after controlling for cohabitation and other background variables. This finding suggests that the crucial factor with respect to the survival chances of the child is not biological fatherhood, but whether the mother is living with a man who, we assume, provides vital economic support for her and her children.

These findings have important implications for future research on infant mortality. Generally the husband has received less attention than the mother in this field, probably because of the recurrent findings that the mother's characteristics are more important determinants of infant mortality than the father's. Although the latter is undoubtedly true, it overlooks the fact that having a husband present, whether he is the father of the child or not, may be a much better option than having no husband around at all. The results of this study suggest that research on infant and child mortality may improve by focusing on the economic support systems from which children benefit. The man a woman is living with, even if he is not the child's biological father, is likely to make important contributions, directly or indirectly, to the child's economic welfare.

Acknowledgements

This paper has benefited from suggestions and encouragement by Caroline Bledsoe, Monica Das Gupta, Gilles Pison, Samuel Preston, Herbert Smith, and Étienne van de Walle. An earlier version of this chapter was presented as a paper at the annual meeting of the Population Association of America, Toronto, 3–5 May 1990. I gratefully acknowledge receiving access to the data as a member of a project on African marriage directed by Professor van de Walle, for which he received permission to use the World Fertility Survey data tape from the International Statistical Institute. Computer time and facilities were provided by the Population Studies Center at the University of Pennsylvania, where I was a graduate student at the time of writing. Additional support was provided by the Population Research Institute, The Pennsylvania State University, which has core support from a grant for international demographic research from the Andrew W. Mellon Foundation and the National Institute for Child Health and Human Development.

Notes

1. Because these types of indicators are vulnerable to biases resulting from the random imputation methods used in the edited data sets (i.e. the Standard Recode files), all relevant dates were reconstructed on the basis of the raw data file (Meekers, 1990, 1991). I assumed that when the specific month of occurrence was not reported, the event occurred in the middle of the year. Contrary to random imputation, which was used in the edited data file, this procedure preserves the order in which the events were reported to have occurred. Because it was impossible to determine whether a child was born before or after the first union when birth and first union occurred in the same month, I have interpreted these births as marital births.
2. Among these institutions is child fosterage, and even very young children can be fostered by a different household (see Antoine and Herry, 1983; Frank, 1985).
3. The results from a separate hazard model (not shown here) indicated that the effects of the time-varying covariates (i.e. cohabitation and the marriage ceremony) were constant over time.

References

Adegbola, O. (1987*a*), 'An Appraisal of Non-Marital Births in Nigeria', in E. van de Walle (ed.), *The Cultural Roots of African Fertility Regimes*. Proceedings of the Ife Conference, 25 Feb.–1 Mar., Obafemi Awolowo University, Ile-Ife, Nigeria, and University of Pennsylvania, Philadelphia, 77–99.

—— (1987*b*), 'A Comparative Study of Mortality of Children of Formal and Informal Unions', Paper presented at the IUSSP Seminar on Mortality and Society in Sub-Saharan Africa, 19–23 Oct. 1987, Yaoundé, Cameroon.

Akoto, E., and Hill, A. G. (1988), 'Morbidité, malnutrition et mortalité des enfants', in D. Tabutin (ed.), *Population et Sociétés en Afrique au Sud du Sahara* (Paris: L'Harmattan), 309–34.

Antoine, P., and Herry, C. (1983), *Urbanisation et dimension du ménage: Le Cas d'Abidjan*, Cahiers ORSTOM, Série Sciences humaines, 19/3.

Augé, M. (1971), *Traité précoloniale, politique matrimoniale et stratégie sociale dans les sociétes lagunaires de Basse Côte d'Ivoire*, Cahiers ORSTOM, 8/2: 143–52.

Berkov, B., and Sklar, J. (1976), 'Does Illegitimacy Make a Difference? A Study of the Life Chances of Illegitimate Children in California', *Population and Development Review*, 2/2: 201–17.

Cantrelle, P. (1980), 'La Mortalité des enfants en Afrique', in P.-M. Boulanger and D. Tabutin (eds.), *La Mortalité des enfants dans le monde et dans l'histoire* (Liège: Ordina Editions), 197–222.

Étienne, M. (1977), 'Women and Men, Cloth and Colonization: The Transformation of Production–Distribution Relations among the Baoulé (Ivory Coast)', *Cahiers d'études africaines*, 17/1: 51–64.

Étienne, P., and Étienne, M. (1971), '"A qui mieux mieux" ou le mariage chez les Baoulé', Cahiers ORSTOM, Série sciences humaines, 8/2: 165–86.

Fields, M. J. (1940), *Social Organization of the Ga Peoples* (London: Crown Agents for the Colonies).

Frank, O. (1985), 'La Mobilité des enfants et l'autosuffisance économique des femmes dans le milieu patriarcal africain', in *Femmes et politiques alimentaires* (Paris: ORSTOM, CIE).

Gluckman, M. (1969), *Ideas and Procedures in African Customary Law* (London: Oxford University Press).

Guèye, M., and van de Walle, E. (1988), 'Some Joint Determinants of Fertility and Infant and Child Mortality in Sub-Saharan Africa', IUSSP African Population Conference, Dakar, Senegal, ii.

Harrell-Bond, B. E. (1975), *Modern Marriage in Sierra Leone: A Study of the Professional Group* (The Hague: Mouton).

Karanja, W. W. (1987), '"Outside Wives" and "Inside Wives" in Nigeria: A Study of Changing Perceptions in Marriage', in D. Parkin and D. Nyamwaya (eds.), *Transformations of African Marriage* (Manchester: Manchester University Press), 247–61.

Kuria, G. K. (1987), 'The African or Customary Marriage in Kenyan Law Today', in D. Parkin and D. Nyamwaya (eds.), *Transformations of African Marriage* (Manchester: Manchester University Press), 283–306.

Lacombe, B. (1986), 'Les Rapports de couple en milieu urbain africain', in *Les Familles d'aujourd'hui*, AIDELF, 2: 285–8.

Lesthaeghe, R., Kaufmann, G., and Meekers, D. (1989), 'The Nuptiality Regimes in Sub-Saharan Africa', in R. J. Lesthaeghe (ed.), *Reproduction and Social Organization in Sub-Saharan Africa* (Berkeley, Calif.: University of California Press), 238–337.

Levasseur, A. A. (1971), 'The Modernization of Law in Africa with Particular Reference to Family Law in the Ivory Coast', in P. Foster and A. Zolberg (eds.), *Ghana and the Ivory Coast: Perspectives on Modernization* (Chicago: University of Chicago Press), 151–66.

Locoh, T. (1988), 'The Evolution of the Family in Africa', in E. van de Walle, P. O. Ohadike, and M. Sala-Diakanda (eds.), *The State of African Demography* (Liège: International Union for the Scientific Study of Population), 47–66.

Lorimer, F. (1954), *Culture and Human Fertility: A Study of the Relation of Cultural Conditions to Fertility in Non-Industrial and Transitional Societies* (Zurich: UNESCO).

Lystad, R. A. (1975), 'Marriage and Kinship among the Ashanti and the Agni: A Study of Differential Acculturation', in W. R. Bascom and M. J. Herskovits (eds.), *Continuity and Change in African Cultures* (Chicago: Phoenix Books, University of Chicago Press), 187–204.

Mair, L. (1971), *Marriage* (New York: Penguin Books).

Malinowski, B. (1927), *Sex and Repression in Savage Societies* (New York: Routledge & Kegan Paul).

—— (1930), 'Parenthood: The Basis of Social Structure', in V. F. Calverton and S. D. Schmalhausen (eds.), *The New Generation* (London: Allen & Unwin), 113–68.

Manoukian, M. (1950), 'Western Africa, Part 1: Akan and Ga-Adangme Peoples of the Gold Coast', in D. Forde (ed.), *Ethnographic Survey of Africa* (London: International African Institute).

Meekers, D. (1990), 'Marriage and Premarital Childbearing in Côte d'Ivoire', unpublished doctoral dissertation, University of Pennsylvania, Philadelphia (Ann Arbor, Mich.: University Microfilms International).

—— (1991), 'The Effect of Imputation Procedures on First Birth Intervals: Evidence from Five African Fertility Surveys', *Demography*, 28/2: 249–60.

—— (1992), 'The Process of Marriage in African Societies: A Multiple Indicator Approach', *Population and Development Review*, 18/1: 61–78.

Ministère de l'Économie et des Finances (1984), *Enquête ivoirienne sur la fécondité*, 1980–1981, Rapport principal, i. *Abidjan*.

Mosley, H., and Chen, L. (1984), 'An Analytical Framework for the Study of Child Survival in Developing Countries', in H. Mosley and L. Chen (eds.), *Child Survival: Strategies for Research Population and Development Review, Supplement*, 10: 25–48.

Murdock, G. P. (1967), *Ethnographic Atlas* (Pittsburgh: University of Pittsburgh Press).

Nelson, N. (1988), 'Marital Options in Mathare Valley (Nairobi)', Paper presented at the IUSSP Seminar on Nuptiality in Sub-Saharan Africa: Current Changes and Impact on Fertility, 14–17 November, Paris.

Oppong, C. (1988), 'Les Femmes africaines: Des épouses, des mères et des travailleuses', in D. Tabutin (ed.), *Population et Sociétés en Afrique au Sud du Sahara* (Paris: L'Harmattan), 421–40.

Paige, K. E., and Paige, J. M. (1981), *The Politics of Reproductive Ritual* (Berkeley, Calif.: University of California Press).

Paulme, D. (1960), 'Structures sociales traditionelles en Afrique Noire', *Cahiers d'études africaines*, 1/1: 15–27.

Radcliffe-Brown, A. R. (1950), 'Introduction', in A. R. Radcliffe-Brown and D. Forde (eds.), *African Systems of Kinship and Marriage* (London: Oxford University Press), 1–85.

Schwartz, A. (1975), *La vie quotidienne dans un village Guéré* (Abidjan: INADES).

Sombo, N. (1985), *Enquête Ivoirienne sur la Fécondité, 1980–81*, WFS Scientific Reports, 70 (Voorburg: International Statistical Institute).

Teichman, J. (1982), *Illegitimacy: A Philosophical Examination* (Oxford: Basil Blackwell).

van de Walle, E. (1968), 'Marriage in African censuses and inquiries', in W. Brass (ed.), *The Demography of Tropical Africa* (Princeton: Princeton University Press), 12–87.

—— (ed.) (1987), *The Cultural Roots of African Fertility Regimes*. Proceedings of the Ife Conference, 25 Feb.–1 Mar., Obafemi Owolowo University, Ile-Ife, Nigeria, and University of Pennsylvania, Philadelphia.

Index